Lecture Notes
in Control and Information Sciences   376

Editors: M. Thoma, M. Morari

Yasumichi Hasegawa

Approximate and Noisy Realization of Discrete-Time Dynamical Systems

 Springer

Author

Prof. Yasumichi Hasegawa

Department of Electronics
Gifu University
Gifu, 501-1193
Japan
E-Mail: yhasega@gifu-u.ac.jp

ISBN 978-3-540-79433-2   e-ISBN 978-3-540-79434-9

DOI 10.1007/978-3-540-79434-9

Lecture Notes in Control and Information Sciences ISSN 0170-8643

Library of Congress Control Number: 2008925333

Typeset & Cover Design: Scientific Publishing Services Pvt. Ltd., Chennai, India.

Printed in acid-free paper

5 4 3 2 1 0

springer.com

Preface

This monograph deals with approximation and noise cancellation of dynamical systems which include linear and nonlinear input/output relations. It will be of special interest to researchers, engineers and graduate students who have specialized in filtering theory and system theory. From noisy or noiseless data, reduction will be made. A new method which reduces noise or models information will be proposed. Using this method will allow model description to be treated as noise reduction or model reduction. As proof of the efficacy, this monograph provides new results and their extensions which can also be applied to nonlinear dynamical systems. To present the effectiveness of our method, many actual examples of noise and model information reduction will also be provided.

Using the analysis of state space approach, the model reduction problem may have become a major theme of technology after 1966 for emphasizing efficiency in the fields of control, economy, numerical analysis, and others.

Noise reduction problems in the analysis of noisy dynamical systems may have become a major theme of technology after 1974 for emphasizing efficiency in control. However, the subjects of these researches have been mainly concentrated in linear systems.

In common model reduction of linear systems in use today, a singular value decomposition of a Hankel matrix is used to find a reduced order model. However, the existence of the conditions of the reduced order model are derived without evaluation of the resultant model. In the common typical noise reduction of linear systems in use today, the order and parameters of the systems are determined by minimizing information criterion.

Approximate and noisy realization problems for input/output relations can be roughly stated as follows:

A. The approximate realization problem.
For any input/output map, find one mathematical model such that it is similar to the input/output map and has a lower dimension than the given minimal state space of a dynamical system which has the same behavior to the input/output map.
B. The noisy realization problem.
For any input/output map which includes noises in output, find one mathematical model which has the same input/output map.

Based on these parameters, we have been able to demonstrate that our new method for nonlinear dynamical systems, fully discussed in this monograph, are effective.

It is worth remembering that the development of approximate and noisy realization has been strongly stimulated by linear system theory and is well-connected to related mathematics, such as for example, matrix theory or mathematical programming. However, such development of nonlinear dynamical systems has not occurred yet because there have been no suitable mathematical method for nonlinear systems.

In this monograph, in related to the approximate quantity of noiseless data as being the noisy part of noisy data, we have identified the approximate realization problem as the noisy realization problem in the sense of how to unify the treatment of these problems. Our method intensively takes a positive attitude toward using computers. We will introduce a new method called the Constrained Least Square method. This method can be unified to solve both an approximate and a noisy realization problem.

The proposed method seeks to find the coefficients of a linear combination without the notion of orthogonal projection, and it can be applied to both approximate and noisy realization problems, namely, in the sense of a unified manner for both approximate and noisy realization problems, a new method will be proposed that provides effective results.

As already mentioned, common approximate and noisy realization problems have been mainly discussed in linear systems. On the other hand, there have been few developments for nonlinear systems. Our recent monograph, *Realization Theory of Discrete-Time Dynamical Systems* (T. Matsuo and Y. Hasegawa, Lecture Notes in Control and Information Science, Vol. 296, Springer, 2003), indicated that any input/output map of nonlinear dynamical systems can be characterized by the Hankel matrix or the Input/output matrix. The monograph also demonstrated that obtaining a dynamical system which describes a given input/output map is equal to determining the rank of the matrix of the input/output map and the coefficients of a linear combination of column vectors in the matrix. This new insight leads to the ability of discussing fruitful approximate and noisy realization problems.

We wish to acknowledge Professor Tsuyoshi Matsuo, who established the foundation for realization theory of continuous and discrete-time dynamical systems, and who taught me much on realization theory for discrete-time non-linear systems. He would have been an author of this monograph, but in April fifteen years ago he sadly passed away. We gratefully consider him one of the authors of this manuscript in spirit.

We also wish to thank Professor R. E. Kalman for his suggestions. He stimulated us to research these realization problems directly as well as through his works.

We also thank Professor Gary B. White for making the first manuscript into a more readable and elegant one.

March 2008   Yasumichi Hasegawa

Contents

1 Introduction

The approximate and noisy realization problems for discrete-time dynamical systems that we will state here can be stated as the following two problems Ⓐ and Ⓑ. The following notations are used in the problem description. I/O is the set of input/output maps which are partial data of given observed objects. DS is the category of mathematical models with a behavior which is an input/output relation.

Ⓐ Approximate realization problem.

For any input/output map $a \in I/O$ without noise, find one mathematical model $\sigma \in DS$ such that its behavior is approximately equal to the input/output map a and has a lower dimension than the given minimal state space of a dynamical system which has the behavior a.

Ⓑ Noisy realization problem.

For any input/output map $a \in I/O$ which includes noises in output, find one mathematical model $\sigma \in DS$ which has the same behavior as the given a.

In 1960, R. E. Kalman stated the realization problem for dynamical systems, that is, systems with input and output mechanisms, and he also established the realization theory for linear systems in an algebraic sense. Based on his ideas, we have solved a realization problem for a very wide class of discrete-time nonlinear systems [Matsuo and Hasegawa, 2003]. In the monograph, we derived fundamental results of realization theory for nonlinear dynamical systems. In particular, proposing some nonlinear dynamical systems, we could obtain when dynamical systems are characterized by their finite dimensionality through introducing a Hankel matrix suited for these dynamical systems. On the basis of these ideas, we will propose an approximate and noisy realization problem for discrete-time dynamical systems which include any nonlinear systems.

Discrete-time dynamical systems have become ever more important synchronously with the development of computers and the establishment of

Y. Hasegawa: Approxi. & Noisy Reali. of Discrete-Time Dyn. Sys., LNCIS 376, pp. 1–6, 2008.
springerlink.com

mathematical programming. Discrete-time linear systems have provided material for many fruitful contributions, as well as for discrete-time nonlinear dynamical systems. R. E. Kalman developed his linear system theory by using algebraic theory. Since then, algebraic theory has provided significant resources for the development of nonlinear dynamical system theory [Matsuo and Hasegawa, 2003] as well. Our processing methods for approximate and noisy realization of discrete-time dynamical systems are the first ones to be proposed in the case of discrete-time nonlinear systems.

In the case of control system design, model reduction method is needed for simplifying models or designing a controller for higher ordered dynamical systems. The usual model reduction (or aproximation) for dynamical systems are restricted to linear systems. The model reduction method can also be used for filtering of signal processing. As one of the main model reduction methods, the Hankel norm method and balanced truncation method have been proposed. Alternately, there is a method of minimum principle and parameter optimization under a certain quadratic performance index.

Our new approach of an approximate problem for nonlinear systems can be stated by using the fact that any input/output relation with causality conditions can be expressed in a sort of Hankel matrix. These relations are discussed in the reference [Matsuo and Hasegawa, 2003]. On this basis, a new method of approximate realization problem for discrete-time dynamical systems including nonlinear input/output maps is possible, and it can be solved by using the Hankel matrix or Input/output matrix. Using a matrix norm for determining the dimension of a given system, we also introduce new methods of determining coefficients for a linear combination of the columns of the Hankel matrix or Input/output matrix. The method is called the Constrained Least Square method. This approximate realization problem is a first trial for both linear and nonlinear dynamical systems.

As a method of modelling for dynamical systems with noise, one typical method is based on AIC (Akaike's Information Criterion). The method has been proposed based on the idea of statistical and probabilistic theory. However, its usage is also restricted to linear systems.

For both linear and nonlinear dynamical systems with noise, we will consider a noisy realization problem while presenting the same method to solve the problem of approximate realization, i.e., we will use the matrix norm and the Constrained Least Square method for our purpose. This noisy realization method is a first trial for nonlinear input/output relations with a noisy case.

In current approximate realization of discrete-time linear systems, once a Hankel matrix expressed in z-transformation is given with its singular value decomposition, a reduction model is constructed by proving the existence of the conditions of the recostructed Hankel matrix from the decomposition. The model reduction method can only be executed without any evaluations which means we don't know how much information have been taken off from the original matrix.

The common approximate realization method is put into execution under prior conditions which means that the ideal input/output value is given, signifying that the infinite sequence of the impulse response is already known.

Since this condition never occurs, the method is clearly not practical.

In current noisy realization of discrete-time linear systems, the problem is discussed using knowledge of statistics and probability distribution function in the typical method, AIC. Therefore, this method can only be applied to linear systems.

We wish to stress that in almost all cases, the methods used to treat approximate and noisy realization need finite data, demonstraing that our situation for our problems is practical.

In this monograph, regarding the approximate quantity of noiseless data as the noisy part of noisy data, we can identify the approximate realization problem as the noisy realization problem in the sense of how to unify the treatment of these problems. Our method intentionally takes a positive attitude toward using computers. We will introduce a new method called the Constrained Least Square method. This method can be unified to solve both an approximate and a noisy realization problem.

The proposed method seeks to find the coefficients of a linear combination without the notion of orthogonal projection which can be applied to both approximate and noisy realization problem. In other words, in the sense of a unified manner for both approximate and noisy realization problem, a new method will be proposed with effective results.

The general plan of this monograph is to propose the method of how to solve approximate and noisy realization problems in a unified way.

In our monograph [Matsuo and Hasegawa, 2003], we proposed the following realization problems A, B and C of nonlinear dynamical systems and solved them by constructing a new and very wide inclusion relation for various nonlinear dynamical systems:

A. The existence and uniqueness in an algebraic sense.
   For any input/output map $a \in I/O$, find at least one dynamical system $\sigma \in CD$ such that its behavior is a.
   Also, prove that any two dynamical systems that have the same behavior a are isomorphic in the sense of the category CD.
B. The finite dimensionality of the dynamical systems.
   Clarify when a dynamical system $\sigma \in CD$ is finite dimensional.
   Because finite dimensional dynamical systems are actually appearing by linear (or nonlinear) circuits or computer programs, it is very important that these conditions become clear.
C. Deriving the dynamical systems from finite data.
   Partial realization problems seek to find the minimal dynamical system fit to a given finite input/output's data and to clarify when the minimal dynamical systems are isomorphic.

In our monograph, we introduced General Dynamical Systems, Linear Representation Systems, Affine Dynamical Systems, Pseudo Linear Systems, Almost Linear Systems and So-called Linear Systems.

Their proposed inclusion relation and usual dynamical systems are shown in the right figure, where arrows imply that the above system includes the below system as a subclass.

We will discuss their approximate and noisy realization except for General Dynamical Systems.

Our realization theory stated here provides a new basis for treating both approximate and noisy realization of each system. Therefore, after two initial chapters regarding basic matters, this monograph is organized into balanced sections of one chapter for each dynamical system.

Each Chapter from 3 to 8 deals with our problem for one dynamical system. The Chapter number and the name of the dynamical system treated in the Chapter are given as follows:

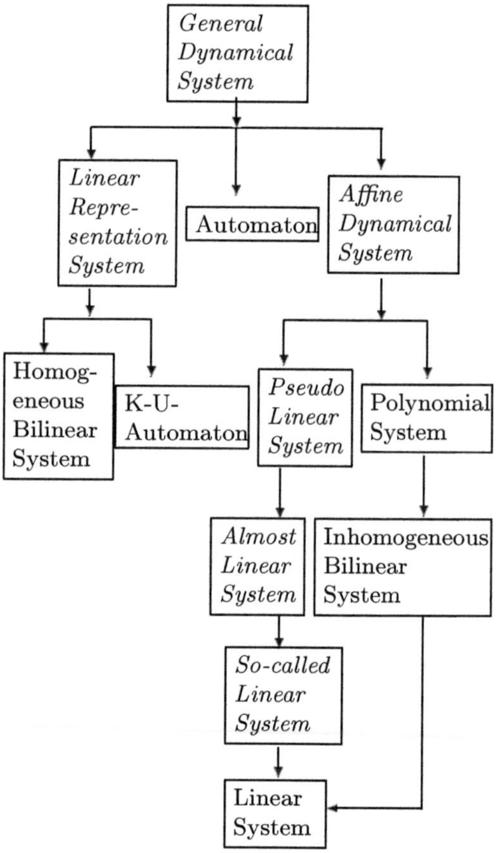

Chapter 3	Linear systems
Chapter 4	So-called linear systems
Chapter 5	Almost linear systems
Chapter 6	Pseudo linear systems
Chapter 7	Affine dynamical systems
Chapter 8	Linear representation systems

Let us preview each chapter in somewhat more detail.

In Chapter 2, we will describe input/output relations and the methods used in this monograph. Using the methods, we will provide a clear connection between signal and approximated signal and we will provide a clear distinction between signal and noise. The methods proposed in this chapter are called Constrained Least Square method abbreviated to the CLS method.

In Chapter 3, we will treat approximate and noisy realization problems for linear systems. Firstly, we will state the facts and establish facts regarding linear systems which are needed for our discussion. Both problems are solved first by decision of order and next by determination of the parameters of the given linear systems. The decision of order will be performed using singular value decomposition. The determination of parameters will be performed using our ConstrainedLleast Square method. Singular value decomposition can be used for reduction of information by cutting a part of space for lower eigenvalues. However, normally the degree of information loss is not made clear in proportion to the part of lower eigenvalues. Hence, in this chapter, good approximation and noisy processing will be tested with the ratio of matrix norm. After selecting a very small ratio, it will be shown that the Constrained Least Square method can produce a good linear system for a given linear system.

In the noisy case, there is a method which can find good systems, which is called AIC, that is, Akaike's Information Criterion. We will compare our method with the AIC method using some examples.

In Chapter 4, we will discuss approximate and noisy realization problems for so-called linear systems which are nonlinear. Such a treatment for problems of nonlinear systems appears for the first time ever in this chapter. Since the characteristic of the systems can be represented by two modified impulse responses, our purpose is to find these two modified impulse responses from the approximated or noisy input/output data. Previously described facts obtained in the monograph [Matsuo and Hasegawa, 2003] and new facts needed in this chapter will be discussed first.

Firstly, we will discuss the approximate realization problem, and then we will prove that our proposed method is effective by illustrating some examples.

Secondly, discussing the noisy realization problem, we will ascertain that our method has considerable merit.

In Chapter 5, we will discuss approximate and noisy realization problems for almost linear systems, which are nonlinear systems. Since the characterisic of almost linear systems can be represented by two modified impulse responses, our purpose is to find these two modified impulse responses from the approximated or noisy input/output data of the systems. Previously described facts obtained in the monograph [Matsuo and Hasegawa, 2003] and new facts needed in our discussion will be discussed first also in this chapter.

Firstly, we will discuss the approximate realization problem, and we will prove that our proposed method is effective by illustrating some examples.

Secondly, discussing the noisy realization problem, some examples will ascertain that our method has considerable merit.

In Chapter 6, we will discuss approximate and noisy realization problems for pseudo linear systems, which are nonlinear systems. Since the characterisic of pseudo linear systems can be represented by some modified impulse responses, our purpose is to find these modified impulse responses from the approximate or noisy input/output data of the given pseudo linear systems. Previously described facts obtained in the monograph [Matsuo and Hasegawa, 2003] and new facts needed in our discussion will be discussed first also in this chapter.

Firstly, we will discuss the approximate realization problem and secondly discuss the noisy realization problem. Some examples will illustrate that our proposed method is effective.

In Chapter 7, we will discuss approximate and noisy realization problems for affine dynamical systems, which are general nonlinear systems and include inhomogeneous bilinear systems as a subclass. Since the characterisic of affine dynamical systems can be represented by an input response map, our purpose is to find this input response map from the approximate or noisy input/output data of the affine dynamical systems. Previously described facts obtained in the monograph [Matsuo and Hasegawa, 2003] and new facts needed in our discussion will be discussed first also in this chapter.

We will discuss the approximate realization problem and then discuss the noisy realization problem. Affine dynamical systems are general nonlinear systems, nevertheless some examples will illustrate that our proposed method is effective for both cases.

In Chapter 8, we will discuss approximate and noisy realization problems for linear representation systems, which are general nonlinear systems and include homogeneous bilinear systems as a subclass. Since the characterisic of linear representation systems can be completely represented by an input response map, our purpose is to find its input response map from the approximate or noisy input/output data of the linear representation systems. Previously described facts obtained in the monograph [Matsuo and Hasegawa, 2003] will be shown first also in this chapter.

We will discuss the approximate realization problem and then discuss the noisy realization problem. The systems are general nonlinear systems, nevertheless some examples will illustrate that our proposed method is effective.

Notations

R : the real number field.

N : the set of non-negative integers.

$F(X, Y)$: the set of all functions from X to Y.

$L(X, Y)$: the set of all linear maps from X to Y.

$L(X)$: the set of all linear maps from X to X.

R^n : an n-dimensional coordinate space over the field R.

$R^{m \times n}$: the set of all $m \times n$-matrices.

im f : the image of a map f.

ker f : the kernel of a map f.

$\ll S \gg$: the smallest linear space which contains a set S.

Gr T : the graph of a relation T.

dom T : the domain of a relation T.

Acronyms

AIC : an information criterion (equivalently, Akaike's information
    criterion)

CLS : constrained least square

2 Input/Output Map and Additive Noises

To obtain concrete results, we will consider a case of dynamical systems with input/output mechanism surrounded by free noise or noise.

2.1 Input Response Maps (Input/Output Maps with Causality)

We will consider a notational method for input/output relations of an object to be observed or to be controlled in a discrete-time case, i.e., a black-box to which any element of the concatenation monoid U^* can be applied and whose output values are in a set of output values, where U^* is the free monoid over the input value's set U. Sometimes, Ω may be used in place of U^*, namely $\Omega = U^*$ always holds. Y is the set of output values. The representation theorems for any input/output map with causality have been given by [Matsuo and Hasegawa 2003]. The theorems can be stated as Lemmas (2.1), (2.5) and (2.8).

Lemma 2.1. *Any input/output relation with causality can be represented as $a \in F(U^*, Y)$. Then, any $a \in F(U^*, Y)$ can be represented as the following equation: $\hat{\gamma}(|\omega|) = a(\omega) \in Y$, where $\hat{\gamma}(|\omega|)$ denotes an output value at the time $|\omega|$ for an input ω to have been ended to apply, where $|\omega|$ is the length of the input ω.*

Definition 2.2. *An element a of $F(U^*, Y)$ is said to be an input response map.*

For the convenience of our discussions, we have utilized some kinds of input response maps from [Matsuo and Hasegawa, 2003].

Definition 2.3. *If an input response map $a \in F(U^*, Y)$ satisfies the following time-invariant condition, then a is said to be a time-invariant input response map.*

Time-invariant condition: $a(\omega_1|\omega) - a(\omega_1) = a(\bar{\omega}_1|\omega) - a(\bar{\omega}_1)$ for any $\omega \in U^*$, and $\omega_1, \bar{\omega}_1 \in U^*$ such that $|\omega_1| = |\bar{\omega}_1|$.

Definition 2.4. *For any time-invariant input response map $a \in F(U^*, Y)$, a function $I_a : U \to F(U^*, Y); u \mapsto I_a(u)[; t \mapsto a(u^t) - a(u^{t-1})]$ is said to be a modified impulse response of a, where u^t is given by $u^t(i) = u$ for $i(1 \leq i \leq t)$.*

Y. Hasegawa: Approxi. & Noisy Reali. of Discrete-Time Dyn. Sys., LNCIS 376, pp. 7–13, 2008.
springerlink.com

Lemma 2.5. *Representation Theorem*
For any time-invariant input response map $a \in F(U^*, Y)$, there exist uniquely modified impulse responses represented by the following equation. This correspondence is bijective.

$$a(\omega) = a(1) + \sum_{j=1}^{|\omega|} \{(I_a(\omega(j)))(|\omega| - j)\}.$$

In our case, we consider input/output maps $a \in F(U^*, Y)$ which satisfy the following time-invariant condition and affinity condition. They are said to be time-invariant, affine input response maps, where U is a linear space in this case. We may treat the case where multi-inputs are fed, i.e., $U = \mathbf{R}^m$, but conveniently, we will discuss a case where one-input is fed, i.e., $U = \mathbf{R}$. And Y is a linear space over the real number field $\mathbf{R}$.

Definition 2.6. *If an input response map a satisfies the following time-invariant and affinity condition, then a is said to be a time-invariant, affine input response map.*
Time-invariant condition
$a(\omega_1|\omega) - a(\omega_1) = a(\bar{\omega}_1|\omega) - a(\bar{\omega}_1)$
for any ω, ω_1, $\bar{\omega}_1$ such that $|\omega_1| = |\bar{\omega}_1|$.
Affinity condition
$a : U^* \to Y$ is an affine map, i.e.,
$a(\omega + \bar{\omega}) + a(0^{|\omega|}) = a(\omega) + a(\bar{\omega})$
$a(\lambda\omega) = \lambda a(\omega) + (1 - \lambda)a(0^{|\omega|})$
for any $\omega, \bar{\omega} \in U^*, |\omega| = |\bar{\omega}|$ and $\lambda \in K$.

Definition 2.7. *For any time-invariant, affine input response map $a \in F(U^*, Y)$, a function $I_a : \{0, 1\} \to F(N, Y); u \mapsto Ia(u)[; t \mapsto a(u^t) - a(u^{t-1})]$ is said to be a modified impulse response of a.*

Lemma 2.8. *Representation Theorem*
For any time-invariant, affine input response map $a \in F(U^*, Y)$, there exist uniquely modified impulse responses represented by the following equation. This correspondence is bijective. $a(\omega) = a(1) + \sum_{j=1}^{|\omega|}(\omega(j))(I_a(1)(|\omega| - j + 1)) + (1 - \omega(j))(I_a(0)(|\omega| - j + 1))$ for any $\omega \in U^*$.

The problem of approximation and noisy realization for input/output relation with causality are roughly stated as follows:

Problem 2.9. Problem statement for approximate realization
For any given data of the input/output map, find an input response map which is suitable in the sense of approximation, namely a dynamical system with approximate behavior for the given input response map.

Problem 2.10. Problem statement for noisy realization
For any given data of the input/output map with noises, find an input response map which is suitable in the sense of noisy data, namely a dynamical system with possibility for the same behavior to the given input response map.

2.2 Analysis for Approximate and Noisy Realization

According to our reference [Matsuo and Hasegawa 2003], any input response maps could be combined into a sort of Hankel matrix or Input/output matrix which are respectively suitable for it.

Here, we will mention the norm of the matrix which is needed to discuss our problems.

First, we will list the facts on singular value decomposition from the reference [R. A. Horn and C. A. Johnson, 1985].

Lemma 2.11. *Let $A \in R^{m \times n}$ with $m \leq n$ and rank $A = k \leq m$. There exists a unitary matrix $X \in R^{m \times m}$, a diagonal matrix $\Lambda \in R^{m \times m}$ with nonnegative diagonal entries $\lambda_1 \geq \lambda_2 \geq \cdots \geq \lambda_k \geq \lambda_{k+1} = \cdots = \lambda_m = 0$, and a matrix $Y \in R^{m \times n}$ with orthogonal rows such that $A = X \Lambda Y^T$. The matrix $\Lambda = diag(\lambda_1, \cdots, \lambda_m)$ is always uniquely determined and $\{\lambda_1^2, \cdots, \lambda_m^2\}$ are eigenvalues of AA^T.*

Lemma 2.12. *Let $A \in R^{m \times n}$ and $A = X \Lambda Y^T$ be the same in lemma (2.11). Let $X \in R^{m \times m}$ and $Y \in R^{m \times n}$ be $X = [\mathbf{x}_1, \mathbf{x}_2, \cdots, \mathbf{x}_n]$ and $Y = [\mathbf{y}_1, \mathbf{y}_2, \cdots, \mathbf{y}_m]$ respectively. Then A can be expressed as follows:*
$A = \lambda_1 \mathbf{x}_1 \mathbf{y}_1^T + \lambda_2 \mathbf{x}_2 \mathbf{y}_2^T + \cdots + \lambda_m \mathbf{x}_m \mathbf{y}_m^T.$

And the following holds:
1) $A\mathbf{y}_j = \lambda_j \mathbf{x}_j$ and $A^T \mathbf{x}_j = \lambda_j \mathbf{y}_j$ hold for $j (j = 1, \cdots, m)$.
2) $A^T A \mathbf{y}_j = \lambda_j^2 \mathbf{y}_j$ and $A\mathbf{y}_j = \lambda_j \mathbf{x}_j$ hold for $j (j = 1, \cdots, m)$.
3) $AA^T \mathbf{x}_j = \lambda_j^2 \mathbf{x}_j$ and $A^T \mathbf{x}_j = \lambda_j \mathbf{y}_j$ hold for $j (j = 1, \cdots, m)$.
4) $\mathbf{x}_i^T \cdot \mathbf{x}_i = \mathbf{y}_j^T \cdot \mathbf{y}_j = 1$ for i $(1 \leq i \leq m, 1 \leq j \leq n)$.

Remark: $A = \lambda_1 \mathbf{x}_1 \mathbf{y}_1^T + \lambda_2 \mathbf{x}_2 \mathbf{y}_2^T + \cdots + \lambda_m \mathbf{x}_m \mathbf{y}_m^T$ in lemma (2.12) is called the singular value decomposition.

Let $P_i := \mathbf{x}_i \mathbf{y}_i^T$. Then $P_i^T P_j = P_i P_j^T = \mathbf{0}$ holds for $i \neq j$.

Next, we will discuss the norm of the matrix.

Lemma 2.13. *Let $\| A \|$ be the norm of the matrix $A \in R^{m \times n}$, i.e., $\| A \| = \max_{\|x\|=1} \| Ax \|$. Then $\| A \| = \sqrt{\mu_1(A)}$ holds, where $\mu_1(A)$ is the maximum value of singular values for A.*

[proof]. Let a scalar valued function $f(\mathbf{x}, \lambda)$ be $f(\mathbf{x}, \lambda) = \mathbf{x}^T A^T A \mathbf{x} + \lambda[\mathbf{x}^T \mathbf{x} - 1]$ for a Lagrange multiplier $\lambda \in K$.

Let the small increment $f(\mathbf{x} + \delta \mathbf{x}, \lambda + \delta \lambda)$ from $f(\mathbf{x}, \lambda)$ be $f(\mathbf{x} + \delta \mathbf{x}, \lambda + \delta \lambda) = (\mathbf{x} + \delta \mathbf{x})^T A^T A (\mathbf{x} + \delta \mathbf{x}) + (\lambda + \delta \lambda)((\mathbf{x} + \delta \mathbf{x})^T (\mathbf{x} + \delta \mathbf{x}) - 1)$. From the equation $f(\mathbf{x} + \delta \mathbf{x}, \lambda + \delta \lambda) - f(\mathbf{x}, \lambda) = 0$, we obtain the equation $A^T A \mathbf{x} = -\lambda \mathbf{x}$. Because of $\mathbf{x}^T A^T A \mathbf{x} = -\lambda \mathbf{x}^T \mathbf{x}$, $\| A\mathbf{x} \|^2 = \mathbf{x}^T A^T A \mathbf{x}$ and $\| \mathbf{x} \|^2 = \mathbf{x}^T \mathbf{x}$, $\| A \| = \max_{\|x\|=1} \| Ax \| = \max_{A^T Ax = -\lambda x} \sqrt{-\lambda}$ holds.

Remark : For the singular values decomposition $A = \lambda_1 \mathbf{x}_1 \mathbf{y}_1^T + \lambda_2 \mathbf{x}_2 \mathbf{y}_2^T + \cdots + \lambda_m \mathbf{x}_m \mathbf{y}_m^T$ in lemma (2.12) and $\bar{m} \leq m$, let $B = \lambda_1 \mathbf{x}_1 \mathbf{y}_1^T + \lambda_2 \mathbf{x}_2 \mathbf{y}_2^T + \cdots + \lambda_{\bar{m}} \mathbf{x}_{\bar{m}} \mathbf{y}_{\bar{m}}^T$. Then $\| A - B \| = \sqrt{\lambda_{\bar{m}+1}}$ holds.

2.3 Measurement Data with Noise

In this section, we will discuss the case where noises are added to dynamical systems with an input/output mechanism.

For observed values $\gamma(t) \in K^p$ of time series, a p-dimensional observed signal $\hat{\gamma}(t) \in K^p$ and an additive noise $\bar{\gamma}(t) \in K^p$ can be considered as the following equation:

$$\gamma(t) = \hat{\gamma}(t) \; + \; \bar{\gamma}(t), \text{ where } t \in N.$$

In the sense of noisy case, the data processing problem can be stated roughly as follows:

For any given data $\{\gamma(t) : t \leq T \text{ for some } T \in N\}$, find the signal $\{\hat{\gamma}(t) : t \leq T\}$ which is the output of a dynamical system.

2.4 Analyses for Noisy Data

Let data with noise be a set $\{\mathbf{x}_t \in K^n : t = 1, 2, \cdots, s \in N\}$. Then $\mathbf{x}_t$ is represented by the equation $\mathbf{x}_i = \hat{\mathbf{x}}_i + \bar{\mathbf{x}}_i$, where $\hat{\mathbf{x}}_i$ is exact data and $\bar{\mathbf{x}}_i$ is noise.

Let $\mathbf{x}$ be $\mathbf{x} := [\mathbf{x}_1^T, \mathbf{x}_2^T, \cdots, \mathbf{x}_s^T]^T$, $\hat{\mathbf{x}} := [\hat{\mathbf{x}}_1^T, \hat{\mathbf{x}}_2^T, \cdots, \hat{\mathbf{x}}_s^T]^T$ and $\bar{\mathbf{x}} := [\bar{\mathbf{x}}_1^T, \bar{\mathbf{x}}_2^T, \cdots, \bar{\mathbf{x}}_s^T]^T$ for $\mathbf{x}_i \in K^n$. Then $\mathbf{x} = \hat{\mathbf{x}} + \bar{\mathbf{x}}$ holds. Let a matrix $A \in K^{q \times n}$ with a full rank satify an equation $A\hat{\mathbf{x}}_i = 0$ for any $1 \leq i \leq s$.

2.5 Constrained Least Square Method

Let $\hat{\mathbf{x}}_i \in K^n$, $\mathbf{x}_i = \hat{\mathbf{x}}_i + \bar{\mathbf{x}}_i$ and $\mathbf{x} = [\mathbf{x}_1^T, \mathbf{x}_2^T, \cdots, \mathbf{x}_s^T]^T$.

Lemma 2.14. *If $\bar{\mathbf{x}}^T \bar{\mathbf{x}}$ is minimum under the condition $S\hat{\mathbf{x}} = 0$ for a proper full rank matrix S, then $\bar{\mathbf{x}}$ is given by the equation $\bar{\mathbf{x}} = S^T[SS^T]^{-1}S\mathbf{x}$.*

[proof]. Let a scalar valued function $f(\bar{\mathbf{x}}, \lambda)$ be $f(\bar{\mathbf{x}}, \lambda) = \bar{\mathbf{x}}^T \bar{\mathbf{x}} + \lambda^T S[\mathbf{x} - \bar{\mathbf{x}}] + [\mathbf{x} - \bar{\mathbf{x}}]^T S^T \lambda$ for a Lagrange multiplier matrix $\lambda \in K^{sn \times 1}$.

Let the small increment $f(\bar{\mathbf{x}} + \delta\bar{\mathbf{x}}, \lambda + \delta\lambda)$ from $f(\bar{\mathbf{x}}, \lambda)$ be $f(\bar{\mathbf{x}} + \delta\bar{\mathbf{x}}, \lambda + \delta\lambda) = [\bar{\mathbf{x}} + \delta\bar{\mathbf{x}}]^T[\bar{\mathbf{x}} + \delta\bar{\mathbf{x}}] + [\lambda + \delta\lambda]^T S[\mathbf{x} - \bar{\mathbf{x}} - \delta\bar{\mathbf{x}}] + [\mathbf{x} - \bar{\mathbf{x}} - \delta\bar{\mathbf{x}}]^T S^T[\lambda + \delta\lambda]$. From the equation $f(\bar{\mathbf{x}} + \delta\bar{\mathbf{x}}, \lambda + \delta\lambda) - f(\bar{\mathbf{x}}, \lambda) = [\bar{\mathbf{x}}^T - \lambda^T S]\delta\bar{\mathbf{x}} + \delta\bar{\mathbf{x}}[\bar{\mathbf{x}} - S^T\lambda] = 0$, we obtain the equation $\bar{\mathbf{x}} = S^T\lambda$. Because of $\lambda = [SS^T]^{-1}S\bar{\mathbf{x}}$, $\bar{\mathbf{x}} = S^T[SS^T]^{-1}S\bar{\mathbf{x}}$ holds.

Next, we will discuss a concretely least square of noise $\bar{\mathbf{x}}_i$ for $1 \leq i \leq s$.

Theorem 2.15. *Under the constraint $A\hat{\mathbf{x}}_i = 0$ for i $(1 \leq i \leq s)$, let $\sum_{i=1}^s \bar{\mathbf{x}}_i \cdot \bar{\mathbf{x}}_i$ take a minimum value, where $\bar{\mathbf{x}}_i \cdot \bar{\mathbf{x}}_i$ denotes the inner product of the vectors $\bar{\mathbf{x}}_i$ and $\bar{\mathbf{x}}_i$. Then $\bar{\mathbf{x}}_\mathbf{i} = A^T[AA^T]^{-1}A\mathbf{x}_i$ holds for i $(1 \leq i \leq s)$.*

[proof]. Let $\mathbf{x} \in K^{sn \times 1}$ and $S \in K^{qs \times ns}$ be $\mathbf{x} := [\mathbf{x}_1^T, \mathbf{x}_2^T, \cdots, \mathbf{x}_s^T]^T$ and

$$S = \begin{bmatrix} A^T & 0 & \cdots & 0 \\ 0 & A^T & \cdots & 0 \\ \vdots & \ddots & \ddots & \vdots \\ 0 & 0 & \cdots & A^T \end{bmatrix}, \text{ where } \mathbf{x} = \hat{\mathbf{x}} + \bar{\mathbf{x}}, \ \hat{\mathbf{x}} := [\hat{\mathbf{x}}_1^T, \hat{\mathbf{x}}_2^T, \cdots, \hat{\mathbf{x}}_s^T]^T, \ \bar{\mathbf{x}} :=$$

$[\bar{\mathbf{x}}_1^T, \bar{\mathbf{x}}_2^T, \cdots, \bar{\mathbf{x}}_s^T]^T$.

Let a scalar function $f(\bar{\mathbf{x}}, \lambda)$ be $f(\bar{\mathbf{x}}, \lambda) = \bar{\mathbf{x}} \bullet \bar{\mathbf{x}} + \lambda^T S[\mathbf{x} - \bar{\mathbf{x}}] + [\mathbf{x} - \bar{\mathbf{x}}]^T S^T \lambda$ for a Lagrange multiplier vector $\lambda \in K^{qs \times 1}$.

Let the small increment $f(\bar{\mathbf{x}} + \delta\bar{\mathbf{x}}, \lambda + \delta\lambda)$ from $f(\bar{\mathbf{x}}, \lambda)$ be $f(\bar{\mathbf{x}} + \delta\bar{\mathbf{x}}, \lambda + \delta\lambda) = [\bar{\mathbf{x}} + \delta\bar{\mathbf{x}}] \bullet [\bar{\mathbf{x}} + \delta\bar{\mathbf{x}}] + [\lambda^T + \delta\lambda^T]S[\mathbf{x} - \bar{\mathbf{x}} - \delta\bar{\mathbf{x}}] + [\mathbf{x} - \bar{\mathbf{x}} - \delta\bar{\mathbf{x}}]^T S^T[\lambda + \delta\lambda]$. From the equation $f(\bar{\mathbf{x}} + \delta\bar{\mathbf{x}}, \lambda + \delta\lambda) - f(\bar{\mathbf{x}}, \lambda) = 0$ and Lemma (2.14), we obtain the equation $\bar{\mathbf{x}} = S^T[SS^T]^{-1}S\mathbf{x}$.

Therefore, $\bar{\mathbf{x}}_i = A^T[AA^T]^{-1}A\mathbf{x}_i$ holds for i ($1 \le i \le s$).

Definition 2.16. *The method given in Theorem (2.15) for an analysis of noisy data is called a Constrained Least Square Method which is abbreviated as the CLS method.*

Theorem 2.17. *Under the constraint $A\hat{\mathbf{x}}_i = 0$, $\bar{\mathbf{x}}_i = [\mathbf{0}^T, \bar{\mathbf{x}}_{i,2}^T] \in K^n$, $\mathbf{0}^T \in K^{n_1}$ and $\bar{\mathbf{x}}_{i,2}^T \in K^{n_2}$ for i ($1 \le i \le s$), let $\sum_{i=1}^s \bar{\mathbf{x}}_i \bullet \bar{\mathbf{x}}_i$ take minimum value,*

where $\bar{\mathbf{x}}_i \bullet \bar{\mathbf{x}}_i$ denotes the inner product of the vectors $\bar{\mathbf{x}}_i$ and $\bar{\mathbf{x}}_i$. Then $\begin{bmatrix} \lambda_2^T \\ \bar{\mathbf{x}}_i \end{bmatrix} =$

$$A^T[AA^T]^{-1}A \left(\hat{\mathbf{x}}_i + \begin{bmatrix} \lambda_2^T \\ \bar{\mathbf{x}}_i \end{bmatrix} \right) \text{ holds for } i \ (1 \le i \le s), \text{ where } \lambda_2^T = \mathbf{a}_1[\mathbf{a}_2^T \mathbf{a}_2]^{-1}$$

$\mathbf{a}_2^T \bar{\mathbf{x}}_{i,2}$ for $A := [\mathbf{a}_1^T \mathbf{a}_2^T]$.

[proof]. Let $\mathbf{x} \in K^{sn \times 1}$, S_1 and $S_2 \in K^{qs \times ns}$ be $\mathbf{x} := [\mathbf{x}_1^T, \mathbf{x}_2^T, \cdots, \mathbf{x}_s^T]^T$

$$S_1 = \begin{bmatrix} A^T & 0 & \cdots & 0 \\ 0 & A^T & \cdots & 0 \\ \vdots & \ddots & \ddots & \vdots \\ 0 & 0 & \cdots & A^T \end{bmatrix}, \ S_2 = \begin{bmatrix} B^T & 0 & \cdots & 0 \\ 0 & B^T & \cdots & 0 \\ \vdots & \ddots & \ddots & \vdots \\ 0 & 0 & \cdots & B^T \end{bmatrix}, \text{ where } \mathbf{x} = \hat{\mathbf{x}} + \bar{\mathbf{x}}, \ \hat{\mathbf{x}} :=$$

$[\hat{\mathbf{x}}_1^T, \hat{\mathbf{x}}_2^T, \cdots, \hat{\mathbf{x}}_s^T]^T$, $\bar{\mathbf{x}} := [\bar{\mathbf{x}}_1^T, \bar{\mathbf{x}}_2^T, \cdots, \bar{\mathbf{x}}_s^T]^T$ and $B^T = [1, \cdots, 1, 0, \cdots, 0]$.

Let a scalar function $f(\bar{\mathbf{x}}, \lambda_1, \lambda_2)$ be $f(\bar{\mathbf{x}}, \lambda_1, \lambda_2) = \bar{\mathbf{x}} \bullet \bar{\mathbf{x}} + \lambda_1^T S_1[\mathbf{x} - \bar{\mathbf{x}}] + \lambda_2^T S_2[\mathbf{x} - \bar{\mathbf{x}}] + [\mathbf{x} - \bar{\mathbf{x}}]^T S_1^T \lambda_1 + [\mathbf{x} - \bar{\mathbf{x}}]^T S_2^T \lambda_2$ for a Lagrange multiplier vector $\lambda_1, \lambda_2 \in K^{qs \times 1}$.

Let the small increment $f(\bar{\mathbf{x}} + \delta\bar{\mathbf{x}}, \lambda_1 + \delta\lambda_1, \lambda_2 + \delta\lambda_2)$ from $f(\bar{\mathbf{x}}, \lambda_1, \lambda_2)$ be $f(\bar{\mathbf{x}} + \delta\bar{\mathbf{x}}, \lambda_1 + \delta\lambda_1, \lambda_2 + \delta\lambda_2) = [\bar{\mathbf{x}} + \delta\bar{\mathbf{x}}] \bullet [\bar{\mathbf{x}} + \delta\bar{\mathbf{x}}] + [\lambda_1^T + \delta\lambda_1^T]S_1[\mathbf{x} - \bar{\mathbf{x}} - \delta\bar{\mathbf{x}}] + [\lambda_2^T + \delta\lambda_2^T]S_2[\mathbf{x} - \bar{\mathbf{x}} - \delta\bar{\mathbf{x}}] + [\mathbf{x} - \bar{\mathbf{x}} - \delta\bar{\mathbf{x}}]^T S_1^T[\lambda_1 + \delta\lambda_1] + [\mathbf{x} - \bar{\mathbf{x}} - \delta\bar{\mathbf{x}}]^T S_2^T[\lambda_2 + \delta\lambda_2]$. From the equation $f(\bar{\mathbf{x}} + \delta\bar{\mathbf{x}}, \lambda_1 + \delta\lambda_1, \lambda_2 + \delta\lambda_2) - f(\bar{\mathbf{x}}, \lambda_1, \lambda_2) = 0$, we obtain the equation $\bar{\mathbf{x}} = S_1^T \lambda_1 + S_2^T \lambda_2$.

Hence, $\bar{\mathbf{x}}_{\mathbf{i}} = A\boldsymbol{\lambda}_1^T - [1, \cdots, 1, 0, \cdots, 0]^T \boldsymbol{\lambda}_2^T = \begin{bmatrix} \mathbf{a}_1 \boldsymbol{\lambda}_1^T - \boldsymbol{\lambda}_2 \\ \mathbf{a}_2 \boldsymbol{\lambda}_1^T \end{bmatrix}$.

Therefore, $\mathbf{a}_1 \boldsymbol{\lambda}_1^T = \boldsymbol{\lambda}_2$ and $\mathbf{x}_i = \begin{bmatrix} \mathbf{0} \\ \bar{\mathbf{x}}_{i,2} \end{bmatrix} = \begin{bmatrix} \mathbf{0} \\ \mathbf{a}_2 \boldsymbol{\lambda}_1^T \end{bmatrix}$ hold.

From the equations $\mathbf{a}_1 \boldsymbol{\lambda}_1^T = \boldsymbol{\lambda}_2$ and $\bar{\mathbf{x}}_{i,2} = \mathbf{a}_2 \boldsymbol{\lambda}_1^T$, $\boldsymbol{\lambda}_2 = \mathbf{a}_1 [\mathbf{a}_2^T \mathbf{a}_2]^{-1} \mathbf{a}_2^T \bar{\mathbf{x}}_{i,2}$ holds.

By the relation $\begin{bmatrix} \boldsymbol{\lambda}_2^T \\ \bar{\mathbf{x}}_{i,2} \end{bmatrix} = \begin{bmatrix} \mathbf{a}_1 \boldsymbol{\lambda}_1^T \\ \mathbf{a}_2 \boldsymbol{\lambda}_1^T \end{bmatrix} = A\boldsymbol{\lambda}_1^T$, $A^T \begin{bmatrix} \boldsymbol{\lambda}_2^T \\ \bar{\mathbf{x}}_{i,2} \end{bmatrix} = A^T A \boldsymbol{\lambda}_1^T$ and $\boldsymbol{\lambda}_1^T = [A^T A]^{-1} A^T \begin{bmatrix} \boldsymbol{\lambda}_2^T \\ \bar{\mathbf{x}}_{i,2} \end{bmatrix}$ can be obtained. Using the relation $A^T \hat{\mathbf{x}}_i = 0$, we obtain

$$\begin{bmatrix} \boldsymbol{\lambda}_2^T \\ \bar{\mathbf{x}}_{\mathbf{i}} \end{bmatrix} = A^T [AA^T]^{-1} A \left(\hat{\mathbf{x}}_i + \begin{bmatrix} \boldsymbol{\lambda}_2^T \\ \bar{\mathbf{x}}_{\mathbf{i}} \end{bmatrix} \right),$$ which holds for i $(1 \leq i \leq s)$ with the relation $\boldsymbol{\lambda}_2^T = \mathbf{a}_1 [\mathbf{a}_2^T \mathbf{a}_2]^{-1} \mathbf{a}_2^T \bar{\mathbf{x}}_{i,2}$.

Definition 2.18. *The method given in Theorem (2.17) for an analysis of noisy data is called a Constrained Least Square Method 2 which is abbreviated as the CLS 2 method.*

An angle for the direction of two vectors 2.19
Let a vector space be $\boldsymbol{R}^n$ and let θ be the angle which implies the difference between the directions of two vectors x, $y \in \boldsymbol{R}^n$. Then $\cos\theta$ can be expressed as follows:

$$\cos\theta = \frac{\|\mathbf{x}\|^2 + \|\mathbf{y}\|^2 - \|\mathbf{x} - \mathbf{y}\|^2}{2\|\mathbf{x}\|\|\mathbf{y}\|},$$

where $\| \mathbf{x} \| = \sqrt{\mathbf{x}^T \mathbf{x}}$.

2.6   Historical Notes and Concluding Remarks

In the field of model reduction of discrete-time systems, singular values decomposition and polynomial equations are used as the effective methods [Glover, 1981]. On the other hand, in the field of modeling under a noisy enviroment, various methods such as AIC (Akaike's Information Criterion) have been proposed from the viewpoint of probabilistic sense.

In this monograph, it is shown for the first time that approximate realization (model reduction) and noisy realization can be proposed for nonlinear dynamical systems with an input and output mechanism. Note that the usual methods for approximation and noisy realization are limited to only linear systems. Of course, our methods in this monograph can be applied not only to linear systems but also to nonlinear systems in a unified manner.

It is noteworthy that our methods are quite different from usual methods and unified for any input/output relations with causality condition.

Also note that our methods are geared only toward the linear combination of vectors. Furthermore, It is also noteworthy that we have shown that any input/output relations with causality condition can be expressed in a Hankel matrix or Input/output matrix which can serve as a linear operator.

3 Approximate and Noisy Realization of Linear Systems

Let the set Y of output's values be a linear space over the real number field $\boldsymbol{R}$. It is well known that Linear System Theory was established in the algebraic sense [Kalman, 1969]. The main theorem says that for any causal linear input/output map, there exist at least two canonical (controllable and observable) Linear Systems which realize (faithfully describe) it and any two canonical Linear Systems with the same behavior are isomorphic.

Details of finite dimensional Linear Systems were investigated. The criterion for the canonical finite dimensional Linear Systems and various standard canonical Linear Systems were given.

Their partial realization was also discussed according to the above results. We will state an algorithm to obtain a canonical partial realization from a given partial input/output map.

Based on fundamentally established results, an approximate realization problem and noisy realization problem will be discussed.

3.1 Basic Facts about Linear Systems

We will summarize fundamentally established facts, which are needed for approximate and noisy realization problems.

Definition 3.1. *Linear Systems*

(1) A system represented by the following equations is written as a collection $\sigma = ((X, F), g, h)$ and it is said to be a linear system:

$$\begin{cases} x(t+1) = Fx(t) + g\omega(t+1) \\ x(0) \quad\ = 0 \\ \hat{\gamma}(t) \quad\ = hx(t) \end{cases}$$

for any $t \in N$, $x(t) \in X$, $\gamma(t) \in Y$, where X is a linear space over the field $\boldsymbol{R}$, F is a linear operator on X, $g \in X$ and $h : X \to Y$ is a linear operator.

Y. Hasegawa: Approxi. & Noisy Reali. of Discrete-Time Dyn. Sys., LNCIS 376, pp. 15–53, 2008.
springerlink.com

(2) The input response map $a_\sigma : U^* \to Y; \omega \mapsto h[\sum_{j=1}^{|\omega|} F^{|\omega|-j} g\omega(j)]$ is said to be the behavior of σ. For an input response map $a \in F(U^*, Y)$, σ which satisfies $a_\sigma = a$ is called a realization of a.

(3) For the linear system σ, $I_\sigma(i) = hF^i g$ is said to be an impulse response of σ. Note that there is a one-to-one correspondence between the behavior of σ and the impulse response of σ.

(4) A linear system σ is said to be reachable if the reachable set $\{\sum_{j=1}^{|\omega|} F^{|\omega|-j} g\omega(j); \omega \in U^*\}$ is equal to X.

(5) A linear system σ is called observable if $hF^i x_1 = hF^i x_2$ for any $i \in N$ implies $x_1 = x_2$.

(6) A linear system σ is called canonical if σ is reachable and observable.

Remark 1: It is meant for σ to be a faithful model for the input response map a that σ realizes a.

Remark 2: A canonical linear system $\sigma = ((X, F), g, h)$ is a system that has the most reduced state space X among systems that have the behavior a_σ.

Remark 3: The linear system $\sigma = ((X, F), g, h)$ obtained by the following common linear system equation and a transformation is a canonical linear system with the same behavior.

$$\begin{cases} x(t+1) = Fx(t) + g\omega(t+1), \\ x(0) \quad = 0, \\ \hat{\gamma}(t) \quad = hx(t), \end{cases} \qquad \begin{cases} \underline{x}(t+1) = A\underline{x}(t) + \mathbf{b}\omega(t) \\ \underline{x}(0) \quad = 0 \\ \hat{\gamma}(t) \quad = c\underline{x}(t) \end{cases}$$

The transformation is given as follows:

$$x(t) = \begin{bmatrix} \underline{x}(t) \\ \omega(t) \end{bmatrix}, \quad F = \begin{bmatrix} A & \mathbf{b} \\ 0 & 0 \end{bmatrix}, \quad g = \begin{bmatrix} 0 \\ 1 \end{bmatrix}, \quad h = \begin{bmatrix} c & 0 \end{bmatrix}.$$

Example 3.2. Let $\mathbf{R}[z]$ be a set of polynomials in one variable z over the field $\mathbf{R}$. The variable $z : \mathbf{R}[z] \to \mathbf{R}[z]; \lambda \mapsto z\lambda$ is a linear operator. Let $a \in F(U^*, Y)$ be regarded as a linear operator : $\mathbf{R}[z] \to Y; z^i \mapsto a(i)$. Then $\sigma_I = ((\mathbf{R}[z], z), 1, a)$ is a linear system whch is a realization of a. The impulse response I_{σ_I} of the linear system σ_I is given by $I_{\sigma_I}(i) = a(0| \cdots |0|1)$, where i is the length of an input $0| \cdots |0|1$.

Remark: For $a \in F(U^*, Y)$, an operator $\tilde{a} : \mathbf{R}[z] \to Y; \alpha_i z^i \mapsto \alpha_i a(i)$ is regarded as a linear operator : $N \to Y; i \mapsto a(i)$. This correspondence is one to one.

Example 3.3. Let $i \in N$, $a \in F(U^*, Y)$, $S_l : F(N, Y) \to F(N, Y); \gamma \mapsto S_l\gamma$ $[; t \mapsto \gamma(t+1)]$ and let $0 : F(N, Y) \to Y; \gamma \mapsto \gamma(0)$ be a linear operator. Then $\sigma_F = ((F(N, Y), S_l), a, 0)$ is a linear system which is a realization of $a \in F(U^*, Y)$.

The impulse response I_{σ_F} of the linear system σ_F is given by $I_{\sigma_F}(i) = a(0| \cdots |0|1)$, where i is the length of an input $0| \cdots |0|1$.

Theorem 3.4. *For an input response map $a \in F(U^*, Y)$, the following two linear systems are both canonical realizations of a:*
1) $((\boldsymbol{R}[z]/_{\equiv a}, \dot{z}), [1], \dot{a})$, where $\boldsymbol{R}[z]/_{\equiv a}$ is a quotient space defined by an equivalence relation $\lambda_1 = \sum_i \lambda_1(i)z^i \equiv \lambda_2 = \sum_i \lambda_2(i)z^i \iff a(\lambda_1) = a(\lambda_2)$, $[1]$ is defined as a map $\boldsymbol{R}[z] \to \boldsymbol{R}[z]/_{\equiv a}; 1 \mapsto [1]$. $\dot{a}$ is defined by $\dot{a}([\lambda]) = a(\lambda)$ for any $\lambda \in \boldsymbol{R}[z]$.
2) $((\ll \{S_l^i a : i \in N\} \gg, S_l), a, 0)$, where $\ll S \gg$ is the linear hull generated by the set S.

Definition 3.5. *Let $\sigma_1 = ((X_1, F_1), g_1, h_1)$ and $\sigma_2 = ((X_2, F_2), g_2, h_2)$ be linear systems. Then a linear operator $T : X_1 \to X_2$ is said to be a linear system morphism $T : \sigma_1 \to \sigma_2$ if T satisfies $TF_1 = F_2 T$, $Tg_1 = g_2$ and $h_1 = h_2 T$. If $T : X_1 \to X_2$ is bijective, then $T : \sigma_1 \to \sigma_2$ is said to be an isomorphism.*

Corollary 3.6. *Let T be a linear system morphism $T : \sigma_1 \to \sigma_2$. Then $a_{\sigma_1} = a_{\sigma_2}$ holds.*

[proof]. The definitions of the behavior and linear system morphism lead to this corollary.

Theorem 3.7. *Realization Theorem of linear systems*

Existence: For any input response map $a \in F(U^*, Y)$, there exist at least two canonical linear systems which realize a.
Uniqueness: Let σ_1 and σ_2 be any two canonical linear systems that realize $a \in F(U^*, Y)$. Then there exists an isomorphism $T : \sigma_1 \to \sigma_2$.

3.2 Finite Dimensional Linear Systems

In this section, a canonical form of finite-dimensional linear systems will be treated based on the realization theorem (3.7). Many results of linear systems have been already shown in a reference [Kalman, 1969]. In the following sections, these results have been summarized for this monograph to be self-contained.

At first, the conditions when a finite dimensional linear system is canonical is presented.

Secondly, the canonical form which is suitable for approximate and noisy realization problems is defined. We introduce a standard system as a representative in their equivalence classes.

Thirdly, a criterion for the behavior of finite dimensional linear systems, that is, the rank condition of an infinite Hankel matrix is presented.

Finally, a procedure to obtain the reachable standard system which realizes a given input response map is presented.

There is a fact about finite dimensional linear spaces that an n-dimensional linear space over the field $\boldsymbol{R}$ is isomorphic to $\boldsymbol{R}^n$ and $L(\boldsymbol{R}^n, \boldsymbol{R}^m)$ is isomorphic to $\boldsymbol{R}^{m \times n}$ (See Halmos [1958]). Therefore, without loss of generality, we can consider a n-dimensional linear system as $\sigma = ((\boldsymbol{R}^n, F), g, h)$, where $F \in \boldsymbol{R}^{n \times n}$, $g \in \boldsymbol{R}^n$ and $h \in \boldsymbol{R}^{p \times n}$.

Lemma 3.8. *A linear system $\sigma = ((\mathbf{R}^n, F), g, h)$ is canonical if and only if the following conditions 1) and 2) hold:*
1) rank $[g, Fg, \cdots, F^{n-1}g] = n$.
2) rank $[h^T, (hF)^T, \cdots, (hF^{n-1})^T] = n$.

Definition 3.9. *A canonical linear system $\sigma_s = ((\mathbf{R}^n, F_s), \mathbf{e}_1, h_s)$ is said to be a reachable standard system if $\mathbf{e}_i = F_s^{i-1}\mathbf{e}_1$ and $F_s^n\mathbf{e}_1 = \sum_{i=1}^{n} \alpha_i F_s^{i-1}\mathbf{e}_1$ hold. Such F_s is presented as follows:*

$$F_s = \begin{bmatrix} 0 & \cdots & \cdots & 0 & \alpha_1 \\ 1 & \ddots & & \vdots & \alpha_2 \\ \vdots & \ddots & \ddots & \vdots & \vdots \\ \vdots & \ddots & \ddots & 0 & \vdots \\ 0 & \cdots & 0 & 1 & \alpha_n \end{bmatrix}.$$

Lemma 3.10. *Lemma for equivalence classes*
For any finite dimensional canonical linear system, there exists a uniquely determined isomorphic reachable standard system.

Definition 3.11. *For any input response map $a \in F(U^*, Y)$, the corresponding linear input/output map $A : \mathbf{R}[z] \to F(N, Y)$ satisfies $A(z^i)(j) = a(0| \cdots |0|1) = I_a(i + j)$ for i, $j \in N$ and the length of an input $0| \cdots |0|1$ is $i + j$.*
Hence, A is represented by the following infinite matrix $\hat{H}_a$. This $\hat{H}_a$ is said to be a Hankel matrix of a.

$$\hat{H}_a = \begin{matrix} & & & i \\ & & \begin{pmatrix} & \vdots & & \\ & \vdots & & \\ & \vdots & & \\ & \vdots & & \\ j & \cdots & \cdots & I_a(i+j) \end{pmatrix} \end{matrix}$$

Note that for the linear input/output map $A : \mathbf{R}[z] \to F(N, Y)$, there exists a unique function $I_a : N \to Y$ such that $I_a(i + j) = A(z^i)(j)$ holds.

It is also noted that the column vectors of $\hat{H}_a$ denote $S_l^i I_a$.

Theorem 3.12. *Theorem for existence criterion*
For an input response map $a \in F(U^*, Y)$, the following conditions are equivalent:
1) The input response map $a \in F(U^*, Y)$ has the behavior of a n-dimensional canonical linear system.
2) There exist n-linearly independent vectors and no more than n-linearly independent vectors in a set $\{S_l^i a; i \leq n$ for $i \in N\}$.
3) The rank of the Hankel matrix $\hat{H}_a$ of a is n.

Theorem 3.13. *Theorem for a realization procedure*
*Let $a \in F(U^*, Y)$ be an input response which satisfies the condition of Theorem
(3.12). Then the reachable standard system $\sigma_s = ((\mathbf{R}^n, F_s), \mathbf{e_1}, h_s)$ which realizes
the input response map a is obtained by the following procedure:*
*1) Select the linearly independent vectors $\{S_l^i a; 0 \le i \le n - 1\}$ from the set
$\{S_l^i a; i \in N\}$.*
2) Let the state be $\mathbf{e_1}$, where $\mathbf{e_1} = [1, 0, \cdots, 0]^T \in \mathbf{R}^n$.
3) Let the output map h_s be $h_s = [a(1), a(1), a(0|1), \cdots, a(0|0| \cdots |0|1)]$.
*4) Let F_s be the same as in the reachable standard system defined in Definition
(3.9) for $S_l^n a = \sum_{i=1}^{n} \alpha_i S_l^{i-1} a$.*

3.3 Partial Realization Theory of Linear Systems

In this section, we consider a partial realization problem of linear systems. Let
$\underline{a}$ be an $\underline{N}$-sized input response map$(\in F(U_{\underline{N}}^*, Y))$, where $\underline{N} \in N$ and $U_{\underline{N}}^* :=
\{\omega \in U^*; |\omega| \le \underline{N}\}$. The $\underline{a}$ is said to be a partial input response map. A finite
dimensional linear system $\sigma = ((X, F), g, h)$ is called a partial realization of $\underline{a}$ if
$hF^{|\omega|-1}g = \underline{a}(0| \cdots |0|1)$ holds for any $\omega \in U_{\underline{N}}^*$, $|\omega| = |0| \cdots |0|1|$.
   A partial realization problem of linear systems is roughly stated as follows:
   < For any given $\underline{a} \in F(U_{\underline{N}}^*, Y)$, find a partial realization σ of $\underline{a}$ such that the
dimension of state space X of σ is minimum. Then the σ is said to be a minimal
partial realization of $\underline{a}$. Moreover, show an algorithm to obtain the minimal
partial realization.>
   For a partial input response map $\underline{a} \in F(U_{\underline{N}}^*, Y)$, the following matrix $\hat{H}_{\underline{a}\ (p,\bar{p})}$
is said to be a finite-sized Hankel matrix of $\underline{a}$.

$$
\hat{H}_{\underline{a}\ (p,\bar{p})}_{\ i} = \begin{pmatrix} & & & \bar{i} \\ & & & \vdots \\ & & & \vdots \\ \cdots & \cdots & I_a(i+\bar{i}) & \end{pmatrix},
$$

where $i \le p$ and $\bar{i} \le \bar{p}$.
   Note that the column vectors of $\hat{H}_{\underline{a}\ (p,\bar{p})}$ is represented by $\underline{S}_l^i I_{\underline{a}}$.
   When we actually treat the approximate and noisy realization problem, we
will use a notation $H_{\underline{a}\ (n_1,\underline{N}-n_1)}$ expressed as follows:
$H_{\underline{a}\ (n_1,\underline{N}-n_1)} = [I_{\underline{a}}, \cdots, S_l^{n_1-1} I_{\underline{a}}]$.

Proposition 3.14. *Let the rank of a finite-sized Hankel matrix $\hat{H}_{\underline{a}\ (p,\bar{p})}$ be n.
Then a minimal partial realization $\sigma_{\underline{a}} = ((\mathbf{R}^n, F_s), \mathbf{e_1}, h_s)$ of the impulse response $I_{\underline{a}}$ is obtained by the following algorithm:*
1) Let F_s be the same as F_s in Definition (3.9) for $\underline{S}_l^n I_{\underline{a}} = \sum_{i=1}^{n} \alpha_i \underline{S}_l^{i-1} I_{\underline{a}}$.

2) Let $\mathbf{e}_1$ be $\mathbf{e}_1 = [1, 0, \cdots, 0]^T$.
3) Let h_s be $h_s = [I_{\underline{a}}(1), I_{\underline{a}}(2), \cdots, I_{\underline{a}}(n)]$.

[proof]. It is obvious from the definition of behavior of the system.

3.4 Approximate Realization of Linear Systems

In this section, we discuss an approximate realization problem for linear systems which is stated as follows:

<For any given finite-length impulse response, find a linear system which approximates it.>

In order to make our discussion simple, we assume that the set Y of output is the set $\boldsymbol{R}$ of real numbers, namely 1-output.

Theorem 3.15. *Algorithm for approximate realization*
Let $\underline{a}$ be a considered object which is a linear system. Then an approximate realization $\sigma = ((\boldsymbol{R}^n, F_s), g, h_s)$ of $\underline{a}$ based on the CLS method is given by the following algorithm:

1) Based on the ratio of the square root of eigenvalues for a matrix $H_{\underline{a}\ (p,\bar{p})} H_{\underline{a}\ (p,\bar{p})}^T$, determine the value n of rank for the matrix $H_{\underline{a}\ (p,\bar{p})}$, where $n \leq p$.

Namely, determine the value n of rank for the matrix $H_{\underline{a}\ (p,\bar{p})}$ such that the ratio of the square root of eigenvalues for the covariance matrix becomes very small. The small ratio means the nearness of approximation degree.

2) We use the CLS method as follows:
① Let a matrix $A \in \boldsymbol{R}^{1 \times (n+1)}$ be $A = [\alpha_1, \alpha_2, \cdots, \alpha_n, -1]$.
② Choose the coefficients $\{\alpha_i : 1 \leq i \leq n\}$ such that $\sum_{j=1}^{n+1} \underline{S}_l^{j-1} \bar{I}_{\underline{a}} \cdot \underline{S}_l^{j-1} \bar{I}_{\underline{a}}$ takes a minimum value, where $\{\underline{S}_l^i \bar{I}_{\underline{a}} \in \boldsymbol{R}^{L \times 1} : 0 \leq i \leq n\}$ are given by the equation $[\bar{I}_{\underline{a}}, \underline{S}_l \bar{I}_{\underline{a}}, \cdots, \underline{S}_l^n \bar{I}_{\underline{a}}]^T := A^T [AA^T]^{-1} A H_{\underline{a}\ (n+1,L)}^T$ and $H_{\underline{a}\ (n+1,L)} := [\bar{I}_{\underline{a}}, \cdots, S_l^{n-1} \bar{I}_{\underline{a}}, S_l^n \bar{I}_{\underline{a}}]$. And $\cdot$ denotes the inner product of two vectors.
③ Let $F \in \boldsymbol{R}^{n \times n}$ be given as below. Let g_0 be $g_0 = \mathbf{e}_1$, where $\mathbf{e}_1 = [1, 0, \cdots, 0]^T \in \boldsymbol{R}^n$.
④ Let h_s be $h_s = [I_{\underline{a}}(1) - \bar{I}_{\underline{a}}(1), I_{\underline{a}}(1) - \bar{I}_{\underline{a}}(1), \cdots, I_{\underline{a}}(0^{n_1-2}|1) - \bar{I}_{\underline{a}}(0^{n_1-2}|1)]$.

$$F_0 = \begin{bmatrix} 0 & \cdots & 0 & \alpha_1 \\ 1 & \ddots & & \alpha_2 \\ \vdots & \ddots & 0 & \vdots \\ 0 & & 1 & \alpha_n \end{bmatrix}.$$

[proof]. By 1), the approximate part of the data can be excluded in the sense of the norm of Hankel matrix $H_{\underline{a}\ (p,\bar{p})}$. The matrix A in 2) corresponds to the matrix A in Proposition (2.14). Hence, if we determine the coefficients $\{\alpha_i : 1 \leq i \leq n\}$, we can obtain the approximation of the given linear system by using Proposition (2.14) in the sense of a linear combination.

Therefore, we obtain the approximate Hankel matrices $\hat{H}_{\underline{a}\ (n_1+1,\bar{p})}(n_1+1,0)$. Finally, we apply Proposition (3.14) to the $\hat{H}_{\underline{a}\ (n_1+1,\bar{p})}(n_1+1,0)$.

Example 3.16. Let a signal be the impulse response of the following 3-dimensional linear system: $\sigma = ((\boldsymbol{R}^3, F), g, h)$, where $F = \begin{bmatrix} 0 & 0 & 0.9 \\ 1 & 0 & 0.2 \\ 0 & 1 & -0.41 \end{bmatrix}$, $h = [10, 2, -5], g = [1, 0, 0]^T$. Then the approximate realization problem is solved as follows:

covariance matrix	eigenvalues			
	1	2	3	4
$H^T_{\underline{a}\ (2,50)} H_{\underline{a}\ (2,50)}$	3228	774		
$H^T_{\underline{a}\ (3,50)} H_{\underline{a}\ (3,50)}$	3653	2292	85.6	
$H^T_{\underline{a}\ (4,50)} H_{\underline{a}\ (4,50)}$	4352	3616	92	0
covariance matrix	square root of eigenvalues			
$H^T_{\underline{a}\ (3,50)} H_{\underline{a}\ (3,50)}$	60.4	47.9	9.3	
$H^T_{\underline{a}\ (4,50)} H_{\underline{a}\ (4,50)}$	65.9	60.1	9.59	0

1) Since the ratio $\frac{9.59}{65.9} = 0.15$ obtained by the square root of $H^T_{\underline{a}\ (4,50)} H_{\underline{a}\ (4,50)}$ is large, the approximate linear system obtained by the CLS method may not be a good model for the original system.

2) After determining the number n of dimensions which is 2, we execute the approximate realization algorithm.

In this connection, the approximate linear system obtained by the Constrained Least Square (CLS) method is a two-dimensional linear system

$$\sigma_1 = ((\boldsymbol{R}^2, F_1),\ g_1,\ h_1), \text{ where } F_1 = \begin{bmatrix} 0 & -0.96 \\ 1 & -1.31 \end{bmatrix}, \ h_1 = [8.1,\ -0.597],$$

$g_1 = [1,\ 0]^T$.

On the other hand, the linear system $\sigma_2 = ((\boldsymbol{R}^3,\ F_2),\ \boldsymbol{e}_1,\ h_2)$ obtained by the CLS method is expressed as follows:

$$F_2 = \begin{bmatrix} 0 & 0 & 0.9 \\ 1 & 0 & 0.2 \\ 0 & 1 & -0.41 \end{bmatrix}, \ h_2 = [10,\ 2,\ -5].$$

In this case, the system σ_2 completely represents the original system.

The following table is a comparison with the mean values of the square root for the sum of the square of the original signal, the obtained signal by CLS and the error to the original signal ratio in terms of the selection of the state space dimension. This table indicates that the 2-dimensional linear system reconstructs the original signal with a 43 % error to signal ratio, and the 3-dimensional linear system completely reconstructs the original system.

Just as we thought, the following table and Fig. 3.1 indicate that the 2-dimensional linear system is not a good approximation to the original 3-dimensional linear system.

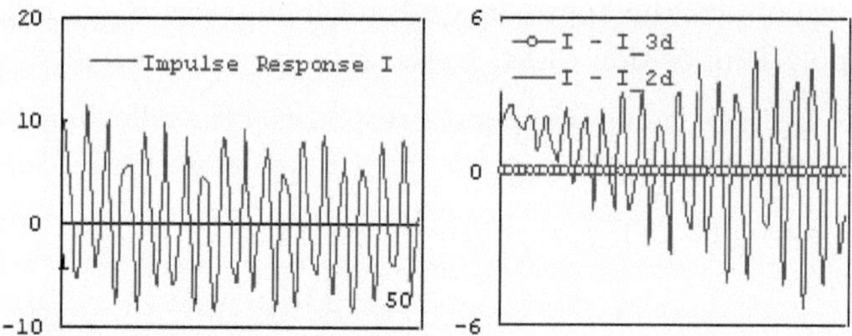

Fig. 3.1. The left is the original impulse response, the right is the difference between the original one and impulse responses approximated by a two or three-dimensional linear system in Example (3.16)

dimen-ion	ratio of matrices	mean values of square root for sum of		error	cosine ① and ②	error ratio
		signal ①	signal by CLS ②	③	$\cos\theta$	③/①
2	0.15	0.9	0.67	0.39	0.918	0.43
3	0	0.9	0.9	0	1	0

Example 3.17. Let a signal be the impulse response of the following 3-dimensional linear system: $\sigma = ((\mathbf{R}^3, F), g, h)$, where $F = \begin{bmatrix} 0 & 0 & -0.8 \\ 1 & 0 & 0.35 \\ 0 & 1 & 0.87 \end{bmatrix}$, $h = [9,\ 9.5,\ 9]$, $g = [1, 0, 0]^T$.

Then the approximate realization problem is solved as follows:

covariance matrix	eigenvalues			
	1	2	3	4
$H_{\underline{a}\ (2,70)}^T H_{\underline{a}\ (2,70)}$	5610	485		
$H_{\underline{a}\ (3,70)}^T H_{\underline{a}\ (3,70)}$	7272	1795	2.5	
$H_{\underline{a}\ (4,70)}^T H_{\underline{a}\ (4,70)}$	8025	4007	2.54	0
covariance matrix	square root of eigenvalues			
$H_{\underline{a}\ (3,70)}^T H_{\underline{a}\ (3,70)}$	85.2	42.3	1.6	
$H_{\underline{a}\ (4,70)}^T H_{\underline{a}\ (4,70)}$	89.6	63.3	1.6	0

1) Since the ratio $\frac{1.6}{89.6} = 0.02$ obtained by the square root of $H_{\underline{a}\ (4,70)}^T H_{\underline{a}\ (4,70)}$ is small, the approximate linear system obtained by the CLS method may be good.
2) After determining the number n of dimensions which is 2, we execute the approximate realization algorithm by the CLS method.

The approximate linear system obtained by the CLS method is a two-dimensional linear system $\sigma_1 = ((\mathbf{R}^2, F_1),\ g_1,\ h_1)$, where $F_1 = \begin{bmatrix} 0 & -0.99 \\ 1 & 1.68 \end{bmatrix}$, $h_1 = [8.58,\ 10.2]$, $g_1 = [1,\ 0]^T$.

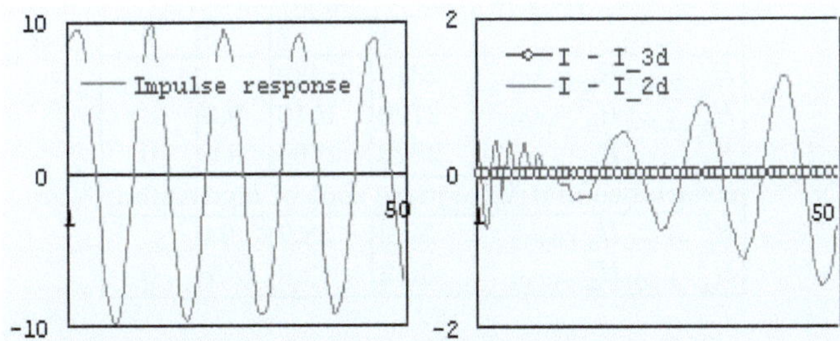

Fig. 3.2. The left is the original impulse response, the right is the difference between the original one and the impulse responses of the two or three-dimensional linear system obtained by the CLS method in Example (3.17)

For reference, a 3-dimensional linear system $\sigma_2 = ((\boldsymbol{R}^3,\ F_2),\ \mathbf{e}_1,\ h_2)$ obtained by the CLS method is expressed as follows:

$$F_2 = \begin{bmatrix} 0 & 0 & -0.8 \\ 1 & 0 & 0.35 \\ 0 & 1 & 0.87 \end{bmatrix}, \ h_2 = [9,\ 9.5,\ 9].$$

In this case, the system σ_2 completely represents the original system.

For reference, in the following table, we list the mean values of the sum of the square for the original, the obtained signal and the error to signal ratio.

This table indicates that the 2-dimensional linear system reconstructs the original signal with a 9 % error to signal ratio, and the 3-dimensional linear system completely reconstructs the original system.

Just as we expected, the following table and Fig. 3.2 indicate that the 2-dimensional linear system is a rather good approximation to the original 3-dimensional linear system.

dimen-ion	ratio of matrices	mean values of the square root for sum of			cosine	error ratio
		signal	signal by CLS	error	① and ②	③/①
		①	②	③	$\cos\theta$	
2	0.02	0.95	0.98	0.09	0.996	0.09
3	0	0.95	0.95	0	1	0

Example 3.18. Let a signal be the impulse response of the following 3-dimensional linear system: $\sigma = ((\boldsymbol{R}^3, F), g, h)$, where $F = \begin{bmatrix} 0 & 0 & 0.8 \\ 1 & 0 & 0.2 \\ 0 & 1 & -0.5 \end{bmatrix}$, $h = [10,\ 2,\ -9]$,

$g = [1, 0, 0]^T$.

Then the approximate realization problem is solved as follows:

covariance matrix	eigenvalues			
	1	2	3	4
$H_{\underline{a}}^T{}_{(2,50)}\,H_{\underline{a}}{}_{(2,50)}$	2375	483		
$H_{\underline{a}}^T{}_{(3,50)}\,H_{\underline{a}}{}_{(3,50)}$	2720	1512	9.3	
$H_{\underline{a}}^T{}_{(4,50)}\,H_{\underline{a}}{}_{(4,50)}$	2828	2705	9.6	0
covariance matrix	square root of eigenvalues			
$H_{\underline{a}}^T{}_{(3,50)}\,H_{\underline{a}}{}_{(3,50)}$	52.2	38.9	3.04	
$H_{\underline{a}}^T{}_{(4,50)}\,H_{\underline{a}}{}_{(4,50)}$	53.2	52.0	3.1	0

1) Since the ratio $\frac{3.1}{53.2} = 0.06$ obtained by the square root of $H_{\underline{a}}^T{}_{(4,50)}\,H_{\underline{a}}{}_{(4,50)}$ is small, the approximate linear system obtained by the CLS method may be good.

2) After determining the number n of dimensions which is 2, we execute the approximate realization algorithm by the CLS method.

The approximate linear system obtained by the CLS method is a two-dimensional linear system $\sigma_1 = ((\boldsymbol{R}^2, F_1),\ g_1,\ h_1)$, where $F_1 = \begin{bmatrix} 0 & -0.92 \\ 1 & -1.34 \end{bmatrix}$,

$h_1 = [9.28,\ 0.95]$, $g_1 = [1,\ 0]^T$.

For reference, a 3-dimensional linear system $\sigma_2 = ((\boldsymbol{R}^3,\ F_2),\ \mathbf{e}_1,\ h_2)$ obtained by the CLS method is expressed as follows:

$$F_2 = \begin{bmatrix} 0 & 0 & 0.8 \\ 1 & 0 & 0.2 \\ 0 & 1 & -0.5 \end{bmatrix},\ h_2 = [10,\ 2,\ -9].$$

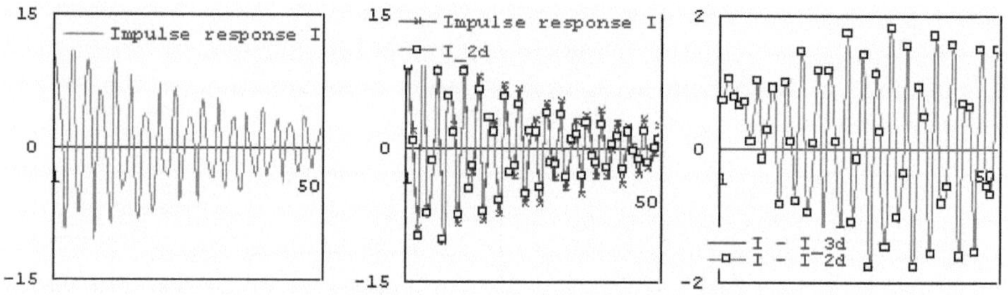

Fig. 3.3. The left is the original impulse response. The middle is the original and obtained one, and the right is the difference between the original one and impulse responses of two or three-dimensional linear systems obtained by the CLS method in Example (3.18)

For reference, we list the mean values of the sum of the square for the original, the obtained signal and the error to signal ratio in terms of the selection of the state space dimension.

This table indicates that the 2-dimensional linear system reconstructs the original signal with a 19 % error to signal ratio, and the 3-dimensional linear system completely reconstructs the original system.

Fig. 3.3 indicates that the 2-dimensional linear system is a fair approximation to the original 3-dimensional linear system except for each peak-value.

dimen-ion	ratio of matrices	mean values of square root for sum of			cosine ① and ②	error ratio
		signal ①	signal by CLS ②	error ③	$\cos\theta$	③/①
2	0.06	0.77	0.67	0.15	0.988	0.19
3	0	0.77	0.77	0	1	0

Example 3.19. Let a signal be the impulse response of the following 4-dimensional linear system: $\sigma = ((\boldsymbol{R}^4, F), g, h)$, where $F = \begin{bmatrix} 0 & 0 & 0 & 0.7 \\ 1 & 0 & 0 & 0.4 \\ 0 & 1 & 0 & -0.2 \\ 0 & 0 & 1 & 0.1 \end{bmatrix}$, $h = [9,\ 15,\ -5,\ 10]$,

$g = [1,\ 0,\ 0,\ 0]^T$.

Then the approximate realization problem is solved as follows:

covariance matrix	eigenvalues				
	1	2	3	4	5
$H_{\underline{a}\ (3,50)}^T H_{\underline{a}\ (3,50)}$	8994	2168	1074		
$H_{\underline{a}\ (4,50)}^T H_{\underline{a}\ (4,50)}$	11724	2306	1982	182	
$H_{\underline{a}\ (5,50)}^T H_{\underline{a}\ (5,50)}$	15062	2774	2172	183	0
covariance matrix	square root of eigenvalues				
$H_{\underline{a}\ (4,50)}^T H_{\underline{a}\ (4,50)}$	108	48	44.5	13.5	
$H_{\underline{a}\ (5,50)}^T H_{\underline{a}\ (5,50)}$	123	53	47	13.5	0

1) Since the ratio $\frac{13.5}{123} = 0.11$ obtained by the square root of $H_{\underline{a}\ (5,50)}^T H_{\underline{a}\ (5,50)}$ is not small, the approximate linear system obtained by the CLS method may not be good.

2) After determining the number n of dimensions which is 3, we execute the approximate realization algorithm by the CLS method.

The approximate linear system obtained by the CLS method is a 3-dimensional linear system $\sigma_1 = ((\boldsymbol{R}^3, F_1),\ g_1,\ h_1)$, where $F_1 = \begin{bmatrix} 0 & 0 & 0.93 \\ 1 & 0 & -0.65 \\ 0 & 1 & 0.74 \end{bmatrix}$,

$h_1 = [13.9,\ 11.5,\ -1.04]$, $g_1 = [1,\ 0,\ 0]^T$.

For reference, a 4-dimensional linear system $\sigma_2 = ((\boldsymbol{R}^4,\ F_2),\ \mathbf{e}_1,\ h_2)$ obtained by the CLS method is expressed as follows:

$F_2 = \begin{bmatrix} 0 & 0 & 0 & 0.7 \\ 1 & 0 & 0 & 0.4 \\ 0 & 1 & 0 & -0.2 \\ 0 & 0 & 1 & 0.1 \end{bmatrix}$, $h_2 = [9,\ 15,\ -5,\ 10]$.

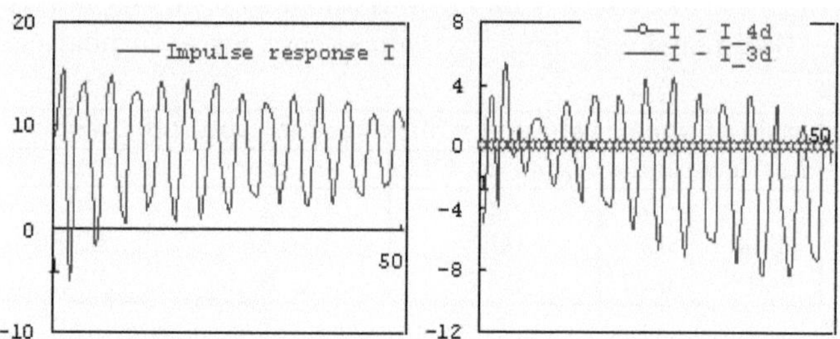

Fig. 3.4. The left is the original impulse response, the right is the difference between the original one and the impulse responses of 3 or 4-dimensional linear systems obtained by the CLS method in Example (3.19)

For reference, in the following table, we list the mean values of the sum of the square for the original signal, the obtained signal and the error to signal ratio.

This table indicates that the 3-dimensional linear system reconstructs the original signal with a 37 % error to signal ratio, and the 4-dimensional linear system completely reconstructs the original system.

Just as we thought, the following table and Fig. 3.4 indicate that the 3-dimensional linear system is not a good approximation to the original 4-dimensional linear system.

dimen-sion	ratio of matrices	mean values of square root for sum of			cosine ① and ②	error ratio
		signal	signal by CLS	error		
		①	②	③	$\cos\theta$	③/①
3	0.11	1.29	1.27	0.48	0.930	0.37
4	0	1.29	1.29	0	1	0

Example 3.20. Let a signal be the impulse response of the following 4-dimensional linear system: $\sigma = ((\boldsymbol{R}^4, F), g, h)$, where $F = \begin{bmatrix} 0 & 0 & 0 & 0.6 \\ 1 & 0 & 0 & 0.55 \\ 0 & 1 & 0 & -0.05 \\ 0 & 0 & 1 & 0.2 \end{bmatrix}$, $h = [9,\ 15,\ -5,\ 10]$, $g = [1,\ 0,\ 0,\ 0]^T$.

Then the approximate realization problem is solved as follows:

covariance matrix	eigenvalues				
	1	2	3	4	5
$H_{\underline{a}\ (3,50)}^T H_{\underline{a}\ (3,50)}$	3.4×10^6	753	530		
$H_{\underline{a}\ (4,50)}^T H_{\underline{a}\ (4,50)}$	5×10^6	804	675	97	
$H_{\underline{a}\ (5,50)}^T H_{\underline{a}\ (5,50)}$	6.9×10^6	896	749	98	0
covariance matrix	square root of eigenvalues				
$H_{\underline{a}\ (4,50)}^T H_{\underline{a}\ (4,50)}$	2236	28.4	26	9.8	
$H_{\underline{a}\ (5,50)}^T H_{\underline{a}\ (5,50)}$	2627	29.9	27.4	9.9	0

1) Since the ratio $\frac{9.9}{2627} = 0.003$ obtained by the square root of $H_{\underline{a}}^{T}{}_{(5,50)} H_{\underline{a}}{}_{(5,50)}$ is very small, the approximate linear system obtained by the CLS method may be good.

2) After determining the number n of dimensions which is 3, we execute the approximate realization algorithm by the CLS method.

The approximate linear system obtained by the CLS method is a 3-dimensional linear system $\sigma_1 = ((\boldsymbol{R}^3, F_1),\ g_1,\ h_1)$, where $F_1 = \begin{bmatrix} 0 & 0 & 0.82 \\ 1 & 0 & -0.24 \\ 0 & 1 & 0.61 \end{bmatrix}$,

$h_1 = [12.6,\ 14,\ -2.3]$, $g_1 = [1,\ 0,\ 0]^T$.

For reference, a 4-dimensional linear system $\sigma_2 = ((\boldsymbol{R}^4,\ F_2),\ \mathbf{e}_1,\ h_2)$ obtained by the CLS method is expressed as follows:

$$F_2 = \begin{bmatrix} 0 & 0 & 0 & 0.6 \\ 1 & 0 & 0 & 0.55 \\ 0 & 1 & 0 & -0.05 \\ 0 & 0 & 1 & 0.2 \end{bmatrix},\ h_2 = [9,\ 15,\ -5,\ 10].$$

This system completely recostructs the original system.

For reference, in the following, we list the mean values of the sum of the square for the original signal, the obtained signal and the error to signal ratio.

This table indicates that the 3-dimensional linear system reconstructs the original signal with a 12 % error to signal ratio, and the 4-dimensional linear system completely reconstructs the original system.

The following table and Fig. 3.5 indicate that the 3-dimensional linear system is not a good approximation to the original 4-dimensional linear system regardless of its very small ratio in cutting the number of dimensions. This result means that our expectations are disappointed. Why this occurs may be caused by the divergence of the impulse response.

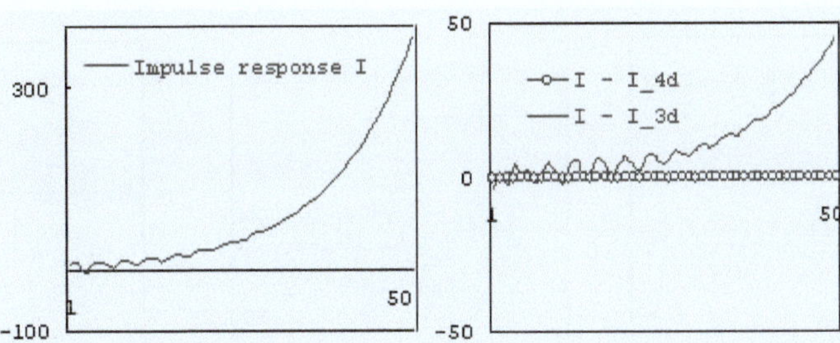

Fig. 3.5. The left is the original impulse response, the right is the difference between the original one and impulse responses of the 3 or 4-dimensional linear systems obtained by the CLS method in Example (3.20)

dimen-ion	ratio of matrices	mean values of square root for sum of			cosine ① and ②	error ratio
		signal ①	signal by CLS ②	error ③	$\cos\theta$	③/①
3	0.003	19.6	17.1	2.50	1	0.12
4	0	19.6	19.6	0	1	0

Example 3.21. Let a signal be the impulse response of the following 5-dimensional linear system: $\sigma = ((\boldsymbol{R}^5, F), g, h)$, where

$$F = \begin{bmatrix} 0 & 0 & 0 & 0 & 0 \\ 1 & 0 & 0 & 0 & -0.04 \\ 0 & 1 & 0 & 0 & -0.1 \\ 0 & 0 & 1 & 0 & 0.5 \\ 0 & 0 & 0 & 1 & -0.4 \end{bmatrix},$$

$h = [10, \ 2, \ -5, \ -1, \ 3]$, $g = [1, \ 0, \ 0, \ 0, \ 0]^T$.

Then the approximate realization problem is solved as follows:

covariance matrix	eigenvalues					
	1	2	3	4	5	6
$H^T_{\underline{a}\ (4,50)} H_{\underline{a}\ (4,50)}$	649	163	37	3		
$H^T_{\underline{a}\ (5,50)} H_{\underline{a}\ (5,50)}$	812	163	43	6.2	1.8	
$H^T_{\underline{a}\ (6,50)} H_{\underline{a}\ (6,50)}$	971	164	49	6.3	1.8	0
covariance matrix	square root of eigenvalues					
$H^T_{\underline{a}\ (5,50)} H_{\underline{a}\ (5,50)}$	28.5	12.8	6.6	2.5	1.3	
$H^T_{\underline{a}\ (6,50)} H_{\underline{a}\ (6,50)}$	31	12.8	7	2.5	1.3	0

1) Since the ratio $\frac{1.3}{31} = 0.04$ obtained by the square root of $H^T_{\underline{a}\ (6,50)} H_{\underline{a}\ (6,50)}$ is not so small, the approximate linear system obtained by the CLS method may be not so good.

2) After determining the number n of dimensions which is 4, we execute the approximate realization algorithm by the CLS method.

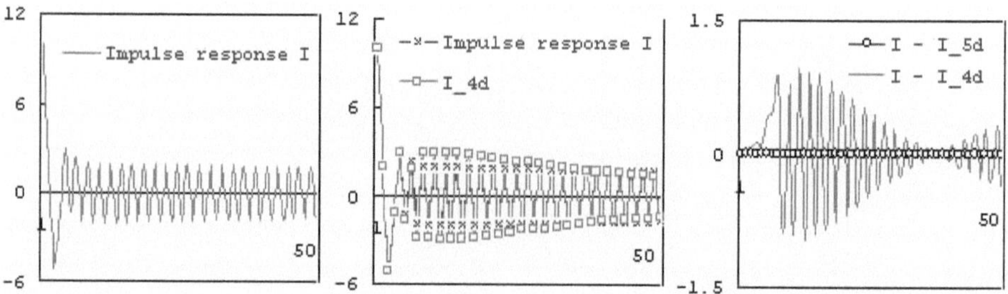

Fig. 3.6. The left is the original impulse response, the middle is the original one and the impulse responses of 4 -dimensional linear system obtained by the CLS method, and the right is the difference between the original one and impulse responses of the 4 or 5-dimensional linear systems obtained by the CLS method in Example (3.21)

The approximate linear system obtained by the CLS method is a 4-dimensional

linear system $\sigma_1 = ((\mathbf{R}^4, F_1),\ g_1,\ h_1)$, where $F_1 = \begin{bmatrix} 0 & 0 & 0 & -0.24 \\ 1 & 0 & 0 & -0.5 \\ 0 & 1 & 0 & -0.9 \\ 0 & 0 & 1 & -1.61 \end{bmatrix}$,

$h_1 = [10,\ 1.96,\ -5.07,\ -1.13],\ g_1 = [1,\ 0,\ 0,\ ,0]^T$.

For reference, a 5-dimensional linear system $\sigma_2 = ((\mathbf{R}^5,\ F_2),\ \mathbf{e}_1,\ h_2)$ obtained by the CLS method is expressed as follows:

$$F_2 = \begin{bmatrix} 0 & 0 & 0 & 0 & 0 \\ 1 & 0 & 0 & 0 & -0.04 \\ 0 & 1 & 0 & 0 & -0.1 \\ 0 & 0 & 1 & 0 & 0.5 \\ 0 & 0 & 0 & 1 & -0.4 \end{bmatrix},\ h_2 = [10,\ 2,\ -5,\ -1,\ 3].$$

This system completely reconstructs the original 5-dimensional linear system.

For reference, in the following table, we list the mean values of the sum of the square for the original signal, the obtained signal and the error to signal ratio.

This table indicates that the 4-dimensional linear system reconstructs the original signal with a 26 % error to signal ratio, and the 5-dimensional linear system completely reconstructs the original system.

Just as we thought, the following table and Fig. 3.6 indicate that the 4-dimensional linear system is not so good, but is a fair approximation to the original 5-dimensional linear system except for each peak-value.

dimen-sion	ratio of matrices	mean values of square root for sum of			cosine	error ratio
		signal	signal by CLS	error	① and ②	
		①	②	③	$\cos\theta$	③/①
4	0.04	0.34	0.4	0.09	0.99	0.23
5	0	0.34	0.34	0	1	0

Example 3.22. Let a signal be the impulse response of the following 6-dimensional

linear system: $\sigma = ((\mathbf{R}^6, F), g, h)$, where $F = \begin{bmatrix} 0 & 0 & 0 & 0 & 0 & 0 \\ 1 & 0 & 0 & 0 & 0 & -0.04 \\ 0 & 1 & 0 & 0 & 0 & -0.03 \\ 0 & 0 & 1 & 0 & 0 & 0.2 \\ 0 & 0 & 0 & 1 & 0 & 0.5 \\ 0 & 0 & 0 & 0 & 1 & -0.5 \end{bmatrix}$,

$h = [10,\ 2,\ -5,\ -1,\ 3,\ -2],\ g = [1,\ 0,\ 0,\ 0,\ 0,\ 0]^T$.

Then the approximate realization problem is solved as follows:

covariance matrix	eigenvalues						
	1	2	3	4	5	6	7
$H_{\underline{a}\ (4,50)}^T H_{\underline{a}\ (4,50)}$	175	87	34	2.8			
$H_{\underline{a}\ (5,50)}^T H_{\underline{a}\ (5,50)}$	179	92	48	5	0.9		
$H_{\underline{a}\ (6,50)}^T H_{\underline{a}\ (6,50)}$	183	99	57	4.9	0.93	0.002	
$H_{\underline{a}\ (7,50)}^T H_{\underline{a}\ (7,50)}$	188	104	61	5.5	0.99	0.002	0
covariance matrix	square root of eigenvalues						
$H_{\underline{a}\ (4,50)}^T H_{\underline{a}\ (4,50)}$	13.2	9.3	5.8	1.7			
$H_{\underline{a}\ (5,50)}^T H_{\underline{a}\ (5,50)}$	13.4	9.6	6.9	2.2	0.9		
$H_{\underline{a}\ (6,50)}^T H_{\underline{a}\ (6,50)}$	13.5	9.9	7.5	2.2	0.96	0.04	
$H_{\underline{a}\ (7,50)}^T H_{\underline{a}\ (7,50)}$	13.7	10.1	7.8	2.3	0.99	0.04	0

1) Since the ratio $\frac{1.7}{13.2} = 0.13$ obtained by the square root of $H_{\underline{a}\ (4,50)}^T H_{\underline{a}\ (4,50)}$ is not small, the approximate linear system obtained by the CLS method may not be good.

2) After determining the number n of dimensions which is 3, we execute the approximate realization algorithm by the CLS method.

The approximate linear system obtained by the CLS method is a 3-dimensional linear system $\sigma_1 = ((\boldsymbol{R}^3, F_1),\ g_1,\ h_1)$, where $F_1 = \begin{bmatrix} 0 & 0 & -0.46 \\ 1 & 0 & -0.64 \\ 0 & 1 & -1.18 \end{bmatrix}$,

$h_1 = [10.15,\ 2.21,\ -4.6]$, $g_1 = [1,\ 0,\ 0]^T$.

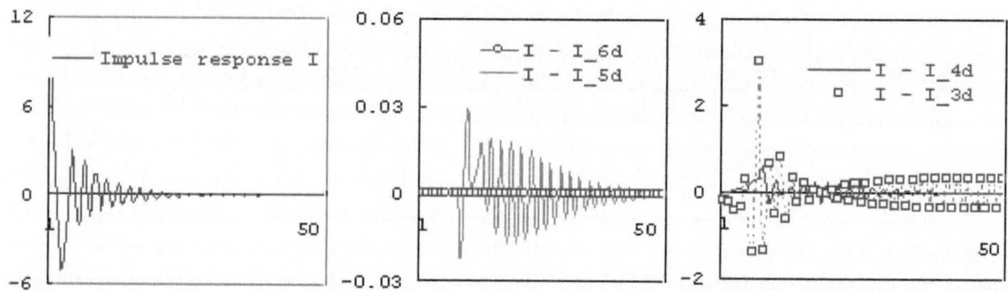

Fig. 3.7. The left is the original impulse response, the middle is the difference between the original one and impulse responses of the 5 or 6-dimensional linear systems by obtained by the CLS method, and the right is the difference between the original one and impulse responses of the 3 or 4-dimensional linear systems obtained by the CLS method in Example (3.22)

3) Since the ratio $\frac{0.9}{13.5} = 0.07$ obtained by the square root of $H_{\underline{a}\ (5,50)}^T H_{\underline{a}\ (5,50)}$ is not so small, the approximate linear system obtained by the CLS method may not be so good.

4) After determining the number n of dimensions which is 4, we execute the approximate realization algorithm by the CLS method.

The approximate linear system obtained by the CLS method is a 4-dimensional linear system $\sigma_2 = ((\boldsymbol{R}^4, F_2),\ g_2,\ h_2)$, where $F_2 = \begin{bmatrix} 0 & 0 & 0 & -0.1 \\ 1 & 0 & 0 & -0.3 \\ 0 & 1 & 0 & -0.6 \\ 0 & 0 & 1 & -1.4 \end{bmatrix}$,

$h_2 = [10,\ 1.99,\ -5.03,\ -1.06]$, $g_2 = [1,\ 0,\ 0,\ ,0]^T$.

5) Since the ratio $\frac{0.04}{13.5} = 0.003$ obtained by the square root of $H_{\underline{a}\ (6,50)}^T H_{\underline{a}\ (6,50)}$ is small, the approximate linear system obtained by the CLS method may be good.

6) After determining the number n of dimensions which is 5, we execute the approximate realization algorithm by the CLS method.

The linear system $\sigma_3 = ((\boldsymbol{R}^5,\ F_3),\ \mathbf{e}_1,\ h_3)$ obtained by the CLS method is expressed as follows:

$$F_3 = \begin{bmatrix} 0 & 0 & 0 & 0 & -0.04 \\ 1 & 0 & 0 & 0 & -0.05 \\ 0 & 1 & 0 & 0 & 0.03 \\ 0 & 0 & 1 & 0 & 0.5 \\ 0 & 0 & 0 & 1 & -0.3 \end{bmatrix},\ h_3 = [10,\ 2,\ -5,\ -1,\ 3].$$

For reference, a 6-dimensional linear system $\sigma_4 = ((\boldsymbol{R}^6,\ F_4),\ \mathbf{e}_1,\ h_4)$ obtained by the CLS method is expressed as follows:

$$F_4 = \begin{bmatrix} 0 & 0 & 0 & 0 & 0 & 0 \\ 1 & 0 & 0 & 0 & 0 & -0.04 \\ 0 & 1 & 0 & 0 & 0 & -0.03 \\ 0 & 0 & 1 & 0 & 0 & -0.2 \\ 0 & 0 & 0 & 1 & 0 & 0.5 \\ 0 & 0 & 0 & 0 & 1 & -0.5 \end{bmatrix},\ h_4 = [10,\ 2,\ -5,\ -1,\ 3,\ -2].$$

We note that this system σ_4 completely reconstructs the original 6-dimensional linear system σ.

For reference, in the following table, we list the mean values of the sum of the square for the original signal, the obtained signal and the error to signal ratio.

This table indicates that the 4-dimensional linear system reconstructs the original signal with an 8 % error to signal ratio, and the 5-dimensional linear system almost reconstructs the original system.

Fig. 3.7 and this table indicate that the 4-dimensional linear system is a very good approximation to the original 6-dimensional linear system, and the 5-dimensional linear system is a fair approximation to the original linear system.

dimen-ion	ratio of matrices	mean values of square root for sum of			cosine ① and ②	error ratio
		signal ①	signal by CLS ②	error ③	$\cos\theta$	③/①
3	0.13	0.25	0.24	0.085	0.940	0.34
4	0.07	0.25	0.25	0.02	0.997	0.08
5	0.003	0.25	0.25	0.001	0.99999	0.004
6	0	0.25	0.25	0	1	0

3.5 Noisy Realization of Linear Systems

In this section, we discuss the noisy realization problem of linear systems. Firstly, we must refer to the information criterion method AIC in noisy case of linear systems which is more commonly used.

We will compare our algorithm by the CLS method with the AIC method.

In order to make our discussion simple, we assume that the set Y of outout is the set $\boldsymbol{R}$ of real numbers, namely 1-output.

AIC criterion 3.23

The information criterion for linear systems is given by the following equation;

AIC $=$ $(-2)\log(\text{maximum likelihood})$ $+2\times$ (number of unknown parameters).

The $AIC(\underline{N},\ n)$ of n-dimensional linear systems with the data number $\underline{N}$ is concretely expressed by $AIC(\underline{N},\ n) = \underline{N}\log((1/\underline{N}) * (\sum_{i=1}^{N}(da(i) - \hat{da}(i))^2)) + 2*2*n$, where $\{da(i) : i \le \underline{N}\}$ are noisy original data obtained by experiments and $\{\hat{da}(i) : i \le \underline{N}\}$ are cleaned-up signals.

A situation for noisy realization problem 3.24

Let the observed object be a linear system and noise be added to output. Then we will obtain the data $\{\gamma(t) = \hat{\gamma}(t) + \bar{\gamma}(t) : 0 \le t \le \underline{N}\}$ for some integer $\underline{N} \in N$, where $\hat{\gamma}(t)$ is the exact signal which come from the observed linear system and $\bar{\gamma}(t)$ is the noise added at observation.

For noise $\{\bar{\gamma}(t) : t \in N\}$ added to the unknown linear system a, we will obtain the observed data $\{\hat{\gamma}(|\omega|) + \bar{\gamma}(|\omega|) : \omega \in U^*\}$. For given $\{\hat{\gamma}(|\omega|) + \bar{\gamma}(|\omega|) : \omega \in U^*\}$, σ which satisfies $\{a_\sigma(\omega) \approx \hat{\gamma}(|\omega|) : \omega \in U^*\}$ is called a noisy realization of a.

We can propose the following noisy realization problem:

For given $\{\hat{\gamma}(|\omega|) + \bar{\gamma}(|\omega|) : \omega \in U^*\}$, find a linear system σ which satisfies $a_\sigma(\omega) \approx \hat{\gamma}(|\omega|)$ for any $\omega \in U^*$.

Problem statement of noisy realization for linear systems 3.25

Let $H_{\underline{a}\ (p,\bar{p})}$ be the measured finite-sized Hankel matrix. Then find the cleaned-up signal Hankel matrix $\hat{H}_{\underline{a}\ (p,\bar{p})}$ such that $H_{\underline{a}\ (p,\bar{p})} = \hat{H}_{\underline{a}\ (p,\bar{p})} + \bar{H}_{\underline{a}\ (p,\bar{p})}$ holds.

Namely, find a minimal dimensional linear system $\sigma = ((\boldsymbol{R}^n, F), g, h))$ which realizes $\hat{H}_{\underline{a}\ (p,\bar{p})}$.

Theorem 3.26. *Algorithm for noisy realization*
Let $\underline{a}$ be a considered object which is a linear system. Then a noisy realization $\sigma = ((\boldsymbol{R}^n, F_s), g, h_s)$ of $\underline{a}$ is given by the following algorithm:

1) Based on the square root of eigenvalues for a matrix $H_{\underline{a}\ (p,\bar{p})} H^T_{\underline{a}\ (p,\bar{p})}$, determine the value n of rank for the matrix $H_{\underline{a}\ (p,\bar{p})}$, where $n \le p$. Namely, determine the value n of rank for the matrix $H_{\underline{a}\ (p,\bar{p})}$ such that a set of the square roots of eigenvalues for the covariance matrix composed of relatively small and equally-sized numbers is excluded, where the signal part effected by the set may be the noisy part.

2) We use the CLS method as follows:
① Let a matrix $A \in \mathbf{R}^{1 \times (n+1)}$ be $A = [\alpha_1, \alpha_2, \cdots, \alpha_n, -1]$.
② Choose the coefficients $\{\alpha_i : 1 \leq i \leq n\}$ such that
$\sum_{j=1}^{n+1} \underline{S}_l^{j-1} \bar{I}_{\underline{a}} \cdot \underline{S}_l^{j-1} \bar{I}_{\underline{a}}$ takes a minimum value, where $\{\underline{S}_l^i \bar{I}_{\underline{a}} \in \mathbf{R}^{L \times 1}$:
$0 \leq i \leq n\}$ is given by the equation $[\bar{I}_{\underline{a}}, \underline{S}_l \bar{I}_{\underline{a}}, \cdots, \underline{S}_l^{n_1} \bar{I}_{\underline{a}}]^T :=$
$A^T [AA^T]^{-1} A H^T_{\underline{a}\ (n_1, L)}$ and $H_{\underline{a}\ (n, L)} :=$
$[I_{\underline{a}}, \cdots, S_l^{n-1} I_{\underline{a}}, S_l^n I_{\underline{a}}]$. And $\cdot$ denotes the inner product of two vectors.
③ Let $F \in \mathbf{R}^{n \times n}$ be given as below. Let g be $g = \mathbf{e}_1$,
where $\mathbf{e}_1 = [1, 0, \cdots, 0]^T \in \mathbf{R}^n$.
④ Let h_s be $h_s = [I_{\underline{a}}(1) - \bar{I}_{\underline{a}}(1), I_{\underline{a}}(1) - \bar{I}_{\underline{a}}(1), \cdots, I_{\underline{a}}(0^{n-2}|1) - \bar{I}_{\underline{a}}(0^{n-2}|1)]$.

$$
F_0 = \begin{bmatrix} 0 & \cdots & 0 & \alpha_1 \\ 1 & \ddots & & \alpha_2 \\ \vdots & \ddots & 0 & \vdots \\ 0 & & 1 & \alpha_n \end{bmatrix}.
$$

[proof]. By 1), the noisy part in the data can be excluded in the sense of the number of dimensions. The matrix A in 2) corresponds to the matrix A in Proposition (2.5). Hence, if we determine the coefficients $\{\alpha_i : 1 \leq i \leq n\}$, we can obtain the noise part of the finite Hankel matrices $H_{\underline{a}\ (n+1, \bar{p})}$ by using Proposition (2.5).

Therefore, we obtain the cleaned-up Hankel matrices $\hat{H}_{\underline{a}\ (n, \bar{p})}$. Finally, we apply Proposition (3.15) to the $\hat{H}_{\underline{a}\ (n+1, \bar{p})}$.

Remark 1: A determination method of the degree n in the linear system $\sigma = ((\mathbf{R}^n, F_s), g, h_s)$ is found in the Principal Component Method. The method is very popular.

Remark 2: Let S and N be the norm of a signal and a noise. Then the selected ratio of matrices in the algorithm may be considered as $\frac{N}{S+N}$.

Remark 3: This noisy realization method is very new.

Remark 4: For a noisy case, the AIC method is famous for determining linear systems including dimensions of the state spaces.

Definition 3.27. *The algorithm for noisy realization (3.19) is called a Constrained Least Square method, abbreaviated, the CLS method.*

We show examples for the CLS method. In addition, we compare the method with the common method for noise processing which is called AIC.

Example 3.28. Let a signal be the impulse response of the following 1-dimensional linear system:
$\sigma = ((\mathbf{R}, -0.8), 1, 10)$. Let added noise be given in Fig. 3.8.

covariance matrix	eigenvalues			
	1	2	3	4
$H_{\underline{a}}^T{}_{(2,50)} H_{\underline{a}}{}_{(2,50)}$	487.3	19.51		
$H_{\underline{a}}^T{}_{(3,50)} H_{\underline{a}}{}_{(3,50)}$	620.8	21.98	14.22	
$H_{\underline{a}}^T{}_{(4,50)} H_{\underline{a}}{}_{(4,50)}$	706.8	24.82	15.32	13.58
covariance matrix	square root of eigenvalues			
$H_{\underline{a}}^T{}_{(2,50)} H_{\underline{a}}{}_{(2,50)}$	22	4.4		
$H_{\underline{a}}^T{}_{(3,50)} H_{\underline{a}}{}_{(3,50)}$	25	4.7	3.8	
$H_{\underline{a}}^T{}_{(4,50)} H_{\underline{a}}{}_{(4,50)}$	26.6	5	3.9	3.7

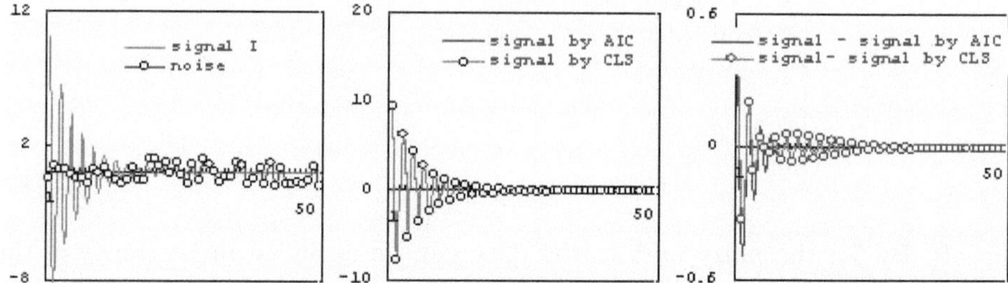

Fig. 3.8. The left is the exact signal of a 1-dimensional linear system and a noise, the middle is the signals obtained by the CLS and AIC methods, and the right is the difference between the original one and signal obtained by CLS or AIC in Example (3.28)

Then the noisy realization problem is solved by the following algorithm:

1) Since a set $\{5, 3.9, 3.7\}$ is composed of relatively small and equally-sized numbers in the square root of $H_{\underline{a}}^T{}_{(4,50)} H_{\underline{a}}{}_{(4,50)}$, the noisy realization of linear system obtained by the CLS method may be good for 1-dimensional space.
2) After determining the number n of dimensions which is 1, we will continue the noisy realization algorithm by the CLS method.

Therefore, the modified impulse response $I(0)$ of a linear system obtained by the CLS method is obtained by a 1-dimensional linear system.

Therefore, the linear system obtained by the CLS method is a 1-dimensional linear system. By the algorithm, the system is given by $\sigma =$ $((\boldsymbol{R}, -0.81839), 1, 9.49675)$.

As a common typical method for signal processing of noisy data, there is the AIC method.

The 2-dimensional linear system $\sigma_a = ((\boldsymbol{R}^2, F_a), \mathbf{e}_1, h_a)$, obtained by AIC method is expressed as follows:

$$F_a = \begin{bmatrix} 0 & 0.42 \\ 1 & -0.28 \end{bmatrix}, \ h_a = [9.63, -7.61].$$

In this example, the original signal is considered as the impulse response of a 1-dimensional linear system and the desirable impulse response is obtained by the two methods, that is, the CLS and AIC methods. For reference, in the following table, we list the mean values of the sum of the square for the original signal, the obtained signal and the error to signal ratio.

This table indicates that the 1-dimensional linear system reconstructs the original signal with a 4 % error to signal ratio and 0.2 noise to signal ratio, please refer to Remark 2 in Theorem 3.26 for the noise to signal ratio.

The model obtained by the CLS method is a 1-dimensional linear system which has the same number of dimensions as the number of the original system. The model obtained by the AIC method is a 2-dimensional linear system.

Nevertheless, Fig. 3.8 indicates that the 2-dimensional linear system obtained by AIC is the model with the same error as in the CLS method.

dimen-sion	ratio of matrices	mean values of square root for sum of			cosine ① and ②	error ratio
		signal	signal by CLS	error		
		①	②	③	$\cos\theta$	③/①
1	0.2	0.34	0.33	0.01	0.999	0.04

Example 3.29. Let a signal be the impulse response of the following 2-dimensional linear system:

$$\sigma = ((\boldsymbol{R}^2, F), \mathbf{e}_1,\ h),$$

where $F = \begin{bmatrix} 0 & -0.08 \\ 1 & -0.81 \end{bmatrix}$, $h = [10, 2]$.

Let an added noise be given in Fig. 3.9.

Then the noisy realization problem is solved as follows:

covariance matrix	eigenvalues				
	1	2	3	4	5
$H_{\underline{a}}^T{}_{(3,50)} H_{\underline{a}\,(3,50)}$	120.9	31.04	6.768		
$H_{\underline{a}}^T{}_{(4,50)} H_{\underline{a}\,(4,50)}$	122.3	36.61	7.075	5.818	
$H_{\underline{a}}^T{}_{(5,50)} H_{\underline{a}\,(5,50)}$	122.8	39.51	8.245	6.210	5.308
covariance matrix	square root of eigenvalues				
$H_{\underline{a}}^T{}_{(3,50)} H_{\underline{a}\,(3,50)}$	11	5.6	2.6		
$H_{\underline{a}}^T{}_{(4,50)} H_{\underline{a}\,(4,50)}$	11	6	2.7	2.4	
$H_{\underline{a}}^T{}_{(5,50)} H_{\underline{a}\,(5,50)}$	11	6.3	2.8	2.5	2.3

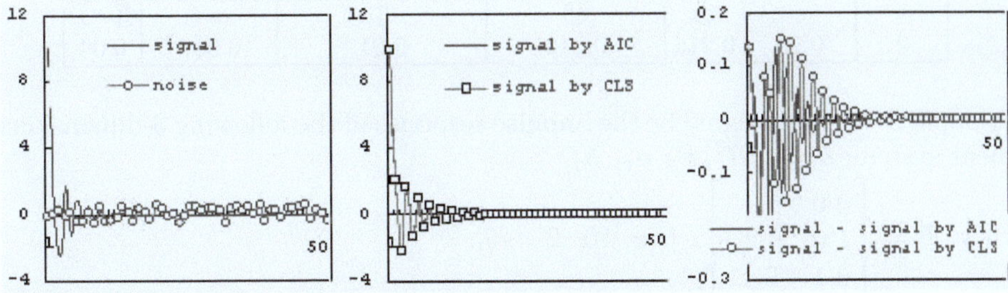

Fig. 3.9. The left is the exact signal of 2-dimensional linear system and a noise, the middle is signals obtained by CLS and AIC method, and the right is the difference between exact signal and signal obtained by the CLS or AIC method in Example (3.29)

1) Since a set $\{2.8,\ 2.5,\ 2.3\}$ is composed of relatively small and equally-sized numbers in the square root of $H_{\underline{a}\ (5,50)}^T H_{\underline{a}\ (5,50)}$, the noisy realization of a linear system obtained by the CLS method may be good for a 2-dimensional space.
2) After determining the number n of dimensions which is 2, we will continue the noisy realization algorithm by the CLS method.

Therefore, the modified impulse response $I(0)$ of a linear system obtained by the CLS method is realized by a 2-dimensional linear system.

Therefore, the linear system obtained by the CLS method is a 2-dimensional linear system. The system is given by $\sigma_c = ((\boldsymbol{R}^2,\ F_c),\ \mathbf{e}_1,\ h_c)$, where $F_c = \begin{bmatrix} 0 & -0.053 \\ 1 & -0.83 \end{bmatrix}$, $h_c = [9.86, 2]$.

Moreover, the 4-dimensional linear system $\sigma_a = ((\boldsymbol{R}^4,\ F_a),\ \mathbf{e}_1,\ h_a)$, obtained by the AIC method is expressed as follows:

$$F_a = \begin{bmatrix} 0 & 0 & 0 & -0.0037 \\ 1 & 0 & 0 & -0.046 \\ 0 & 1 & 0 & 0.31 \\ 0 & 0 & 1 & -0.12 \end{bmatrix},\quad h_a = [9.86,\ 1.98,\ -2.23,\ 1.7].$$

In this example, the original signal is considered as the impulse response of a 2-dimensional linear system and the desirable impulse response is obtained by the two methods, that is, the CLS and AIC methods.

For reference, in the following table, we list the mean values of the sum of the square for the original signal, the obtained signal and the error to signal ratio.

This table indicates that a 2-dimensional linear system reconstructs the original signal with a 4 % error to signal ratio and 0.23 noise to signal ratio, please refer to Remark 2 in Theorem 3.26 for the noise to signal ratio.

The model obtained by the CLS method is the 2-dimensional linear system which has the same number of dimensions as the number of the original system. The model obtained by the AIC method is the 4-dimensional linear system.

Nevertheless, Fig. 3.9 indicates that the 4-dimensional linear system obtained by AIC is the system with the same error as in the CLS method.

dimen-ion	ratio of matrices	mean values of square root for sum of		cosine ① and ②	error ratio
		signal	signal by CLS	error	
		①	②	③	$\cos\theta$	③/①
2	0.23	0.215	0.212	0.01	0.999	0.04

Example 3.30. Let a signal be the impulse response of the following 3-dimensional linear system: $\sigma = ((\boldsymbol{R}^3, F), \mathbf{e}_1,\ h)$,

where $F = \begin{bmatrix} 0 & 0 & 0.9 \\ 1 & 0 & 0.2 \\ 0 & 1 & -0.41 \end{bmatrix}$, $h = [10,\ 2,\ -5]$.

Let an added noise be given in Fig. 3.10.

Then the noisy realization problem is solved as follows:

covariance matrix	eigenvalues					
	1	2	3	4	5	6
$H_{\underline{a}\ (4,50)}^{T} H_{\underline{a}\ (4,50)}$	4263	3565	106.7	10.79		
$H_{\underline{a}\ (5,50)}^{T} H_{\underline{a}\ (5,50)}$	5870	3784	133.3	12.42	9.710	
$H_{\underline{a}\ (6,50)}^{T} H_{\underline{a}\ (6,50)}$	6787	4729	151.9	12.51	11.46	7.537
covariance matrix	square root of eigenvalues					
$H_{\underline{a}\ (4,50)}^{T} H_{\underline{a}\ (4,50)}$	65.3	60	10.3	3.3		
$H_{\underline{a}\ (5,50)}^{T} H_{\underline{a}\ (5,50)}$	76.6	61.5	11.5	3.5	3.1	
$H_{\underline{a}\ (6,50)}^{T} H_{\underline{a}\ (6,50)}$	82.4	68.8	12.3	3.5	3.4	2.7

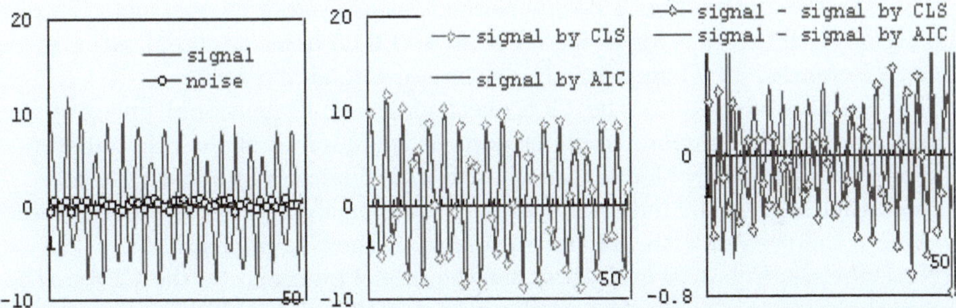

Fig. 3.10. The left is the exact signal of a 3-dimensional linear system and noise, the middle is signals by the CLS and AIC methods, and the right is the difference between the original signal and the signal obtained by the CLS or AIC methods in Example (3.30)

1) Since a set {3.5, 3.4, 2.7} is composed of relatively small and equally-sized numbers in the square root of $H_{\underline{a}\ (6,50)}^{T} H_{\underline{a}\ (6,50)}$, the noisy realization of a linear system obtained by the CLS method may be good for 3-dimensional space.
2) After determining the number n of dimensions which is 3, we will continue the noisy realization algorithm by the CLS method.

Therefore, the modified impulse response $I(0)$ of a linear system obtained by the CLS method is realized by a 3-dimensional linear system.

Therefore, the linear system obtained by the CLS method is a 3-dimensional linear system and it is expressed as follows:

$$\sigma_c = ((\boldsymbol{R}^3,\ F_c),\ \mathbf{e}_1,\ h_c),\ \text{where } F_c = \begin{bmatrix} 0 & 0 & 0.92 \\ 1 & 0 & 0.22 \\ 0 & 1 & -0.4 \end{bmatrix},\ h_c = [9.7,\ 2.46,\ -5.36].$$

Moreover, the linear system obtained by the AIC method is a 7-dimensional linear system $\sigma_a = ((\boldsymbol{R}^7,\ F_a),\ \mathbf{e}_1,\ h_a)$ given as follows:

$$F_a = \begin{bmatrix} 0\ 0\ 0\ 0\ 0\ 0 & -0.33 \\ 1\ 0\ 0\ 0\ 0\ 0 & 0.29 \\ 0\ 1\ 0\ 0\ 0\ 0 & 0.5 \\ 0\ 0\ 1\ 0\ 0\ 0 & 0.02 \\ 0\ 0\ 0\ 1\ 0\ 0 & 0.32 \\ 0\ 0\ 0\ 0\ 1\ 0 & -0.008 \\ 0\ 0\ 0\ 0\ 0\ 1 & -0.15 \end{bmatrix}, \quad h_a = [9.31,\ 2.37,\ -5.19,\ 12,\ -4.6,\ -0.04,\ 9.55].$$

In this example, the original signal is considered as the impulse response of a 3-dimensional linear system and the desirable impulse response is obtained by the two methods, that is, the CLS and AIC methods.

For reference, in the following table, we list the mean values of the sum of the square for the original signal, the obtained signal and the error to signal ratio.

This table indicates that a 3-dimensional linear system reconstructs the original signal with a 5 % error to signal ratio and 0.05 noise to signal ratio, please refer to Remark 2 in Theorem 3.26 for the noise to signal ratio.

The model obtained by the CLS method is a 3-dimensional linear system which has the same number of dimensions as the number of the original system. The model obtained by the AIC method is a 7-dimensional linear system.

Nevertheless, Fig. 3.9 indicates that the 7-dimensional linear system obtained by AIC is the system with the same error as in the CLS method.

Nevertheless, Fig. 3.10 indicates that the model obtained by the CLS method causes the same degree of error as the model obtained by AIC.

dimen- sion	ratio of matrices	mean values of square root for sum of			cosine	error ratio
		signal	signal by CLS	error	① and ②	
		①	②	③	$\cos\theta$	③/①
3	0.05	0.902	0.928	0.05	0.999	0.05

Example 3.31. Let a signal be the impulse response of the following 3-dimensional linear system: $\sigma = ((\boldsymbol{R}^3, F), \mathbf{e}_1,\ h)$,

$$\text{where } F = \begin{bmatrix} 0\ 0\ -0.8 \\ 1\ 0\ 0.35 \\ 0\ 1\ 0.87 \end{bmatrix}, \quad h = [9,\ 9.5,\ 5.9].$$

Let an added noise be given in Fig. 3.11.

Then the noisy realization problem is solved by the following algorithm:

covariance matrix	eigenvalues					
	1	2	3	4	5	6
$H_{\underline{a}\ (4,50)}^{T} H_{\underline{a}\ (4,50)}$	6127	2917	3.041	0.53		
$H_{\underline{a}\ (5,50)}^{T} H_{\underline{a}\ (5,50)}$	6200	5031	4.053	0.54	0.42	
$H_{\underline{a}\ (6,50)}^{T} H_{\underline{a}\ (6,50)}$	7402	6043	4.11	0.76	0.44	0.39
covariance matrix	square root of eigenvalues					
$H_{\underline{a}\ (4,50)}^{T} H_{\underline{a}\ (4,50)}$	78.3	54	1.7	0.7		
$H_{\underline{a}\ (5,50)}^{T} H_{\underline{a}\ (5,50)}$	78.7	71	2.01	0.73	0.64	
$H_{\underline{a}\ (6,50)}^{T} H_{\underline{a}\ (6,50)}$	86	77.7	2.02	0.87	0.66	0.62

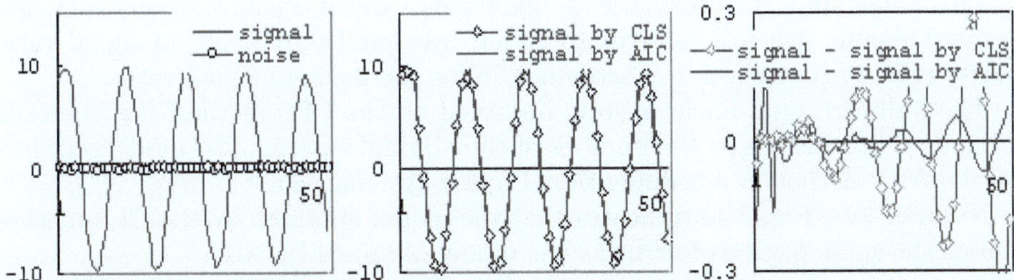

Fig. 3.11. The left is the signal of a 3-dimensional linear system and noise, the middle are signals of a 3-dimensional linear system obtained by the CLS method and a 5-dimensional linear system obtained by the AIC method, and the right is the difference between the exact signal and signal obtained by CLS or AIC in Example (3.31)

1) Since a set $\{0.87,\ 0.66,\ 0.62\}$ is composed of relatively small and equally-sized numbers in the square root of $H_{\underline{a}\ (6,50)}^{T}\, H_{\underline{a}\ (6,50)}$, the noisy realization of a linear system obtained by the CLS method may be good for 3-dimensional space.
2) After determining the number n of dimensions which is 3, we will continue the noisy realization algorithm by the CLS method.

Therefore, the modified impulse response $I(0)$ of a linear system obtained by the CLS method is realized by a 3-dimensional linear system.

Therefore, the linear system obtained by the CLS method is a 3-dimensional linear system and the system is given by $\sigma_c = ((\boldsymbol{R}^3,\ F_c),\ \mathbf{e}_1,\ h_c)$, where

$$F_c = \begin{bmatrix} 0 & 0 & -0.79 \\ 1 & 0 & 0.33 \\ 0 & 1 & 0.88 \end{bmatrix},\ h_c = [9.1,\ 9.5,\ 9].$$

Moreover, the linear system obtained by the AIC method is a 5-dimensional linear system $\sigma_a = ((\boldsymbol{R}^5,\ F_a),\ \mathbf{e}_1,\ h_a)$ given as follows:

$$F_a = \begin{bmatrix} 0 & 0 & 0 & 0 & -0.3 \\ 1 & 0 & 0 & 0 & -0.29 \\ 0 & 1 & 0 & 0 & -0.32 \\ 0 & 0 & 1 & 0 & 0.48 \\ 0 & 0 & 0 & 1 & 0.31 \end{bmatrix},\ h_a = [9.2,\ 9.4,\ 8.8,\ 4.1,\ -1].$$

In this example, the original signal is considered as the impulse response of a 3-dimensional linear system and the desirable impulse response is obtained by the two methods, that is, the CLS and AIC methods. The model obtained by the CLS method is a 3-dimensional linear system which has the same number of dimensions as the number of the original system.

For reference, in the following table, we list the mean values of the sum of the square for the original signal, the obtained signal and the error to signal ratio.

This table indicates that the 3-dimensional linear system reconstructs the original signal with a 2 % error to signal ratio and 0.01 noise to signal ratio, please refer to Remark 2 in Theorem 3.26 for the noise to signal ratio.

The 3-dimensional linear system obtained by the CLS method has the same number of dimensions as the number of the original system. The model obtained by the AIC method is a 5-dimensional linear system.

Nevertheless, Fig. 3.11 indicates that the model obtained by the CLS method causes the same degree of error as the model obtained by AIC.

dimen-ion	ratio of matrices	mean values of square root for sum of			cosine ① and ②	error ratio
		signal	signal by CLS	error		
		①	②	③	$\cos\theta$	③/①
3	0.01	0.958	0.96	0.02	0.999	0.02

Example 3.32. Let a signal be the impulse response of the following 3-dimensional linear system: $\sigma = ((\boldsymbol{R}^3, F), \mathbf{e}_1, \ h)$, where $F = \begin{bmatrix} 0 & 0 & -0.8 \\ 1 & 0 & 0.35 \\ 0 & 1 & 0.87 \end{bmatrix}$, $h = [9, \ 9.5, \ 5.9]$.

Let an added noise be given in Fig. 3.12.

Then the noisy realization problem is solved as follows:

covariance matrix	eigenvalues					
	1	2	3	4	5	6
$H_{\underline{a}\,(4,50)}^T H_{\underline{a}\,(4,50)}$	6108	2912	3.70	0.71		
$H_{\underline{a}\,(5,50)}^T H_{\underline{a}\,(5,50)}$	6181	5020	4.53	0.72	0.71	
$H_{\underline{a}\,(6,50)}^T H_{\underline{a}\,(6,50)}$	7384	6029	4.56	0.76	0.72	0.70
covariance matrix	square root of eigenvalues					
$H_{\underline{a}\,(4,50)}^T H_{\underline{a}\,(4,50)}$	78.2	54	1.9	0.8		
$H_{\underline{a}\,(5,50)}^T H_{\underline{a}\,(5,50)}$	78.6	70.9	2.1	0.8	0.8	
$H_{\underline{a}\,(6,50)}^T H_{\underline{a}\,(6,50)}$	86	77.6	2.1	0.87	0.84	0.84

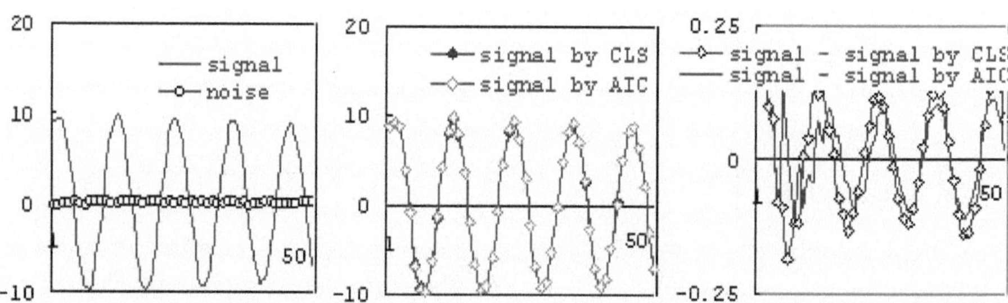

Fig. 3.12. The left is the original signal of a 3-dimensional linear system and noise, the middle are signals of a 3-dimensional linear system obtained by the CLS method and a 6-dimensional linear system obtained by the AIC method, and the right is the difference between the original signal and the signal obtained by the CLS or AIC methods in Example (3.32)

1) Since a set $\{0.87,\ 0.84,\ 0.84\}$ is composed of relatively small and equally-sized numbers in the square root of $H_{\underline{a}\ (6,50)}^{T} H_{\underline{a}\ (6,50)}$, the noisy realization of a linear system obtained by the CLS method may be good for a 3-dimensional space.
2) After determining the number n of dimensions which is 3, we will continue the noisy realization algorithm by the CLS method.

Therefore, the modified impulse response $I(0)$ of a linear system obtained by the CLS method is obtained by a 3-dimensional linear system.

Therefore, the linear system obtained by the CLS method is a 3-dimensional linear system. The system is given by $\sigma_c = ((\boldsymbol{R}^3,\ F_c),\ \mathbf{e}_1,\ h_c)$, where $F_c =$

$$\begin{bmatrix} 0 & 0 & -0.82 \\ 1 & 0 & 0.38 \\ 0 & 1 & 0.85 \end{bmatrix},\ h_c = [8.84,\ 9.3,\ 8.9].$$

Moreover, the linear system obtained by the AIC method is a 6-dimensional linear system $\sigma_a = ((\boldsymbol{R}^6,\ F_a),\ \mathbf{e}_1,\ h_a)$ given as follows:

$$F_a = \begin{bmatrix} 0 & 0 & 0 & 0 & 0 & -0.05 \\ 1 & 0 & 0 & 0 & 0 & -0.4 \\ 0 & 1 & 0 & 0 & 0 & -0.26 \\ 0 & 0 & 1 & 0 & 0 & -0.09 \\ 0 & 0 & 0 & 1 & 0 & 0.2 \\ 0 & 0 & 0 & 0 & 1 & 0.3 \end{bmatrix},\ h_a = [8.8,\ 9.3,\ 8.9,\ 3.9,\ -0.92,\ -6.8].$$

In this example, the original signal is considered as the impulse response of a 3-dimensional linear system and the desirable impulse response is obtained by the two methods, that is, the CLS and AIC methods.

For reference, in the following table, we list the mean values of the sum of the square for the original signal, the obtained signal and the error to signal ratio.

This table indicates that the 3-dimensional linear system obtained by the CLS method reconstructs the original signal with a 2 % error to signal ratio and 0.01 noise to signal ratio, please refer to Remark 2 in Theorem 3.26 for the noise to signal ratio.

The 3-dimensional linear system obtained by the CLS method has the same number of dimensions as the number of the original system.

The AIC method produces a 6-dimensional linear system. Nevertheless, Fig. 3.12 indicates that the model obtained by the CLS method causes the same degree of error as the model obtained by AIC.

dimen-ion	ratio of matrices	mean values of square root for sum of			cosine	error ratio
		signal	signal by CLS	error	① and ②	
		①	②	③	$\cos\theta$	③/①
3	0.01	0.958	0.948	0.01	0.999	0.02

Example 3.33. Let a signal be the impulse response of the following 4-dimensional linear system: $\sigma = ((\boldsymbol{R}^4, F), \mathbf{e}_1, h)$,

where $F = \begin{bmatrix} 0 & 0 & 0 & 0.7 \\ 1 & 0 & 0 & 0.4 \\ 0 & 1 & 0 & -0.2 \\ 0 & 0 & 1 & 0.1 \end{bmatrix}$, $h = [9,\ 15,\ -5,\ 10]$.

Let an added noise be given in Fig. 3.13.

Then the noisy realization problem is solved as follows:

covariance matrix	eigenvalues						
	1	2	3	4	5	6	7
$H^T_{\underline{a}\ (5,50)} H_{\underline{a}\ (5,50)}$	14860	2759	2163	187.6	9.91		
$H^T_{\underline{a}\ (6,50)} H_{\underline{a}\ (6,50)}$	17730	3360	2514	187.9	12.07	7.28	
$H^T_{\underline{a}\ (7,50)} H_{\underline{a}\ (7,50)}$	20260	4011	3098	187.9	12.57	10.41	5.28
covariance matrix	square root of eigenvalues						
$H^T_{\underline{a}\ (5,50)} H_{\underline{a}\ (5,50)}$	122	52.5	46.5	13.7	3.01		
$H^T_{\underline{a}\ (6,50)} H_{\underline{a}\ (6,50)}$	133	58	50.1	13.7	3.5	2.7	
$H^T_{\underline{a}\ (7,50)} H_{\underline{a}\ (7,50)}$	142	63	55.7	13.7	3.5	3.2	2.3

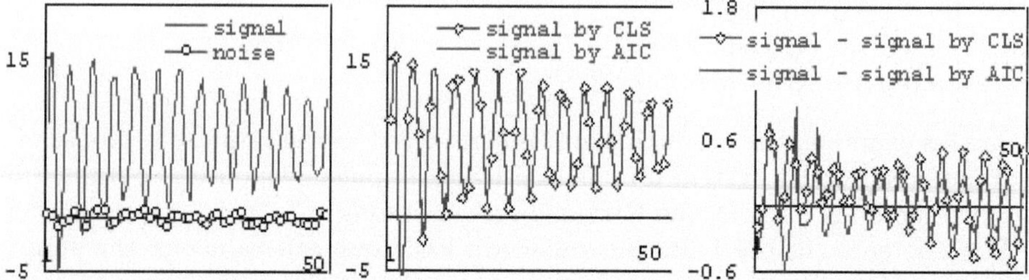

Fig. 3.13. The left is the exact signal of a 4-dimensional linear system and noise, the middle are the signals of a 4-dimensional linear system obtained by the CLS method and the signal of a 7-dimensional linear system obtained by the AIC method, and the right is the difference between the original signal and the siganl by the CLS and AIC methods in Example (3.33)

1) Since a set $\{3.5,\ 3.2,\ 2.3\}$ is composed of relatively small and equally-sized numbers in the square root of $H^T_{\underline{a}\ (7,50)} H_{\underline{a}\ (7,50)}$, the noisy realization of a linear system obtained by the CLS method may be good for a 4-dimensional space.
2) After determining the number n of dimensions which is 4, we will continue the noisy realization algorithm by the CLS method.

Therefore, the modified impulse response $I(0)$ of a linear system obtained by the CLS method is realized by a 4-dimensional linear system.

Therefore, the linear system obtained by the CLS method is a 4-dimensional linear system. The system is given by $\sigma_c = ((\boldsymbol{R}^4,\ F_c),\ \mathbf{e}_1,\ h_c)$, where $F_c = \begin{bmatrix} 0 & 0 & 0 & 0.7 \\ 1 & 0 & 0 & 0.4 \\ 0 & 1 & 0 & -0.19 \\ 0 & 0 & 1 & 0.1 \end{bmatrix}$, $h_c = [9.2,\ 15,\ -5.65,\ 9.5]$.

Moreover, the linear system obtained by the AIC method is a 7-dimensional linear system $\sigma_a = ((\boldsymbol{R}^7, \ F_a), \ \mathbf{e}_1, \ h_a)$ given as follows:

$$F_a = \begin{bmatrix} 0 & 0 & 0 & 0 & 0 & 0 & 0.19 \\ 1 & 0 & 0 & 0 & 0 & 0 & 0.09 \\ 0 & 1 & 0 & 0 & 0 & 0 & -0.23 \\ 0 & 0 & 1 & 0 & 0 & 0 & 0.5 \\ 0 & 0 & 0 & 1 & 0 & 0 & 0.3 \\ 0 & 0 & 0 & 0 & 1 & 0 & -0.3 \\ 0 & 0 & 0 & 0 & 0 & 1 & 0.48 \end{bmatrix} \ , \ h_a = [9.4 \ 15.2, \ -5.7, \ 9.5, \ 14.1, \ 8.5, \ -1.2].$$

In this example, the original signal is considered as the impulse response of a 4-dimensional linear system and the desirable impulse response is obtained by the two methods, that is, the CLS and AIC methods.

For reference, in the following table, we list the mean values of the sum of the square for the original signal, the obtained signal and the error to signal ratio.

This table indicates that the 4-dimensional linear system obtained by the CLS method reconstructs the original signal with a 4 % error to signal ratio and 0.02 noise to signal ratio, please refer to Remark 2 in Theorem 3.26 for the noise to signal ratio.

The 4-dimensional linear system obtained by the CLS method has the same number of dimensions as the number of the original system.

The AIC method produces a 7-dimensional linear system.

Nevertheless, Fig. 3.13 indicates that the model obtained by the CLS method causes the same degree of error as the model obtained by AIC.

dimen-	ratio of	mean values of square root for sum of			cosine	error
ion	matrices	signal	signal by CLS	error	① and ②	ratio
		①	②	③	$\cos\theta$	③/①
4	0.02	1.287	1.268	0.04	0.999	0.04

Example 3.34. Let a signal be the impulse response of the following 4-dimensional linear system: $\sigma = ((\boldsymbol{R}^4, F), \mathbf{e}_1, \ h)$,

$$\text{where } F = \begin{bmatrix} 0 & 0 & 0 & 0.6 \\ 1 & 0 & 0 & 0.55 \\ 0 & 1 & 0 & -0.05 \\ 0 & 0 & 1 & 0.2 \end{bmatrix} \ , \ h = [9, \ 15, \ -5, \ 10].$$

Let an added noise be given in Fig. 3.14.

Then the noisy realization problem is solved as follows:

covariance matrix	eigenvalues						
	1	2	3	4	5	6	7
$H_{\underline{a}\,(5,120)}^T H_{\underline{a}\,(5,120)}$	8.0×10^{11}	920	784	145	20.5		
$H_{\underline{a}\,(6,120)}^T H_{\underline{a}\,(6,120)}$	1.0×10^{12}	1082	839	148	21.6	19.7	
$H_{\underline{a}\,(7,120)}^T H_{\underline{a}\,(7,120)}$	1.4×10^{12}	1178	994	149	22.0	20.7	19.5
covariance matrix	square root of eigenvalues						
$H_{\underline{a}\,(5,120)}^T H_{\underline{a}\,(5,120)}$	8.9×10^5	30.3	28	12	4.5		
$H_{\underline{a}\,(6,120)}^T H_{\underline{a}\,(6,120)}$	1.0×10^6	32.9	29	12.2	4.6	4.4	
$H_{\underline{a}\,(7,120)}^T H_{\underline{a}\,(7,120)}$	1.2×10^6	34.3	31.5	12.2	4.7	4.5	4.4

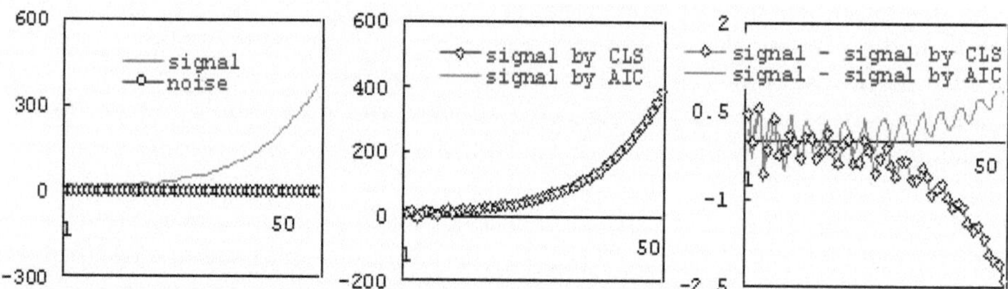

Fig. 3.14. The right is the exact signal of a 4-dimensional linear system and noise, the middle are signals of a 4-dimensional linear system obtained by the CLS method and a 9-dimensional linear system obtained by the AIC method, and the right is the difference between the exact signal and signal obtained by the CLS or AIC methods in Example (3.34)

1) Since a set $\{4.7,\ 4.5,\ 4.4\}$ is composed of relatively small and equally-sized numbers in the square root of $H_{\underline{a}\,(7,120)}^T H_{\underline{a}\,(7,120)}$, the noisy realization of a linear system obtained by the CLS method may be good for a 4-dimensional space.

2) After determining the number n of dimensions which is 4, we will continue the noisy realization algorithm by the CLS method.

Therefore, the modified impulse response $I(0)$ of a linear system obtained by the CLS method is realized by a 4-dimensional linear system.

Therefore, the linear system obtained by the CLS method is a 4-dimensional linear system. The system is given by $\sigma_c = ((\boldsymbol{R}^4,\ F_c),\ \mathbf{e}_1,\ h_c)$, where $F_c =$

$$\begin{bmatrix} 0 & 0 & 0 & 0.56 \\ 1 & 0 & 0 & 0.55 \\ 0 & 1 & 0 & -0.07 \\ 0 & 0 & 1 & 0.24 \end{bmatrix},\ h_c = [8.5\ 15,\ -5.5,\ 10.6].$$

Moreover, the linear system obtained by the AIC method is a 9-dimensional linear system $\sigma_a = ((\boldsymbol{R}^9,\ F_a),\ \mathbf{e}_1,\ h_a)$ given as follows:

$$F_a = \begin{bmatrix} 0\ 0\ 0\ 0\ 0\ 0\ 0\ 0 & 0.08 \\ 1\ 0\ 0\ 0\ 0\ 0\ 0\ 0 & 0.22 \\ 0\ 1\ 0\ 0\ 0\ 0\ 0\ 0 & 0.25 \\ 0\ 0\ 1\ 0\ 0\ 0\ 0\ 0 & -0.01 \\ 0\ 0\ 0\ 1\ 0\ 0\ 0\ 0 & -0.054 \\ 0\ 0\ 0\ 0\ 1\ 0\ 0\ 0 & 0.35 \\ 0\ 0\ 0\ 0\ 0\ 1\ 0\ 0 & 0.39 \\ 0\ 0\ 0\ 0\ 0\ 0\ 1\ 0 & 0.05 \\ 0\ 0\ 0\ 0\ 0\ 0\ 0\ 1 & 0.17 \end{bmatrix},$$

$$h_a = [8.8,\ 15.3,\ -5.6,\ 10.7,\ 15.6,\ 8.8,\ 3.4,\ 15.5,\ 16.9].$$

In this example, the original signal is considered as the impulse response of a 4-dimensional linear system and the desirable impulse response is obtained by the two methods, that is, the CLS and AIC methods.

For reference, in the following table, we list the mean values of the sum of the square for the original signal, the obtained signal and the error to signal ratio.

This table indicates that the 4-dimensional linear system obtained by the CLS method reconstructs the original signal with a 1 % error to signal ratio and zero noise to signal ratio, please refer to Remark 2 in Theorem 3.26 for the noise to signal ratio.

The 4-dimensional linear system obtained by the CLS method has the same number of dimensions as the number of the original system.

The AIC method produces a 9-dimensional linear system.

Nevertheless, Fig. 3.14 indicates that the model obtained by the CLS method causes the same degree of error as the model obtained by AIC.

dimen-ion	ratio of matrices	mean values of square root for sum of		error	cosine	error
		signal	signal by CLS	error	① and ②	ratio
		①	②	③	$\cos\theta$	③/①
4	0	19.5	19.7	0.12	0.999	0.01

Example 3.35. Let a signal be the impulse response of the following 4-dimensional linear system: $\sigma = ((\boldsymbol{R}^4, F), \mathbf{e}_1,\ h)$,

$$\text{where } F = \begin{bmatrix} 0\ 0\ 0 & -0.0024 \\ 1\ 0\ 0 & 0.014 \\ 0\ 1\ 0 & 0.1 \\ 0\ 0\ 1 & -0.2 \end{bmatrix},\ h = [10,\ 2,\ -5,\ -1].$$

Let an added noise be given in Fig. 3.15.

Then the noisy realization problem is solved as follows:

covariance matrix	eigenvalues					
	1	2	3	4	5	6
$H_{\underline{a}\,(4,40)}^{T}\,H_{\underline{a}\,(4,40)}$	164.1	31.76	11.22	6.37		
$H_{\underline{a}\,(5,40)}^{T}\,H_{\underline{a}\,(5,40)}$	164.3	31.89	11.49	6.97	6.08	
$H_{\underline{a}\,(6,40)}^{T}\,H_{\underline{a}\,(6,40)}$	164.3	31.89	12.73	7.17	6.08	5.77
covariance matrix	square root of eigenvalues					
$H_{\underline{a}\,(4,40)}^{T}\,H_{\underline{a}\,(4,40)}$	12.8	5.6	3.34	2.5		
$H_{\underline{a}\,(5,40)}^{T}\,H_{\underline{a}\,(5,40)}$	12.8	5.6	3.4	2.6	2.5	
$H_{\underline{a}\,(6,40)}^{T}\,H_{\underline{a}\,(6,40)}$	12.8	5.6	3.6	2.7	2.5	2.4

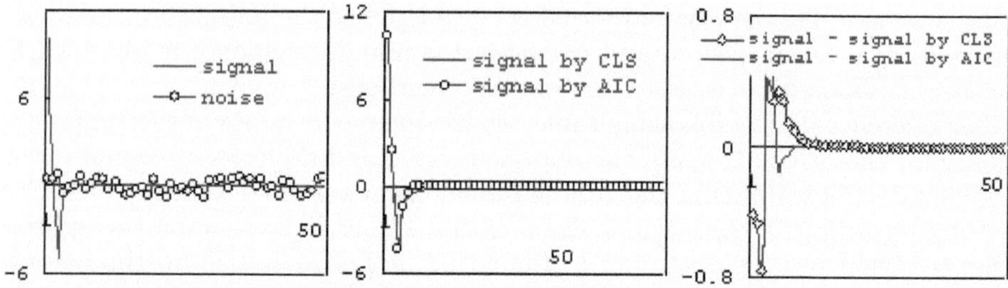

Fig. 3.15. The left is the exact signal of a 4-dimensional linear system and noise, the middle are the signals of a 3-dimensional linear system obtained by the CLS method and a 3-dimensional linear system obtained by the AIC method, and the right is the difference between the exact signal and the signal by the CLS or AIC methods in Example (3.35)

1) Since a set $\{2.7,\ 2.5,\ 2.4\}$ is composed of relatively small and equally-sized numbers in the square root of $H_{\underline{a}\,(6,40)}^{T}\,H_{\underline{a}\,(6,40)}$, the noisy realization of a linear system obtained by the CLS method may be good for a 3-dimensional space.
2) After determining the number 3 for the dimension, we will continue the noisy realization algorithm by the CLS method.

Therefore, the modified impulse response $I(0)$ of a linear system obtained by the CLS method is realized by a 3-dimensional linear system.

Therefore, the linear system obtained by the CLS method is a 3-dimensional linear system. The system is given by $\sigma_c = ((\mathbf{R}^3,\ F_c),\ \mathbf{e}_1,\ h_c)$, where $F_c =$

$$\begin{bmatrix} 0 & 0 & 0.05 \\ 1 & 0 & -0.03 \\ 0 & 1 & 0.5 \end{bmatrix},\ h_c = [10.4,\ 2.5,\ -4.2].$$

Moreover, the linear system obtained by the AIC method is a 3-dimensional linear system $\sigma_a = ((\mathbf{R}^9,\ F_a),\ \mathbf{e}_1,\ h_a)$ given as follows:

$$F_a = \begin{bmatrix} 0 & 0 & -0.05 \\ 1 & 0 & 0.03 \\ 0 & 1 & 0.23 \end{bmatrix},\ h_a = [10.4,\ 2.5,\ -4.3].$$

In this example, the original signal is considered as the impulse response of a 4-dimensional linear system and the desirable impulse response is obtained by the two methods, that is, the CLS and AIC methods.

For reference, in the following table, we list the mean values of the sum of the square for the original signal, the obtained signal and the error to signal ratio.

This table indicates that the 4-dimensional linear system obtained by the CLS method reconstructs the original signal with an 11 % error to signal ratio and 0.19 noise to signal ratio, please refer to Remark 2 in Theorem 3.26 for the noise to signal ratio.

The 4-dimensional linear system obtained by the CLS method has the same number of dimensions as the number of the original system.

The AIC method produces a 3-dimensional linear system.

Fig. 3.15 indicates that the model obtained by the CLS method causes the same degree of error as the model obtained by AIC.

dimen-ion	ratio of matrices	mean values of square root for sum of			cosine ① and ②	error ratio
		signal ①	signal by CLS ②	error ③	$\cos\theta$	③/①
4	0.19	0.228	0.233	0.03	0.994	0.114

Example 3.36. Let a signal be the impulse response of the following 5-dimensional linear system: $\sigma = ((\boldsymbol{R}^5, F), \mathbf{e}_1, h)$,

$$\text{where } F = \begin{bmatrix} 0 & 0 & 0 & 0 & 0 \\ 1 & 0 & 0 & 0 & -0.04 \\ 0 & 1 & 0 & 0 & -0.11 \\ 0 & 0 & 0 & 1 & 0.52 \\ 0 & 0 & 0 & 1 & -0.4 \end{bmatrix}, \ h = [10,\ 2,\ -5,\ -1,\ 3].$$

Let an added noise be given in Fig. 3.16.

Then the noisy realization problem is solved as follows:

covariance matrix	eigenvalues							
	1	2	3	4	5	6	7	8
$H^T_{a\ (6,80)} H_{\underline{a}\ (6,80)}$	1377	196.4	75.44	24.03	19.85	13.38		
$H^T_{a\ (7,80)} H_{\underline{a}\ (7,80)}$	1600	196.6	80.0	24.95	21.12	14.11	12.57	
$H^T_{a\ (8,80)} H_{\underline{a}\ (8,80)}$	1822	196.5	81.58	24.98	21.37	17.52	12.93	11.33
covariance matrix	square root of eigenvalues							
$H^T_{a\ (6,80)} H_{\underline{a}\ (6,80)}$	37	14	8.7	4.9	4.5	3.7		
$H^T_{a\ (7,80)} H_{\underline{a}\ (7,80)}$	40	14	8.9	5	4.6	3.8	3.5	
$H^T_{a\ (8,80)} H_{\underline{a}\ (8,80)}$	42.7	14	9	5	4.6	4.2	3.6	3.4

1) Since a set {4.2, 3.6, 3.4} is composed of relatively small and equally-sized numbers in the square root of $H^T_{\underline{a}\ (8,80)} H_{\underline{a}\ (8,80)}$, the noisy realization of a linear system obtained by the CLS method may be good for a 5-dimensional space.

2) After determining the number n of dimensions which is 5, we will continue the noisy realization algorithm by the CLS method.

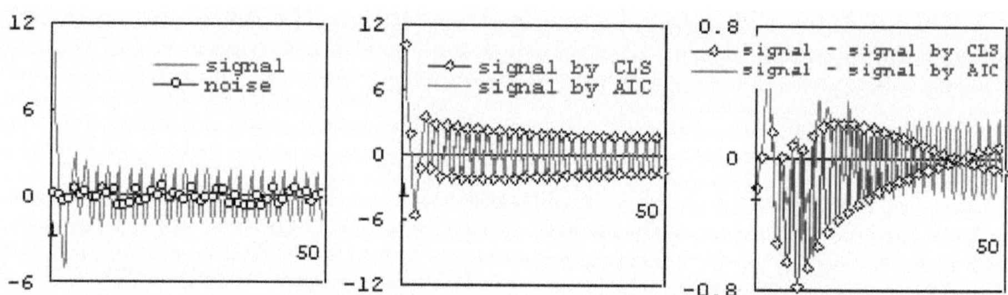

Fig. 3.16. The left is the exact signal of a 5-dimensional linear system and noise, the middle are signals of a 5-dimensional linear system obtained by the CLS method and a 9-dimensional linear system obtained by the AIC method, and the right is the difference between the signal by the CLS or AIC methods in Example (3.36)

Therefore, the linear system obtained by the CLS method is a 5-dimensional linear system. The system is given by $\sigma_c = ((\boldsymbol{R}^5, \ F_c), \ \mathbf{e}_1, \ h_c)$, where $F_c =$

$$\begin{bmatrix} 0\ 0\ 0\ 0 & 0.05 \\ 1\ 0\ 0\ 0 & 0.02 \\ 0\ 1\ 0\ 0 & -0.0001 \\ 0\ 0\ 1\ 0 & 0.77 \\ 0\ 0\ 0\ 1 & -0.25 \end{bmatrix}, \ h_c = [10.2, \ 2, \ -5.5, \ -1.16, \ 3.5].$$

Moreover, the linear system obtained by the AIC method is a 9-dimensional linear system $\sigma_a = ((\boldsymbol{R}^9, \ F_a), \ \mathbf{e}_1, \ h_a)$ given as follows:

$$F_a = \begin{bmatrix} 0\ 0\ 0\ 0\ 0\ 0\ 0\ 0 & -0.1 \\ 1\ 0\ 0\ 0\ 0\ 0\ 0\ 0 & -0.09 \\ 0\ 1\ 0\ 0\ 0\ 0\ 0\ 0 & -0.228 \\ 0\ 0\ 1\ 0\ 0\ 0\ 0\ 0 & 0.2 \\ 0\ 0\ 0\ 1\ 0\ 0\ 0\ 0 & -0.12 \\ 0\ 0\ 0\ 0\ 1\ 0\ 0\ 0 & -0.01 \\ 0\ 0\ 0\ 0\ 0\ 1\ 0\ 0 & 0.03 \\ 0\ 0\ 0\ 0\ 0\ 0\ 1\ 0 & 0.3 \\ 0\ 0\ 0\ 0\ 0\ 0\ 0\ 1 & -0.16 \end{bmatrix},$$

$$h_a = [10.2, \ 2, \ -5.5, \ -1.2, \ 3.5, \ -1.2, \ 2.6, \ -2, \ 1.9].$$

In this example, the original signal is considered as the impulse response of a 4-dimensional linear system and the desirable impulse response is obtained by two methods, that is, the CLS and AIC methods.

For reference, in the following table, we list the mean values of the sum of the square for the original signal, the obtained signal and the error to signal ratio.

This table indicates that the 5-dimensional linear system obtained by the CLS method reconstructs the original signal with a 11 % error to signal ratio and 0.1 noise to signal ratio, please refer to Remark 2 in Theorem 3.26 for the noise to signal ratio.

The 5-dimensional linear system obtained by the CLS method has the same number of dimensions as the number of the original system.

The AIC method produces a 9-dimensional linear system.

Nevertheless, Fig. 3.16 indicates that the model obtained by the CLS method causes the same degree of error as the model obtained by AIC.

dimen- ion	ratio of matrices	mean values of square root for sum of			cosine ① and ②	error ratio
		signal ①	signal by CLS ②	error ③	$\cos\theta$	③/①
5	0.1	0.341	0.366	0.04	0.997	0.11

Example 3.37. Let a signal be the impulse response of the following 6-dimensional linear system: $\sigma = ((\boldsymbol{R}^6, F), \mathbf{e}_1, h)$,

where $F = \begin{bmatrix} 0\,0\,0\,0\,0 & 0 \\ 1\,0\,0\,0\,0 & -0.04 \\ 0\,1\,0\,0\,0 & -0.03 \\ 0\,0\,0\,1\,0 & 0.16 \\ 0\,0\,0\,1\,0 & 0.48 \\ 0\,0\,0\,0\,1 & -0.5 \end{bmatrix}$, $h = [10,\ 2,\ -5,\ -1,\ 3,\ -2]$.

Let an added noise be given in Fig. 3.17.

Then the noisy realization problem is solved as follows:

covariance matrix	eigenvalues							
	1	2	3	4	5	6	7	8
$H_{\underline{a}\,(6,20)}^T H_{\underline{a}\,(6,20)}$	192	96.3	55.3	8.5	4.5	1.6		
$H_{\underline{a}\,(7,20)}^T H_{\underline{a}\,(7,20)}$	197	101	58	8.5	5.1	2.9	1.4	
$H_{\underline{a}\,(8,20)}^T H_{\underline{a}\,(8,20)}$	202	104	58.2	9.7	5.2	3.6	2.7	1
covariance matrix	square root of eigenvalues							
$H_{\underline{a}\,(6,20)}^T H_{\underline{a}\,(6,20)}$	13.9	9.8	7.4	2.9	2.1	1.3		
$H_{\underline{a}\,(7,20)}^T H_{\underline{a}\,(7,20)}$	14	10	7.6	2.9	2.3	1.7	1.2	
$H_{\underline{a}\,(8,20)}^T H_{\underline{a}\,(8,20)}$	14.2	10.2	7.6	3.1	2.3	1.9	1.6	1

1) Since a set $\{1.9,\ 1.6,\ 1\}$ is composed of relatively small and equally-sized numbers in the square root of $H_{\underline{a}\,(8,20)}^T H_{\underline{a}\,(8,20)}$, the noisy realization of a linear system obtained by the CLS method may be good for a 5-dimensional space.

2) After determining the number n of dimensions which is 5, we will continue the noisy realization algorithm by the CLS method.

Therefore, the linear system obtained by the CLS method is a 5-dimensional linear system. The system is given by $\sigma_c = ((\boldsymbol{R}^5,\ F_c),\ \mathbf{e}_1,\ h_c)$, where $F_c =$

$\begin{bmatrix} 0\,0\,0\,0 & -0.28 \\ 1\,0\,0\,0 & -0.48 \\ 0\,1\,0\,0 & -0.86 \\ 0\,0\,1\,0 & -0.73 \\ 0\,0\,0\,1 & -0.91 \end{bmatrix}$, $h_c = [10.3,\ 1.5,\ -4.5,\ -1.1,\ 3.6]$.

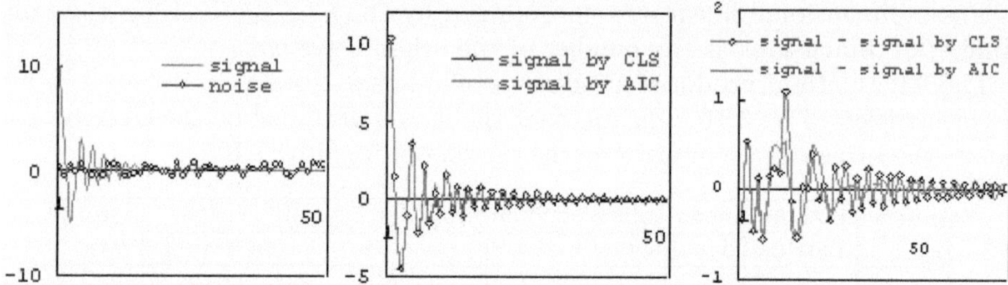

Fig. 3.17. The left is the exact signal of a 6-dimensional linear system and noise, the middle are signals of a 5-dimensional linear system obtained by the CLS method and a 9-dimensional linear system obtained by the AIC method, and the right is the difference between the exact signal and the signal by the CLS or AIC methods in Example (3.37)

Moreover, the linear system obtained by the AIC method is a 9-dimensional linear system $\sigma_a = ((\boldsymbol{R}^9, \ F_a), \ \mathbf{e}_1, \ h_a)$ given as follows:

$$F_a = \begin{bmatrix} 0\ 0\ 0\ 0\ 0\ 0\ 0\ 0 & -0.17 \\ 1\ 0\ 0\ 0\ 0\ 0\ 0\ 0 & -0.24 \\ 0\ 1\ 0\ 0\ 0\ 0\ 0\ 0 & -0.35 \\ 0\ 0\ 1\ 0\ 0\ 0\ 0\ 0 & 0.15 \\ 0\ 0\ 0\ 1\ 0\ 0\ 0\ 0 & -0.43 \\ 0\ 0\ 0\ 0\ 1\ 0\ 0\ 0 & -0.81 \\ 0\ 0\ 0\ 0\ 0\ 1\ 0\ 0 & -0.09 \\ 0\ 0\ 0\ 0\ 0\ 0\ 1\ 0 & -0.13 \\ 0\ 0\ 0\ 0\ 0\ 0\ 0\ 1 & -0.43 \end{bmatrix},$$

$$h_a = [10.3, \ 1.45, \ -4.6, \ -1.2, \ 3.5, \ -2.2, \ 1.8, \ -1.9, \ 0.94].$$

In this example, the original signal is considered as the impulse response of a 6-dimensional linear system and the desirable impulse response is obtained by the two methods, that is, the CLS and AIC methods.

For reference, in the following table, we list the mean values of the sum of the square for the original signal, the obtained signal and the error to signal ratio.

This table indicates that the 6-dimensional linear system obtained by the CLS method reconstructs the original signal with a 15 % error to signal ratio and 0.09 noise to signal ratio, please refer to Remark 2 in Theorem 3.26 for the noise to signal ratio.

The 6-dimensional linear system obtained by the CLS method has the same number of dimensions as the number of the original system.

The AIC method produces a 9-dimensional linear system.

Nevertheless, Fig. 3.17 indicates that the model obtained by the CLS method causes the same degree of error as the model obtained by AIC.

dimen-ion	ratio of matrices	mean values of square root for sum of			cosine	error ratio
		signal	signal by CLS	error	① and ②	
		①	②	③	$\cos \theta$	③/①
6	0.09	0.251	0.256	0.04	0.989	0.15

3.5.1   Comparative Table of CLS and AIC Method

We have proposed a noisy realization problem to clean up noise from actual observed data. To solve the problem, up to now, we could only use the AIC method. Here we have introduced a new method which is the CLS method. But we know that the AIC method is only applied to linear systems.

Here, we list the difference between the CLS and AIC methods through our examples.

The error in the next table means the mean value of the square root of the following value: $(1/V) * \sum_{i=1}^{V} (\text{signal}(i) - \text{obtained signal}(i))^2$.

No.	28	29	30	31	32	33	34	35	36	37
dim.	1	2	3	3	3	4	4	4	5	6
CLS	1	2	3	3	3	4	4	3	5	5
error	0.01	0.01	0.05	0.02	0.01	0.04	41	0.02	0.04	0.04
AIC	2	4	7	5	6	7	9	3	9	9
error	0.01	0.01	0.04	0.01	0.01	0.04	16	0.02	0.05	0.03

'No.' denotes the number of examples in this chapter.
'dim.' denotes the number of dimensions of the
original systems.
Numbers in the upper stand denote the number
of dimensions of the obtained ones.
Numbers in the lower stand denote
the root mean square error.

3.6   Historical Notes and Concluding Remarks

Approximate realization and noisy realization problems of linear systems were studied with the notion of the ratio of Hankel norm and the CLS method. The ratio of Hankel norm is used for determining the dimension of a state space and the CLS method is used for determining the parameters of linear systems.

For our treatement of the approximate and noisy realization problems, there may be a point for using singular value decomposition and Constrained Least Square (CLS) in Kalman [1997]. In this reference, Kalman pointed out that the identification problem from noisy data should be treated without any prejudice, hence, should be approached in a statistical sense, not in a probabilistic sense. Here, we only insist that the signal and the noise are not correlated. Allowing for this, we could discuss approximate and noisy realization problems for linear systems with a unified method.

Since our determination method of the dimensions for linear systems is directly executed without any restrictions, our method is very useful and convenient for both approximate realization, equivalency, model reduction, and noisy realization problems.

However, we cannot fully apply the approximate realization algorithm to impulse responses which increase in numerical value, which can be seen in Example

3.20. For the noisy realization problem, we cannot fully apply the noisy realization algorithm to impulse responses which have rapid damping or values near zero, which can be seen in Examples 3.28, 3.29, 3.35 and 3.37.

In multivariable analysis which is a traditional method for analysis in economic, biology, psychology and others, it is known that the factor number is determined by the number of eigenvalues of the covariance matrix which are greater than one. Our determination for the dimension of linear systems is based on the ratio of the Hankel norm. This direction is presented by showing examples.

For normal noisy realization, we can easily perform AIC as a typical example. It is known that this method has been proposed with the notion of both ideas of statistical and probabilistic view points. Therefore, the AIC method is considered as a very technical idea. Since we only stress statistical idea for our treatment of noisy data, we can easily connect the idea of the Hankel norm and Constrained Least Square method.

In order to show that our method for approximate and noisy realization are effective, we provided several examples. Based on the result of these examples, we demonstrated that the ratio of the square root of singular values imply the degree of approximation. For our noisy realization problems, we have demonstrated that we can determine the dimension of linear systems when the set of equally-sized numbers of the square root of singular values can be found.

Roughly speaking, our several examples for the approximate realization problem suggest that the smaller the ratio of matrix norm is, the smaller the error to signal ratio is. The change of ratio ranges from one to six percent for the error to signal ratio per 0.01 ratio of matrix norm.

The several examples suggest that two unique features can be expressed as follows:

(1) The ratio of the matrix norm determines the degree of the crossed angle between directions of the approximated signal and the original signal.
(2) The CLS method determines the coefficients of linearly dependent vectors such that the error between the approximate signal and the original signal has a minimum value in the sense of a square norm while conserving the crossed angle.

Intuitively, our several examples for the noisy realization problem show that the smaller the ratio of matrices is, the smaller the error to signal ratio is. The change ratio ranges from one to two percent by the error to signal ratio per 0.01 ratio of matrix norm.

The several examples suggest that two unique features can be expressed as follows:

(1) The ratio of matrices determines the degree of the crossed angle between directions of the obtained signal and the original signal.
(2) The CLS method determines the coefficients of linearly dependent vectors such that the error between the obtained signal and the original signal has a minimum value in the sense of a square norm while conseving the crossed angle.

In subsection (3.5.1), we compared our CLS method with the AIC method and we showed that the CLS method is more useful than the AIC method in the sense of noisy realization because the CLS method result in less dimensional state space than the AIC method.

We want to say that our approximate and noisy realizations were developed by our realization procedure for obtaining the reachable standard system from a given input response map and a partial realization algorithm.

4 Approximate and Noisy Realization of So-called Linear Systems

Let the set Y of output's values be a linear space over the field $\boldsymbol{R}$. Almost linear systems were introduced in the monograph [Matsuo and Hasegawa, 2003], and it was also shown that the systems contain so-called linear systems as a sub-class, where so-called linear systems are linear systems with a non-zero initial state.

It is well known that a common method to obtain so-called linear systems is solved through two problems.

One is the realization problem to obtain linear systems with a zero initial state and the other is the state estimation problem for systems with a non-zero initial state. Based upon the prejudice that so-called linear systems are completely the same as linear systems, so-called linear systems were treated separately.

In the monograph, it was also shown that so-called linear systems can be obtained from input/output data from a single experiment.

In this chapter, based on the results regarding so-called linear systems, we will discuss approximate and noisy realization of the systems. For our discussion, we will present for the first time a concrete method to discuss approximate and noisy realization problems from partial data, equivalently, i.e., data obtained in finite real time. Hence, this new method is very useful and practical.

Note that because of the system's nonlinearity, these problems were never discussed before.

For self-contained, we will list the main results needed for our discussion from our monograph.

In order to solve our problems, we will use singular value decomposition and Constrained Least Square method, which is abbreviated to the CLS method which has been discussed in Chapter 2. The singular value decomposition is used to determine the dimension of so-called linear systems and the CLS method is used to determine parameters of a so-called linear system.

At first, we will discuss approximate realization problems and give many example to ascertain the effectiveness of our algorithm. Next, we will discuss noisy realization problems and give several examples to ascertain the effectiveness of our noisy realization algorithm.

Y. Hasegawa: Approxi. & Noisy Reali. of Discrete-Time Dyn. Sys., LNCIS 376, pp. 55–94, 2008.
springerlink.com

4.1 Basic Facts about So-called Linear Systems

Definition 4.1. *So-called Linear Systems*

1) A system given by the following system equation is said to be a so-called linear system $\sigma = ((X, F), x^0, g, h)$. This system is a linear system with a non-zero initial state.

$$\begin{cases} x(t+1) = Fx(t) + g\omega(t+1) \\ x(0) \quad = x^0 \\ \gamma(t) \quad = hx(t) \end{cases},$$

where $F \in L(X)$, $\omega(t+1) \in U$, g, $x^0 \in X$. In addition, h is a linear operator: $X \to Y$ for any $t \in N$, $\gamma(t) \in Y$.

2) The input response map $a_\sigma : U^* \to Y; \omega \mapsto h(\sum_{j=1}^{|\omega|} F^{|\omega|-j}(Fx^0 + g\omega(j)))$ is said to be the behavior of σ.

3) For the so-called linear system σ and any $i \geq 1$,
$I_\sigma(1)(i) := a_\sigma(0^i|1) - a_\sigma(0^i) = hF^i(g^0 + g)$ and
$I_\sigma(0)(i) := a_\sigma(0^{i+1}) - a_\sigma(0^i) = hF^i g^0$ are said to be modified impulse responses of σ, where $0^0 := 1$, $g^0 := Fx^0 - x^0$.
Note that there is a one-to-one correspondence between the behavior of σ and the modified impulse responses $I_\sigma(0)$ and $I_\sigma(1) \in F(N, Y)$ of σ by the relations $a_\sigma(\omega) = (\sum_{j=1}^{|\omega|}(I_\sigma(0)(|\omega| - j + 1) + I_\sigma(1)(|\omega| - j + 1) \times \omega(j))$.

4) A so-called linear system σ is said to be reachable if the reachable set $\{\sum_{j=1}^{|\omega|} F^{|\omega|-j}(g^0 + g\omega(j)); \omega \in U^*\}$ is equal to X and the system σ is called to be observable if $hF^i x_1 = hF^i x_2$ for any $i \in N$ implies $x_1 = x_2$,
where $g^0 := Fx^0 - x^0$.

5) A so-called linear system σ is called canonical if σ is reachable and observable.

Remark 1: It is meant for σ to be a faithful model for the input response map a that σ realizes a.

Remark 2: Notice that a canonical so-called linear system $\sigma = ((X, F), x^0, g, h)$ is a system that has the most reduced state space X among systems that have the behavior a_σ.

Proposition 4.2. *For any so-called linear system $\tilde{\sigma} = ((X, F), x^0, g, h)$, there exists an almost linear system $\sigma = ((X, F), g^0, g, h, h^0)$ with the same input/output relation which satisfies $g^0 = Fx^0 - x^0$ and $h^0 = hx^0$.*

Remark: For details of almost linear systems, see Definition (5.1) in Chapter 5.

Lemma 4.3. *Let $\sigma = ((X, F), x^0, g, h)$ be a canonical (controllable and observable) so-called linear system, then the almost linear system σ obtained by Proposition (4.2) is intrinsically canonical.*

Conversely, let $\sigma = ((X, F), g^0, g, h, h^0)$ be an intrinsically canonical almost linear system, then so-called linear system $\tilde{\sigma}$ obtained by Proposition (4.2) is canonical.

Example 4.4. Let $F(N, Y) := \{$ any function $f : N \to Y\}$. Let $S_l\gamma(t) = \gamma(t+1)$ for any $\gamma \in F(N, Y)$ and $t \in N$, then $S_l \in L(F(N, Y))$. Let a map $\chi^0 \in F(N, Y)$ be $(\chi^0)(t) := a(\omega|0) - a(\omega)$ and $\chi \in F(N, Y)$ be $\chi(t) := a(\omega|1) - a(\omega)$ for any $t \in N$, a time-invariant, affine input response map $a \in F(U^*, Y)$ and $\omega \in U^*$ such that $|\omega| = t$. Moreover, let a linear map 0 be $F(N, Y) \to Y; \gamma \mapsto \gamma(0)$. Then a collection $((F(N, Y), S_l), \chi^0, \chi, 0, a(1))$ is an observable almost linear system that realizes a.

Theorem 4.5. *The following almost linear system is the canonical realizations of any time-invariant, affine input response map $a \in F(U^*, Y)$.*
$((\ll S_l^N(\chi(U)) \gg, S_l), \chi^0, \chi, 0, a(1))$,
where $\ll S_l^N(\chi(U)) \gg$ is the smallest linear space that contains $S_l^N(\chi(U)) :=$
$\{S_l^i(\chi^0 + \chi \times u); u \in \boldsymbol{R}, i \in N, S_l^i(\chi^0 + \chi u)(t) = (\chi(u)(t+1) = a(\omega|u) - a(\omega),$
$\omega \in U^*\}$.

Proposition 4.6. *Let $\sigma = ((\ll S_l^N(\chi(U)) \gg, S_l), \chi^0, \chi, 0, a(1))$ be the intrinsically canonical almost linear system which is given in Theorem (4.5).*
The so-called linear system $((\ll S_l^N(\chi(U)) \gg, S_l), x^0, \chi, 0))$ is given by σ if and only if there exists a $x^0 \in \ll S_l^N(\chi(U)) \gg$ such that $\chi^0 = S_l x^0 - x^0$ and $a(1) = 0x^0$.

Definition 4.7. *Let $\sigma_1 = ((X_1, F_1), g_1^0, g_1, h_1, h^0)$ and*
$\sigma_2 = ((X_2, F_2), g_2^0, g_2, h_2, h^0)$ *be almost linear systems. Then a linear operator $T : X_1 \to X_2$ is said to be an almost linear system morphism $T : \sigma_1 \to \sigma_2$ if T satisfies $TF_1 = F_2T$, $Tg_1^0 = g_2^0$, $Tg_1 = g_2$ and $h_1 = h_2T$.*
If $T : X_1 \to X_2$ is bijective, then $T : \sigma_1 \to \sigma_2$ is said to be an isomorphism.

Corollary 4.8. *Let T be an almost linear system morphism $T : \sigma_1 \to \sigma_2$. Then $a_{\sigma_1} = a_{\sigma_2}$ holds.*

4.2   Finite Dimensional So-called Linear Systems

We will state facts regarding finite-dimensional so-called linear systems in this section. Since many results of so-called linear systems have been shown in a monograph [Matsuo and Hasegawa, 2003], the main results are cited from the monograph.

Firstly, we introduce conditions in which a finite dimensional so-called linear system is canonical.

Secondly, we introduce a canonical form which is suitable for approximate and noisy realization problems.

Namely, we introduce a standard system as a representative in their equivalence classes.

Thirdly, we introduce a criterion for the behavior of finite dimensional so-called linear systems, i.e., a rank condition of infinite Input/output matrix.

Lastly, we introduce a procedure to obtain a real-time standard system which realizes a given input response map.

There is a fact regarding finite dimensional linear spaces that an n-dimensional linear space over the field $\boldsymbol{R}$ is isomorphic to $\boldsymbol{R}^n$ and $L(\boldsymbol{R}^n, \boldsymbol{R}^m)$ is isomorphic to $\boldsymbol{R}^{m \times n}$ (See Halmos [1958]). Therefore, without loss of generality, we can consider a n-dimensional linear system as $\sigma = ((\boldsymbol{R}^n, F), g, h)$, where $F \in \boldsymbol{R}^{n \times n}$, $g \in \boldsymbol{R}^n$ and $h \in \boldsymbol{R}^{p \times n}$.

Definition 4.9. *For any time-invariant, affine input response map $a \in F(U^*, Y)$, the corresponding linear input/output map $A : (A(N \times \{0,1\}), S_r) \to (F(N, Y), S_l)$ satisfies $A(\mathbf{e}_{(\mathbf{s},\mathbf{u})})(t) = a(u^{s+t+1}) - a(u^{s+t})$ for any $u \in \{0,1\}$. Therefore, A is represented by the next infinite matrix $(I/O)_a$. This $(I/O)_a$ is said to be an Input/output matrix of a. For the $(A(N \times \{0,1\}), S_r)$, see Example (5.2) in chapter 5.*

$$
\begin{array}{c}
\\
(I/O)_a = \\
t
\end{array}
\begin{array}{c}
(s, u) \\
\left(
\begin{array}{ccc}
 & \vdots & \\
 & \vdots & \\
 & \vdots & \\
 & \vdots & \\
\cdots \quad \cdots & a(u^{s+t+1}) - a(u^{s+t}) & \\
\end{array}
\right)
\end{array}
$$

Note that for the linear input/output map $A : A(N \times \{0,1\}) \to F(N,Y)$, there exists a unique function $I_a : \{0,1\} \to F(N,Y)$ such that $I_a(u)(i+j) = A(\mathbf{e}_{(\mathbf{i},\mathbf{u})})(j) = a(u^{i+j+1}) - a(u^{i+j})$ holds for $u \in \{0,1\}$.
Also note that column vectors of $(I/O)_a$ denote $S_l^i I_a(u)$.

Theorem 4.10. *Theorem for existence criterion*
For a time-invariant, affine input response map $a \in F(U^*, Y)$, the following conditions are equivalent:
1) The input response map $a \in F(U^*, Y)$ has the behavior as a n-dimensional canonical almost linear system.
2) There exist n linearly independent vectors and no more than n linearly independent vectors in a set $\{S_l^i I_a(u) \in \ll S_l^N(\chi(U)) \gg; i \leq n$ for $i \in N, u \in \{0,1\}\}$.
3) The rank of the Input/output matrix $(I/O)_a$ of a is n.

Definition 4.11. *Let $\sigma_r = ((\boldsymbol{R}^n, F_r), g_r^0, g_r, h_r, h_r^0)$ be a canonical almost linear system. The σ_r which satisfies the following conditions is called a real time standard system.*
1) $g_r^0 = \mathbf{e}_1$, $F_r^{i-1} g_r^0 = \mathbf{e}_i$, $1 \leq i \leq n_1$ and $F_r^{n_1} g_r^0 = \sum_{i=1}^{n_1} \alpha_{1i} F_r^{i-1} g_r^0$, $\alpha_{1i} \in \boldsymbol{R}$ hold.
2) $g_r^0 + g_r = \mathbf{e}_{n_1+1}$, $F_r^{i-1}(g_r^0 + g_r) = \mathbf{e}_{n_1+i}$, $1 \leq i \leq n_2$ and
$F_r^{n_2}(g_r^0 + g_r) = \sum_{i=1}^{n_1} \alpha_{2i} F_r^{i-1} g_r^0 + \sum_{i=n_1+1}^{n_1+n_2} \alpha_{2i} F_r^{i-1} g_r$, $\alpha_{1i}, \alpha_{2i} \in \boldsymbol{R}$ hold.

3) $n = n_1 + n_2$ holds.
4) F_r is given as follows:

$$
F_r =
\begin{bmatrix}
0 \cdots 0 & \alpha_{11} & 0 \cdots \cdots 0 & \alpha_{21} \\
1 \ddots & \alpha_{12} & 0 \cdots \qquad 0 & \alpha_{22} \\
\vdots \ddots \ 0 & \vdots \quad \vdots & & \vdots \quad \vdots \\
0 \qquad 1 & \alpha_{1n_1} \ \vdots & & \vdots \ \alpha_{2n_1} \\
0 \ 0 \ \cdots \ 0 & 0 \cdots \cdots 0 & \alpha_{2n_1+1} \\
0 \ 0 \ \cdots \ \vdots & 1 \ \ddots & \vdots \ \alpha_{2n_1+2} \\
0 \ 0 \ \cdots \ \vdots & 0 \ \ddots \ \ddots \ \vdots & \vdots \\
0 \ \ddots \cdots \ \vdots & \vdots \ \ddots \ 1 \ 0 & \alpha_{2n-1} \\
0 \ 0 \ \cdots \ 0 & 0 \cdots 0 \ 1 & \alpha_{2n}
\end{bmatrix} \ .
$$

For the real time standard system $\sigma_r = ((\mathbf{R}^n, F_r), g_r^0, g_r, h_r, h_r^0)$, its modified impulse responses $I(0)(i) := h_r F_r^i g_r^0$ and $I(1)(i) := h_r F_r^i (g_r^0 + g_r)$ may be written as $I(0)\text{-}(n_1, n_2)$ and $I(1)\text{-}(n_1, n_2)$ respectively.

Theorem 4.12. *Representation Theorem for equivalence classes*
   For any finite dimensional canonical almost linear system, there exists a uniquely determined isomorphic real time standard system.

[proof]. Let $\sigma = ((\mathbf{R}^n, F), g^0, g, h, h^0)$ be any finite dimensional canonical almost linear system. We select the set of linearly n independent vectors $\{g^0, Fg^0, F^2 g^0, \cdots, F^{n_1-1} g^0, g^0 + g, F(g^0 + g), F^2(g^0 + g), \cdots, F^{n_2-1}(g^0 + g); n = n_1 + n_2\}$ among $\{F^i g^0, F^j g; 1 \le i \le n, \ 1 \le j \le n\}$ in the order of a set $\{g^0, Fg^0, F^2 g^0, \cdots, F^{n_1-1} g^0, g^0 + g, F(g^0 + g), F^2(g^0 + g), \cdots, F^{n_2-1}(g^0 + g)\}$. Then we introduce a linear operator $T : \mathbf{R}^n \to \mathbf{R}^n$ by setting $TF^{i-1} g^0 = \mathbf{e}_i$ for $i(1 \le i \le n_1)$ and $TF^{j-1}(g^0 + g) = \mathbf{e}_{n_1+j}$ for $j \ (1 \le j \le n_2)$, then T is a regular matrix. Let $F_r := TFT^{-1}$. Then $F_r \in \mathbf{R}^{n \times n}$. Since T is a regular matrix, $T : \mathbf{R}^n \to \mathbf{R}^n$ preserves linear independence and dependence. Also, T satisfies the equations $F_r T = TF$, $Tg^0 = g_r^0$ and $Tg = g_r$ by the construction of T. Let $h_r = hT^{-1}$. Then T is an almost linear system morphism : $\sigma = ((K^n, F), g^0, g, h, h^0) \to \sigma_r = ((K^n, F_r), g_r^0, g_r, h_r, h^0)$. T is bijective and σ_r is the only real time standard system by the selection of T. By Corollary (4.9), the behaviors of σ and σ_r are the same.
   Moreover, we can show that its uniqueness comes from the selection of $\{F^i g^0, F^j(g^0 + g) \ 1 \le n_1, \ n_2 \le n, \ n = n_1 + n_2\}$.

Theorem 4.13. *Theorem for a realization procedure*
   Let a time-invariant, affine input response map $a \in F(U^*, Y)$ satisfy rank $(I/O)_a = n$. Then the real time standard system $\sigma_r = ((\mathbf{R}^n, F_r), g_r^0, g_r, h_r, a(1))$ which realizes a is obtained by the following procedure:

1) Select n_1 independent vectors on the vectors $\{S_l^s I_a(0) : 0 \leq s \leq n\}$. Next select n_2 independent vectors in $\{S_l^s I_a(1) : 0 \leq s \leq n\}$.
(2) Let the state space be $\mathbf{R}^n$. And let g_r^0 and g_r be as follows: $g_r^0 = \mathbf{e}_1$,

$g_r^0 + g_r = \mathbf{e}_{n_1+1}$. Moreover, $n = n_1 + n_2$ and $\mathbf{e}_i = [0, \cdots, 0, \overset{i}{1}, 0, \cdots, 0]^T$ hold.
(3) $F_r \in \mathbf{R}^{n \times n}$ is given in Definition (4.11),
where $S_l^{n_1} I_a(0) = \sum_{i=1}^{n_1} \alpha_{1i} S_l^{i-1} I_a(0)$,
$S_l^{n_2} I_a(1) = \sum_{i=1}^{n_1} \alpha_{2i} S_l^{i-1} I_a(0) + \sum_{j=1}^{n_2} \alpha_{2n_1+j} S_l^{j-1} I_a(1)$.
(4) Let h_r be $h_r = [a(0) - a(1), a(0^2) - a(0), \cdots, a(0^{n_1}) - a(0^{n_1-1})$,
* $a(1) - a(1), a(0|1) - a(0), \cdots, a(0^{n_1-1}|1) - a(0^{n_1-1})]$.*
(5) Let h^0 be $h^0 = a(1)D$

[proof]. Since a time-invariant, affine input response map $a \in F(U^*, Y)$ satisfies rank $(I/O)_a = n$, the system $((\ll S_l^N(\chi(U)) \gg, S_l), \chi^0, \chi, 0, a(1))$ which realizes a is a canonical n-dimensional almost linear system by theorem (4.5).

Select the linearly independent vectors $\{S_l^{i-1} \chi^0; 1 \leq i \leq n_1\}$ and select the linearly independent vectors $\{S_l^{j-1}(\chi^0 + \chi); 1 \leq j \leq n_2\}$ from the linearly independent vectors $\{S_l^i(\chi^0 + \chi u); u \in \{0,1\}, i \in N, 1 \leq i \leq n\}$. Let a linear map $T :\ll S_l^N(\chi(U)) \gg \to \mathbf{R}^n$ be $TS_l^{i-1}\chi^0 = \mathbf{e}_i$, $1 \leq i \leq n_1$ and $TS_l^{j-1}(\chi^0 + \chi) = \mathbf{e}_{n_1+j}$, $1 \leq j \leq n_2$. Then, by step 2), $T\chi^0 = g_r^0$ and $T\chi = g_r$ hold and by step 3), $h_r T = 0$ holds. By step 4), $F_r T = T S_l$ holds. Consequently, T is bijective and an almost linear system morphism : $((\ll R(\chi) \gg, S_l), \chi^0, \chi, 0, a(1)) \to \sigma_r = ((\mathbf{R}^n, F_r), g_r^0, g_r, h_r, a(1))$. By Corollary (4.9), the behavior of σ_s is a. It follows from the choice of $\{S_l^{i-1}\chi^0, S_l^{j-1}\chi; 1 \leq i \leq n_1, 1 \leq j \leq n_2\}$ and the determination of map T that σ_r is the real time standard system.

4.3 Partial Realization of So-called Linear Systems

Here we consider a partial realization problem by multi-experiment. Let $\underline{a}$ be an $\underline{N}$ sized time-invariant, affine input response map $(\underline{a} \in F(U_{\underline{N}}^*, Y))$, where $\underline{N} \in N$ and $U_{\underline{N}}^* := \{\omega \in U^*; |\omega| \leq \underline{N}\}$. The $\underline{a}$ is said to be a partial time-invariant, affine input response map.

A finite dimensional so-called linear system $\sigma = ((K^n, F), x^0, g, h)$ is said to be a partial realization of $\underline{a}$ if $h(\sum_{j=1}^{|\omega|} F^{|\omega|-j}(Fx^0 + g\omega(j))) = \underline{a}(\omega)$ holds for any $\omega \in U_{\underline{N}}^*$.

A partial realization problem of so-called linear systems is stated as follows:

$<$ For any given partial time-invariant, affine input response $\underline{a} \in F(U_{\underline{N}}^*, Y)$, find a partial realization σ of $\underline{a}$ such that the dimensions of state space X of σ is minimum, where the σ is said to be a minimal partial realization of $\underline{a}$. Moreover, show when the minimal realizations are isomorphic.$>$

We have noted the representation for the time-invariant, affine input response maps. The representation says that any time-invariant, affine input response map can be characterized by the modified impulse response in Definition (4.1).

Note that the modified impulse response $I : \{0,1\} \to F(N,Y)$ can be represented by $(I(u)(t)) = a(u^{t+1}) - a(u^t)$ for $u \in \{0,1\}$, $t \in N$ and the time-invariant, affine input response map $a \in F(U^*, Y)$.

For any given partial time-invariant, affine input response $\underline{a} \in F(U_{\underline{N}}^*, Y)$, this correspondence can determine a partial modified impulse response $\underline{I} : \{0,1\} \to F(N_{\underline{N}}, Y); u \mapsto [t \mapsto (\underline{I}(u))(t) = \underline{a}(u^{t+1}) - \underline{a}(u^t)$, where $N_{\underline{N}} := \{1, 2, \cdots, \underline{N}$; for some $\underline{N} \in N\}$.

For a partial time-invariant, affine input response map $\underline{a} \in F(U_{\underline{N}}^*, Y)$, the following matrix $(I/O)_{\underline{a}\ (p,\underline{N}-p)}$ is said to be a finite-sized Input/output matrix of $\underline{a}$.

$$
(I/O)_{\underline{a}\ (p,\underline{N}-p)} =
\begin{array}{c} (s,u) \\ \left(\begin{array}{ccc} & & \vdots \\ & & \vdots \\ & & \vdots \\ & & \vdots \\ \cdots & \cdots & a(u^{s+t+1}) - a(u^{s+t}) \end{array} \right) \end{array},
$$

where the row index is t.

where $0 \le s \le p$, $0 \le t \le \underline{N} - p$ and $u \in \{0,1\}$.

Since $I_{\underline{a}}(u)(i+j) = \underline{a}(u^{i+j+1}) - \underline{a}(u^{i+j})$ holds for $u \in \{0,1\}$, column vectors of $(I/O)_{\underline{a}}$ denote $S_l^i I_{\underline{a}}(u)$.

Let a matrix $(I/O)_{\underline{a}\ (p,\underline{N}-p)}(v,w)$ denote $(I/O)_{\underline{a}\ (p,\underline{N}-p)}(v,w) = [I_{\underline{a}}(0), S_l I_{\underline{a}}(0), \cdots, S_l^{v-1} I_{\underline{a}}(0), I_{\underline{a}}(1), S_l I_{\underline{a}}(1), \cdots, S_l^{w-1} I_{\underline{a}}(1)]$.

When we actually treat approximate and noisy realization problems, we will use a notation $H_{\underline{a}\ (n_1+n_2,\underline{N}-n_1-n_2)}(n_1, n_2)$ expressed as follows:
$H_{\underline{a}\ (n_1+n_2,\underline{N}-n_1-n_2)}(n_1,n_2) = [I_{\underline{a}}(0), \cdots, S_l^{n_1-1} I_{\underline{a}}(0), I_{\underline{a}}(1), \cdots, S_l^{n_2-1} I_{\underline{a}}(1)]$.

Theorem 4.14. *Let a time-invariant, affine input response map $\underline{a} \in F(U_{\underline{N}}^*, Y)$ satisfy rank $(I/O)_{\underline{a}\ (p,\underline{N}-p)} = n$ such that the rank value becomes the maximum value for some $p \in N$. Then the real time standard system $\sigma_r = ((\boldsymbol{R}^n, F_r), g_r^0, g_r, h_r, a(1))$ which realizes $\underline{a}$ is obtained by the following procedure:*
1) Select n_1 independent vectors on the vectors $\{S_l^s I_{\underline{a}}(0) : 0 \le s \le n\}$.
Select n_2 independent vectors in $\{S_l^s I_{\underline{a}}(1) : 0 \le s \le n\}$.
(2) Let the state space be $\boldsymbol{R}^n$. And let g_r^0 and g_r be as follows: $g_r^0 = \mathbf{e}_1$,

$g_r = \mathbf{e}_{n_1+1}$. Moreover, $n = n_1 + n_2$ and $\mathbf{e}_i = [0, \cdots, 0, \overset{i}{1}, 0, \cdots, 0]^T$ hold.
(3) $F_r \in \boldsymbol{R}^{n \times n}$ is given in Definition (4.11),
where $S_l^{n_1} I_{\underline{a}}(0) = \sum_{i=1}^{n_1} \alpha_{1i} S_l^{i-1} I_{\underline{a}}(0)$,
$S_l^{n_2} I_{\underline{a}}(1) = \sum_{i=1}^{n_1} \alpha_{2i} S_l^{i-1} I_{\underline{a}}(0) + \sum_{j=1}^{n_2} \alpha_{2n_1+j} S_l^{j-1} I_{\underline{a}}(1)$.
(4) Let h_r be $h_r = [\underline{a}(0) - \underline{a}(1), \underline{a}(0^2) - \underline{a}(0), \cdots, \underline{a}(0^{n_1}) - \underline{a}(0^{n_1-1}),$
$\underline{a}(1) - \underline{a}(1), \underline{a}(0|1) - \underline{a}(0), \cdots, \underline{a}(0^{n_1-1}|1) - \underline{a}(0^{n_1-1})]$.
(5) Let h^0 be $h^0 = a(1)D$

[proof]. For the selected value $p \in N$, a linear combination obtained by $S_l^{n_1} I_{\underline{a}}(0)$ and $S_l^{n_2} I_{\underline{a}}(1)$ produces $a \in F(U^*, Y)$ obtained from the linear combination. Then the obtained time-invariant, affine input response map $a \in F(U^*, Y)$

satisfies rank $(I/O)_a = n$. Hence, the almost linear system $((\ll S_l^N(\chi(U)) \gg, S_l), \chi^0, \chi, 0, a(1))$ which realizes a is a canonical n-dimensional almost linear system by theorem.

Therefore, based on the $a \in F(U^*, Y)$, we obtain the real time standard system $\sigma_r = ((\boldsymbol{R}^n, F_r), g_r^0, \ g_r, \ h_r, \ a(1))$ which realizes a is obtained by the same procudure in theorem (4.13).

4.4 Real-Time Partial Realization of Almost Linear Systems

In general, it is well known that non-linear systems can only be determined by multi-experiments. The condition that a single experiment may pretend to produce the same effects is very hard for us to find. However, we can look for special single-experiments that simulate multi-experiments for any almost linear system.

In this section, based on the results of partial realization theory in the reference [Matsuo and Hasegawa, 2003], we will state a single-experiment for so-called linear systems.

Problem 4.15. Real time partial realization problem

Let a physical object, that is, $\underline{a} \in F(U_{\underline{N}}^*, Y)$ be a finite dimensional so-called linear system. Then, for any given finite data $\{a(\bar{\omega}); \bar{\omega}$ is a finite length input $\}$, find a so-called linear system $\sigma = ((K^n, F), x^0, g, h)$ and an input $\bar{\omega} \in U^*$ such that $a_\sigma(\bar{\omega}) = a(\bar{\omega})$ for any $\omega \in U^*$.

Definition 4.16. *For finite dimensional almost linear system, if there exists a solution of the real time partial realization problem, then an input $\bar{\omega} \in U^*$ of the solution is said to be a (real time partial) realization signal.*

Lemma 4.17. *Let a given time-invariant, affine input response map $a \in F (U^*, Y)$ have the behavior of an almost linear system whose state space is less than L-dimensional. Then there exists an input of finite length $\bar{\omega} \in U^*$ such that the following algorithm provides a finite Input/output matrix, where $p := max\{L_1, L_2\}$.*

1) Find an integer L_1 such that row vectors $\{\underline{S_l}^i \chi^0 \in K^L; 0 \leq i \leq L_1 - 1\}$ are linearly independent and $\{\underline{S_l}^i \chi^0 \in K^L; 0 \leq i \leq L_1\}$ are linearly dependent. Namely, feed an input $\omega_1 := 0^{L_1+L+1}$ into the plant, where $\underline{S_l}^i \chi^0 = [a(0^{n+1}) - a(0^n), a(0^n) - a(0^{n-1}), \cdots, a(0^{L+i+1}) - a(0^{L+i})]^T$.

2) Find an integer L_2 such that row vectors $\{\underline{S_l}^i \chi^0, \underline{S_l}^i(\chi^0 + \bar{\chi}) \in K^L; 0 \leq i \leq L_j - 1, 1 \leq j \leq 2\}$ are linearly independent and $\{\underline{S_l}^i \cdot \chi^0, \underline{S_l}^i(\chi^0 + \bar{\chi} \cdot u \in K^L; 0 \leq i \leq L_j, 1 \leq j \leq 2\}$ are linearly dependent. Namely, feed a further input $\omega_2 := 0^{L_1+L-1}|1$ into the plant. Let $\bar{\omega} = \omega_2|\omega_1$.

Making the row vectors of a matrix from the row vectors
$\{\underline{S_l}^i(\chi^0 + \bar{\chi} \cdot u)) \in K_L; 0 \leq i \leq L_j, 1 \leq j \leq 2, u \in \{0,1\}\}$ *obtained by the*
above iterations, we will obtain a finite-sized Input/output matrix
$(I/O)_{a\ (L-1,p)}$*, where* $\underline{S_l}^i \bar{\chi}$
$= [a(0^i|1) - a(0^{i+1}), a(0^{i+1}|1) - a(0^{i+2}), \cdots, a(0^{i+L}|1) - a(0^{i+L+1})]^T$.
And $a(0^i|1)$ *is given by* $a(0^j|1) = a(0^{i+1}|1|0^t) - a(0^{t+1}) + a(0^t)$
for any $i, t \in N$.

Theorem 4.18. *Let a given time-invariant, affine input response map* $a \in F$
(U^*, Y) *have the behavior of an almost linear system whose state space is less*
than L-dimensional. Then there exists a realization signal such that the real-time
standard system $\sigma_s = ((K^n, F_s), g_s^0, g_s, h_s, h^0)$ *which realizes a is obtained by the*
following algorithm:
1) Find a finite Input/output matrix $(I/O)_{a\ (L-1,p)}$ *based on the algorithm*
   given in Lemma (4.17).
2) Apply the algorithm given in Theorem (4.14) to the above finite
   Input/output matrix $(I/O)_{a\ (L-1,p)}$.

4.5 Approximate Realization of So-called Linear Systems

Here, we will discuss the approximate realization problem of so-called linear
systems, which is stated as follows:
<For any given finite-length modified impulse response, find a so-called linear
system which approximates it.>
The approximate realization of non-linear systems is presented here for the
first time.

In order to make our discussion simple, we assume that the set Y of outout
is the set R of real numbers, namely 1-output.

Theorem 4.19. *Algorithm for approximate realization*
Let a partial input response map $\underline{a}$ *be considered an object which is a so-called*
linear system. Then an approximate realization $\sigma = ((R^n, F_r), g_r^0, g_r, h_r)$ *of* $\underline{a}$ *is*
given by the following algorithm:
1) Based on the ratio of the square root of eigenvalues for a matrix
   $H_{\underline{a}\ (p,\bar{p})}(p,0)H_{\underline{a}\ (p,\bar{p})}(p,0)^T$*, determine the value* n_1 *of rank for the Input/output*
   matrix $H_{\underline{a}\ (p,\bar{p})}(p,0)$*, where* $n_1 \leq p$.
   Namely, determine the value n_1 *of rank for the matrix* $H_{\underline{a}\ (p,\bar{p})}(p,0)$ *such*
   that the ratio of the square root of eigenvalues for the covariance matrix
   becomes very small. The small ratio defines the nearness of approximation
   degree.
2) The CLS method is used as follows:
   ① Let a matrix $A_1 \in R^{1 \times (n_1+1)}$ *be* $A_1 = [\alpha_{11}, \alpha_{12}, \cdots, \alpha_{1n_1}, -1]$.
   ② Choose the coefficients $\{\alpha_{1i} : 1 \leq i \leq n_1\}$ *such that*

$\sum_{j=1}^{n_1+1} \underline{S}_l^{j-1} \bar{I}_{\underline{a}} \cdot \underline{S}_l^{j-1} \bar{I}_{\underline{a}}$ *takes a minimum value, where* $\{\underline{S}_l^i \bar{I}_{\underline{a}} \in \mathbf{R}^{L \times 1} :$
$0 \leq i \leq n_1\}$ *are given by the equation* $[\bar{I}_{\underline{a}}(0), \underline{S}_l \bar{I}_{\underline{a}}(0), \cdots, \underline{S}_l^{n_1} \bar{I}_{\underline{a}}(0)]^T :=$
$A_1^T [A_1 A_1^T]^{-1} A_1 H_{\underline{a} \ (n_1+1,L)}^T (n_1+1, 0)$ *and* $H_{\underline{a} \ (n_1,L)}^T (n_1, 0) :=$
$[I_{\underline{a}}(0), \cdots, S_l^{n_1-1} I_{\underline{a}}(0), S_l^{n_1} I_{\underline{a}}(0)]$ *and* **.** *denotes the inner product of two*
vectors.
③ *Let* $h_{1r} \in \mathbf{R}^{1 \times n_1}$ *be* h_{1r}
$= [(I_{\underline{a}}(0))(0) - (\bar{I}_{\underline{a}}(0))(0), (S_l I_{\underline{a}}(0))(0) - (S_l \bar{I}_{\underline{a}}(0))(0), \cdots ,$
$(S_l^{n_1-1} I_{\underline{a}}(0))(0) - (S_l^{n_1-1} \bar{I}_{\underline{a}}(0))(0)]$.

3) Based on the ratio of the square root of eigenvalues for a matrix
$H_{\underline{a} \ (n_1+p,\bar{p})}(n_1, p) H_{\underline{a} \ (n_1+p,\bar{p})}(n_1, p)^T$, *determine the value* n_2 *of rank for the*
matrix $H_{\underline{a} \ (n_1+p,\bar{p})}(n_1, p)$, *where* $n_2 \leq p$.
Namely, determine the value n_2 *of rank for the matrix* $H_{\underline{a} \ (p,\bar{p})}(p, 0)$ *such*
that the ratio of the square root of eigenvalues for the covariance matrix
becomes very small. The small ratio defines the nearness of approximation
degree.

4) The CLS method is used as follows:
① *Let a matrix* $A_2 \in \mathbf{R}^{1 \times (n_1+n_2+1)}$ *be* $A_2 = [\alpha_{21}, \alpha_{22}, \cdots, \alpha_{2n_1+n_2}, -1]$.
② *Choose the coefficients* $\{\alpha_{2i} : 1 \leq i \leq n_1 + n_2\}$ *such that*
$\sum_{j=1}^{n_1+n_2+1} \underline{S}_l^{j-1} \bar{I}_{\underline{a}} \cdot \underline{S}_l^{j-1} \bar{I}_{\underline{a}}$ *takes a minimum value, where* $\{\underline{S}_l^i \bar{I}_{\underline{a}} \in \mathbf{R}^{L \times 1} :$
$0 \leq i \leq n_1 + n_2\}$ *are given by the equation* $[\bar{I}_{\underline{a}}, \underline{S}_l \bar{I}_{\underline{a}}, \cdots, \underline{S}_l^{n_1} \bar{I}_{\underline{a}}]^T :=$
$A_2^T [A_2 A_2^T]^{-1} A_2 H_{\underline{a} \ (n_1+n_2,L)}^T (n_1, n_2+1)$ *and* $H_{\underline{a} \ (n_1,L)}^T (n_1, n_2+1) :=$
$[I_{\underline{a}}(0), \cdots, S_l^{n_1-1} I_{\underline{a}}(0), I_{\underline{a}}(1), \cdots, S_l^{n_2-1} I_{\underline{a}}(1), S_l^{n_2} I_{\underline{a}}(1)]$. *And* **.** *denotes the*
inner product of two vectors.
③ *Let* $F_r \in \mathbf{R}^{(n_1+n_2) \times (n_1+n_2)}$ *be given as the same as in Definition (4.11).*
Let g_r^0 *be* $g_r^0 = \mathbf{e}_1$ *and* g_r *be* $g_r = \mathbf{e}_{n_1+1} - \mathbf{e}_1$,
where $\mathbf{e}_i = [0, \cdots, 0, \overset{i}{1}, 0, \cdots, 0]^T \in \mathbf{R}^{n_1+n_2}$.
④ *Let* $h_r \in \mathbf{R}^{1 \times (n_1+n_2)}$ *be* h_r
$= [h_{1r}, (I_{\underline{a}}(1))(0) - (\bar{I}_{\underline{a}}(1))(0), (S_l I_{\underline{a}}(1))(0) - (S_l \bar{I}_{\underline{a}}(1))(0), \cdots ,$
$(S_l^{n_2-1} I_{\underline{a}}(1))(0) - (S_l^{n_2-1} \bar{I}_{\underline{a}}(1))(0)]$.

[proof]. By 1) and 3), the reduction part in the data can be excluded in the sense
of the number of dimensions by using the ratio of matrix norm, which produces
a degree of information loss. The matrices A_1 in 2) and A_2 in 4) correspond
to the matrix A in Proposition (2.14). Hence, if we determine the coefficients
$\{\alpha_{ij} : i \leq i \leq 2, \ 1 \leq j \leq n_i\}$, we can obtain the approximate part of the finite-
sizes Input/output matrices $H_{\underline{a} \ (n_1+1,\bar{p})}(n_1+1, 0)$ and $H_{\underline{a} \ (n_1+n_2+1,\bar{p})}(n_1, n_2+1)$
by using Proposition (2.14).

Therefore, we obtain the approximate Input/output matrices $\hat{H}_{\underline{a} \ (n_1+1,\bar{p})}(n_1+$
$1, 0)$ and $\hat{H}_{\underline{a} \ (n_1+n_2+1,\bar{p})}(n_1, n_2+1)$. Finally, we apply Proposition (3.15) to the
$\hat{H}_{\underline{a} \ (n_1+1,\bar{p})}(n_1+1, 0)$ and $\hat{H}_{\underline{a} \ (n_1+n_2+1,\bar{p})}(n_1, n_2+1)$.

Example 4.20. Let the signals be the modified impulse response of the following 3-dimensional so-called linear system: $\sigma = ((\boldsymbol{R}^3, F), x^0, g, h)$, where $F = \begin{bmatrix} 0 & 0 & -0.7 \\ 1 & 0 & 0.6 \\ 0 & 1 & 0.7 \end{bmatrix}$, $x^0 = [0,\ 1.5,\ -2]^T$, $h = [2,\ 5,\ -3]$, $g = [1, 0, 0]^T$.

Then the approximate realization problem is solved as follows:

covariance matrix	eigenvalues			
	1	2	3	4
$H^T_{\underline{a}\ (2,50)}(2,0) H_{\underline{a}\ (2,50)}(2,0)$	474	269		
$H^T_{\underline{a}\ (3,50)}(3,0) H_{\underline{a}\ (3,50)}(3,0)$	732	288	39	
$H^T_{\underline{a}\ (4,50)}(4,0) H_{\underline{a}\ (4,50)}(4,0)$	786	380	64	0
covariance matrix	square root of eigenvalues			
$H^T_{\underline{a}\ (3,50)}(3,0) H_{\underline{a}\ (3,50)}(3,0)$	27	17	6.2	
$H^T_{\underline{a}\ (4,50)}(4,0) H_{\underline{a}\ (4,50)}(4,0)$	28	19	8	0
covariance matrix	eigenvalues			
	1	2	3	4
$H^T_{\underline{a}\ (3,50)}(2,1) H_{\underline{a}\ (3,50)}(2,1)$	2020	354	151	
$H^T_{\underline{a}\ (4,50)}(3,1) H_{\underline{a}\ (4,50)}(3,1)$	2088	489	264	0
covariance matrix	square root of eigenvalues			
$H^T_{\underline{a}\ (3,50)}(2,1) H_{\underline{a}\ (3,50)}(2,1)$	45	19	12.2	
$H^T_{\underline{a}\ (4,50)}(3,1) H_{\underline{a}\ (4,50)}(3,1)$	45.7	22	16.2	0

1) Since the ratio $\frac{8}{28} = 0.28$ obtained by the square root of $H^T_{\underline{a}\ (4,50)}(4,0) \times H_{\underline{a}\ (4,50)}(4,0)$ is large, the approximate linear system obtained by the CLS method may not be good.

2) After determining the numbers n_1 and n_2 of dimensions which are 2 and 0, we execute the approximate realization algorithm by the CLS method.

A so-called linear system $\sigma = ((\boldsymbol{R}^3,\ F_2),\ x^0,\ g,\ h_2)$ obtained by the CLS method is expressed as follows:

$$F_2 = \begin{bmatrix} 0 & 0 & -0.7 \\ 1 & 0 & 0.6 \\ 0 & 1 & 0.7 \end{bmatrix}, h_2 = [10.3,\ 1.7,\ -12],\ g = [1,\ 0,\ 0]^T, g^0 = [3.3,\ 0.35,\ -3.6]^T.$$

In this example, the original signals are considered as modified impulse responses of a 3-dimensional so-called linear system and the desirable modified impulse responses are obtained by the CLS method. The model obtained by the CLS method is a 3-dimensional so-called linear system. Therefore, a good approximate realization could not be obtained. For reference, a two-dimensional so-called linear system obtained by the CLS method is also given.

This table indicates that the 2-dimensional so-called linear system reconstructs the original signal with 63 and 129 % error to signal ratio and 0.28 and 0.27 ratio of matrices, and the 3-dimensional linear system completely reconstructs the original system.

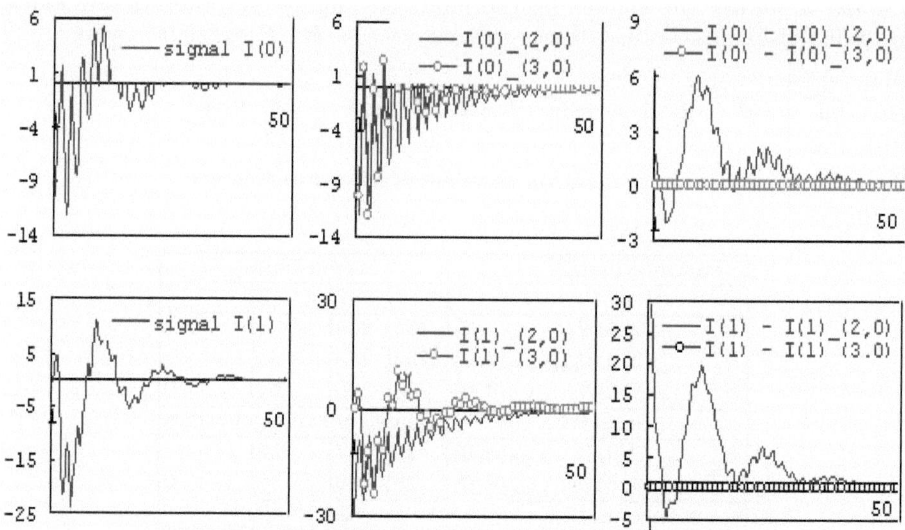

Fig. 4.1. The left are the original modified impulse responses $I(0)$ and $I(1)$. The middle are the obtained modified impulse responses $I(0)\_(2,0)$, $I(0)\_(3,0)$, $I(1)\_(2,0)$ and $I(1)\_(3,0)$ by the CLS method. The right are the difference between $I(0)$ and $I(0)\_(2,0)$ or $I(0)\_(3,0)$ and the difference between $I(1)$ and $I(1)\_(2,0)$ or $I(1)\_(3,0)$ in Example (4.20).

Just as we thought, the following table and Fig. 4.1 truly indicate that the two-dimensional so-called linear system is not a good approximation for the given system.

dimen-sion	ratio of matrices	mean values of square root for sum of			cosine	error ratio
		signal	signal by CLS	error	① and ②	
		①	②	③	$\cos\theta$	③/①
$I(0)\_(2,0)$	0.28	0.41	0.45	0.26	0.82	0.63
$I(1)\_(2,0)$	0.27	0.84	1.2	1.1	0.48	1.29
$I(0)\_(3,0)$	0	0.41	0.41	0	1	0
$I(1)\_(3,0)$	0	0.84	0.84	0	1	0

For the notations $I(0)\_(n_1, n_2)$ and $I(1)\_(n_1, n_2)$, see Definition (4.11).

Example 4.21. Let the signals be the modified impulse responses of the following 3-dimensional so-called linear system: $\sigma = ((\boldsymbol{R}^3, F), x^0, g, h)$, where $F =$

$$\begin{bmatrix} 0 & 0 & 0.8 \\ 1 & 0 & 0.2 \\ 0 & 1 & -0.5 \end{bmatrix}, \ x^0 = [1, \ 0, \ 0]^T, \ h = [10, \ 2, \ -9], \ g = [3, \ 2, \ 2]^T.$$

Then the approximate realization problem is solved as follows:

covariance matrix	eigenvalues			
	1	2	3	4
$H^T_{\underline{a}\ (2,50)}(2,0)H_{\underline{a}\ (2,50)}(2,0)$	7313	1390		
$H^T_{\underline{a}\ (3,50)}(3,0)H_{\underline{a}\ (3,50)}(3,0)$	8490	4427	0.2	
$H^T_{\underline{a}\ (4,50)}(4,0)H_{\underline{a}\ (4,50)}(4,0)$	8502	8206	0.22	0
covariance matrix	square root of eigenvalues			
$H^T_{\underline{a}\ (3,50)}(3,0)H_{\underline{a}\ (3,50)}(3,0)$	92	66.5	0.4	
$H^T_{\underline{a}\ (4,50)}(4,0)H_{\underline{a}\ (4,50)}(4,0)$	92	91	0.47	0
covariance matrix	eigenvalues			
	1	2	3	4
$H^T_{\underline{a}\ (3,50)}(2,2)H_{\underline{a}\ (3,50)}(2,2)$	8030	1581	269	0
$H^T_{\underline{a}\ (4,50)}(3,1)H_{\underline{a}\ (4,50)}(3,1)$	9318	4809	156	0
covariance matrix	square root of eigenvalues			
$H^T_{\underline{a}\ (3,50)}(2,2)H_{\underline{a}\ (3,50)}(2,2)$	89.6	39.8	16.4	0
$H^T_{\underline{a}\ (4,50)}(3,1)H_{\underline{a}\ (4,50)}(3,1)$	96.5	69.3	12.5	0

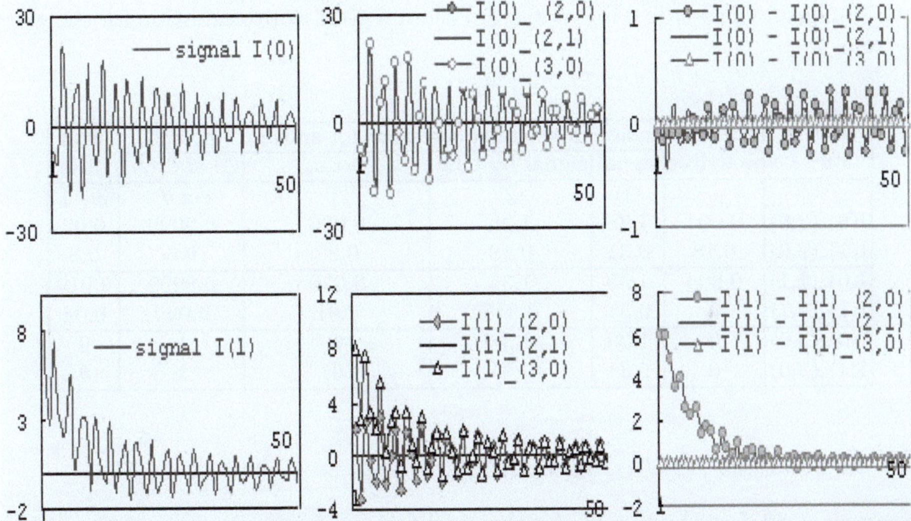

Fig. 4.2. The left are original modified impulse responses $I(0)$ and $I(1)$. The middle are obtained modified impulse responses $I(0)_-(2,0)$, $I(0)_-(2,1)$, $I(0)_-(3,0)$, $I(1)_-(2,0)$, $I(1)_-(2,1)$ and $I(1)_-(3,0)$ by the CLS method, The right is the difference between $I(0)$ and $I(0)_-(2,0)$, $I(0)_-(2,1)$ or $I(0)_-(3,0)$ and the difference between $I(1)$ and $I(1)_-(2,0)$, $I(1)_-(2,1)$ or $I(1)_-(3,0)$ in Example (4.21).

1) Since the ratio $\frac{0.4}{92} = 0.004$ obtained by the square root of $H^T_{\underline{a}\ (3,50)}(3,0) \times H_{\underline{a}\ (3,50)}(3,0)$ is small, but the ratio $\frac{16.4}{89.6} = 0.18$ obtained by the square root of $H^T_{\underline{a}\ (3,50)}(2,2)H_{\underline{a}\ (3,50)}(2,2)$ is a little large, the approximate linear system obtained by the CLS method may not be good.

2) After determining the numbers n_1 and n_2 of dimensions which are 2 and 0, we execute the approximate realization algorithm by the CLS method.

The so-called linear system $\sigma_2 = ((\mathbf{R}^3, F_2), x_2^0, g_2, h_2)$ obtained by the CLS method is expressed as follows:

$$F_2 = \begin{bmatrix} 0 & 0 & 0.8 \\ 1 & 0 & 0.2 \\ 0 & 1 & -0.5 \end{bmatrix}, \ h_2 = [-8, \ -11, \ 22], \ g_2^0 = [1, \ 0, \ 0]^T, g_2 = [-14.2, \ -19, \ -14]^T.$$

In this example, original signals are considered as the modified impulse responses of a 3-dimensional so-called linear system. The following table shows that the 2-dimensional so-called linear system reconstructs the original signal with 2 and 80 % error to signal ratio and 0.004 and 0.18 ratio of matrices. Therefore, an approximate realization could not be obtained. For reference, the two-dimensional so-called linear system obtained by the CLS method is also shown.

Just as we thought, the following table and Fig. 4.2 truly indicate that the two-dimensional so-called linear system is not a good approximation for the given system.

dimen-ion	ratio of matrices	mean values of square root for sum of signal ①	signal by CLS ②	error ③	cosine ① and ② $\cos\theta$	error ratio ③/①
I(0)-(2,0)	0.004	1.38	1.36	0.026	0.9999	0.02
I(1)-(2,0)	0.18	0.31	0.19	0.25	0.6	0.8
I(0)-(2,1)	0.004	1.38	1.36	0.026	0.9999	0.019
I(1)-(2,1)	0	0.31	0.315	0.01	0.999	0.04
I(0)-(3,0)	0	1.38	1.38	0	1	0
I(1)-(3,0)	0	0.31	0.31	0	1	0

For the notations $I(0)_-(n_1, n_2)$ and $I(1)_-(n_1, n_2)$, see Definition (4.11).

Example 4.22. Let the signals be the modified impulse responses of the following 4-dimensional so-called linear system: $\sigma = ((\mathbf{R}^4, F), x^0, g, h)$, where $F =$

$$\begin{bmatrix} 0 & 0 & 0 & 0.7 \\ 1 & 0 & 0 & 0.4 \\ 0 & 1 & 0 & -0.2 \\ 0 & 0 & 1 & 0.1 \end{bmatrix}, \ x^0 = [-1, \ 0, \ 0, \ 0]^T, \ h = [2, \ 3, \ 0, \ -4], \ g = [1, \ 0, \ 0, \ 0]^T.$$

Then the approximate realization problem is solved by the following algorithm:

covariance matrix	eigenvalues				
	1	2	3	4	5
$H_{\underline{a}\,(3,50)}^{T}(3,0)\,H_{\underline{a}\,(2,50)}(3,0)$	723	387	13.1		
$H_{\underline{a}\,(4,50)}^{T}(4,0)\,H_{\underline{a}\,(4,50)}(4,0)$	812	653	24	0	
$H_{\underline{a}\,(5,50)}^{T}(5,0)\,H_{\underline{a}\,(5,50)}(5,0)$	1043	750	24.5	0	0
covariance matrix	square root of eigenvalues				
$H_{\underline{a}\,(4,50)}^{T}(4,0)\,H_{\underline{a}\,(4,50)}(4,0)$	28.5	25.6	4.9	0	
$H_{\underline{a}\,(5,50)}^{T}(5,0)\,H_{\underline{a}\,(5,50)}(5,0)$	32.3	27.4	4.9	0	0
covariance matrix	eigenvalues				
	1	2	3	4	5
$H_{\underline{a}\,(5,50)}^{T}(3,2)\,H_{\underline{a}\,(5,50)}(3,2)$	2039	1508	27	0.7	0
$H_{\underline{a}\,(5,50)}^{T}(4,1)\,H_{\underline{a}\,(5,50)}(4,1)$	1954	896	36	0.5	0
covariance matrix	square root of eigenvalues				
$H_{\underline{a}\,(5,50)}^{T}(3,2)\,H_{\underline{a}\,(5,50)}(3,2)$	45	39	5.2	0.8	0
$H_{\underline{a}\,(5,50)}^{T}(4,1)\,H_{\underline{a}\,(5,50)}(4,1)$	44.2	30	6	0.7	0

1) Since the ratio $\frac{0}{28.5} = 0$ obtained by the square root of $H_{\underline{a}\,(4,50)}^{T}(4,0) \times H_{\underline{a}\,(4,50)}(4,0)$ is very small and the ratio $\frac{0.8}{45} = 0.02$ obtained by the square root of $H_{\underline{a}\,(5,50)}^{T}(3,2)\,H_{\underline{a}\,(5,50)}(3,2)$ is also small, an approximate almost linear system obtained by the CLS method is obtained as follows:

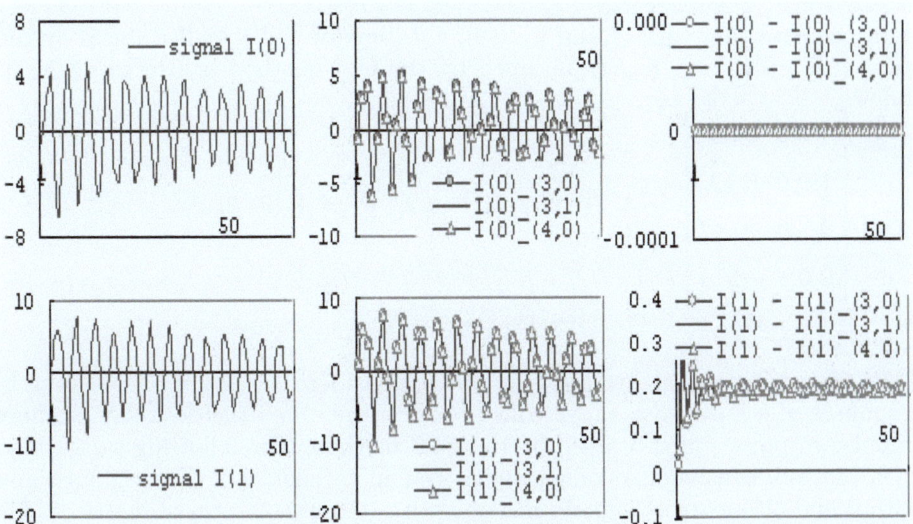

Fig. 4.3. The left are orignal modified impulse responses $I(0)$ and $I(1)$. The middle are the obtained modified impulse responses $I(0)$_$(3,0)$, $I(0)$_$(3,1)$, $I(0)$_$(4,0)$, $I(1)$_$(3,0)$, $I(1)$_$(3,1)$ and $I(1)$_$(4,0)$ by the CLS method. The right are the difference between $I(0)$ and $I(0)$_$(3,0)$, $I(0)$_$(3,1)$ or $I(0)$_$(4,0)$ and the difference between $I(1)$ and $I(1)$_$(3,0)$, $I(1)$_$(3,1)$ or $I(1)$_$(4,0)$ in Example (4.22).

2) After determining the numbers n_1 and n_2 of dimensions which are 3 and 0, we execute the approximate realization algorithm by the CLS method.

$$\sigma_2 = ((\boldsymbol{R}^3,\ F_2),\ g_2^0,\ g_2,\ h_2, h^0),\ \text{where}\ F_2 = \begin{bmatrix} 0 & 0 & -0.7 \\ 1 & 0 & -1.1 \\ 0 & 1 & -0.9 \end{bmatrix},\ h_2 = [-1,\ 3,\ 4],$$

$g_2 = [1,\ 0,\ 0]^T, g_2^0 = [0.86\ 0.52\ 0.32]^T,\ h^0 = -2.$

3) Since the ratio $\frac{3 \times 10^{-6}}{28.5} = 1 \times 10^{-7}$ obtained by the square root of $H^T_{\underline{a}\ (4,50)}(4,0) H_{\underline{a}\ (4,50)}(4,0)$ is very small and the ratio $\frac{0}{45} = 0$ obtained by the square root of $H^T_{\underline{a}\ (5,50)}(3,2) H_{\underline{a}\ (5,50)}(3,2)$ is also very small, the approximate realization obtained by the CLS method may be good.

4) After determining the numbers n_1 and n_2 of dimensions which are 3 and 1, we execute the approximate realization algorithm by the CLS method.

The approximate almost system obtained by the CLS method is obtained as follows:

$$\sigma_3 = ((\boldsymbol{R}^4,\ F_3),\ g_3^0,\ g_3,\ h_3, h^0),\ \text{where}\ F_3 = \begin{bmatrix} 0 & 0 & -0.7 & -2 \\ 1 & 0 & -1.1 & 1 \\ 0 & 1 & -0.9 & 0 \\ 0 & 0 & 0 & 1 \end{bmatrix},\ h_3 = [-1,\ 3,\ 4,\ 1],$$

$g_3 = [-1,\ 0,\ 0,\ 1]^T, g_3^0 = [1,\ 0,\ 0,\ 0]^T,\ h^0 = -2.$

In the case that $n_1 = 4$ and $n_2 = 0$, a 4-dimensional so-called linear system $\sigma_4 = ((\boldsymbol{R}^4,\ F_4),\ x_4^0,\ g,\ h_4)$ obtained by the CLS method is also expressed as follows:

$$F_4 = \begin{bmatrix} 0 & 0 & 0 & 0.53 \\ 1 & 0 & 0 & 0.13 \\ 0 & 1 & 0 & -0.42 \\ 0 & 0 & 1 & -0.14 \end{bmatrix},\ h_4 = [-1,\ 3,\ 4,\ -6.2],\ g_4^0 = [1,\ 0,\ 0,\ 0]^T,$$

$g_4 = [0.42,\ -0.15,\ -0.24,\ -0.61]^T,\ h^0 = -2.$

In this example, the original signals are considered as the modified impulse responses of a 4-dimensional so-called linear system and the desirable modified impulse responses are obtained by the CLS method. The following table shows that the 3-dimensional so-called linear system reconstructs the original signal with 0 and 4 % error to signal ratio and with 0 and 0.02 ratio of matrices. The model obtained by the CLS method is a 3-dimensional so-called linear system. Therefore, a good approximate realization was obtained. For reference, two 4-dimensional so-called linear systems are shown.

Fig. 4.3 also indicates that even the 3-dimensional so-called linear system is a good approximation for the given system.

dimen-ion	ratio of matrices	mean values of square root for sum of		error	cosine ① and ②	error ratio
		signal ①	signal by CLS ②	error ③	$\cos\theta$	③/①
$I(0)\_(3,0)$	0	0.42	0.42	0	1	0
$I(1)\_(3,0)$	0.02	0.66	0.66	0.027	0.999	0.04
$I(0)\_(3,1)$	0	0.42	0.42	0	1	0
$I(1)\_(3,1)$	0	0.66	0.66	0	1	0
$I(0)\_(4,0)$	0	0.42	0.42	0	1	0
$I(1)\_(4,0)$	0	0.66	0.66	0.027	0.999	0.04

For the notations $I(0)\_(n_1, n_2)$ and $I(1)\_(n_1, n_2)$, see Definition (4.11).

Example 4.23. Let the signals be the modified impulse responses of the following 5-dimensional so-called linear system: $\sigma = ((\boldsymbol{R}^5, F), x^0, g, h)$, where $F =$

$$\begin{bmatrix} 0\,0\,0\,0 & 0 \\ 1\,0\,0\,0 & -0.0384 \\ 0\,1\,0\,0 & -0.112 \\ 0\,0\,1\,0 & 0.52 \\ 0\,0\,0\,1 & -0.4 \end{bmatrix}, x^0 = [-10,\ 0,\ 1,\ 0,\ 0]^T,\ h = [1,\ 2,\ -5,\ -1,\ 3],$$

$g = [1,\ 0,\ 0,\ 0,\ 0]^T$.

Then the approximate realization problem is solved by the following algorithm:

covariance matrix	eigenvalues					
	1	2	3	4	5	6
$H^T_{\underline{a}\ (5,50)}(5,0)H_{\underline{a}\ (5,50)}(5,0)$	188750	16785	8583	801	15	
$H^T_{\underline{a}\ (6,50)}(6,0)H_{\underline{a}\ (6,50)}(6,0)$	225857	19651	8750	805	15	0
covariance matrix	square root of eigenvalues					
$H^T_{\underline{a}\ (5,50)}(5,0)H_{\underline{a}\ (5,50)}(5,0)$	434	130	93	28.3	3.8	
$H^T_{\underline{a}\ (6,50)}(6,0)H_{\underline{a}\ (6,50)}(6,0)$	475	140	93.5	28.3	3.8	0

covariance matrix	eigenvalues					
	1	2	3	4	5	6
$H^T_{\underline{a}\ (5,50)}(4,1)H_{\underline{a}\ (5,50)}(4,1)$	2.9×10^5	1.9×10^4	8400	663	0.2	
$H^T_{\underline{a}\ (6,50)}(5,1)H_{\underline{a}\ (6,50)}(5,1)$	3.2×10^5	2.7×10^4	1×10^4	859	15	0
covariance matrix	square root of eigenvalues					
$H^T_{\underline{a}\ (5,50)}(4,1)H_{\underline{a}\ (5,50)}(4,1)$	538	138	92	25.7	0.4	
$H^T_{\underline{a}\ (6,50)}(5,1)H_{\underline{a}\ (6,50)}(5,1)$	565	152	100	29	3.9	0

1) Since the ratio $\frac{3.8}{434} = 0.01$ obtained by the square root of $H^T_{\underline{a}\ (5,50)}(5,0) \times H_{\underline{a}\ (5,50)}(5,0)$ is very small and the ratio $\frac{0.4}{538} \approx 0$ obtained by the square root of $H^T_{\underline{a}\ (5,50)}(4,1)H_{\underline{a}\ (5,50)}(4,1)$ is also very small, the approximation of the original so-called linear system may be good.

2) After determining the numbers n_1 and n_2 of dimensions which are 4 and 0, we execute the approximate realization algorithm by the CLS method.

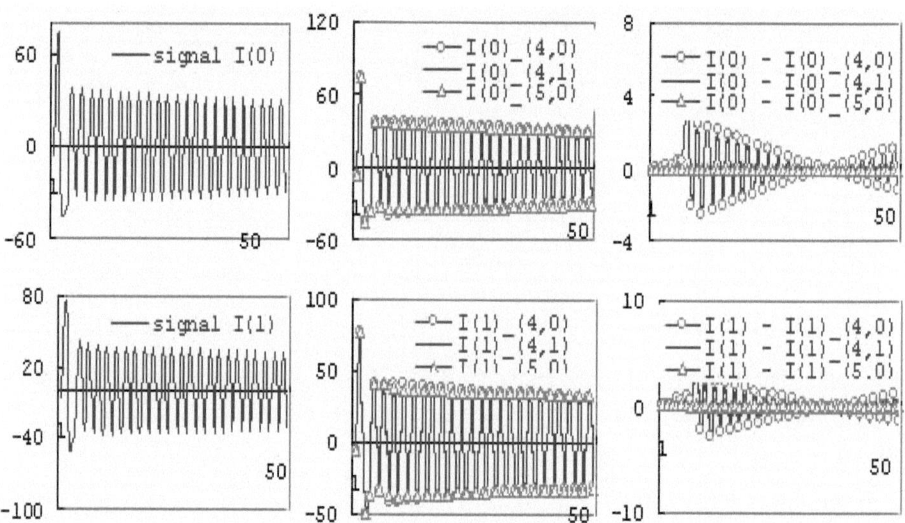

Fig. 4.4. The left are original modified impulse responses $I(0)$ and $I(1)$. The middle are obtained modified impulse responses $I(0)\_(4,0)$, $I(0)\_(4,1)$, $I(0)\_(5,0)$, $I(1)\_(4,0)$, $I(1)\_(4,1)$ and $I(1)\_(5,0)$ by the CLS method, The right are the difference between $I(0)$ and $I(0)\_(4,0)$, $I(0)\_(4,1)$ or $I(0)\_(5,0)$ and the difference between $I(1)$ and $I(1)\_(4,0)$, $I(1)\_(4,1)$ or $I(1)\_(5,0)$ in Example (4.23).

The approximate almost system obtained by the CLS method is obtained as follows:

$$\sigma_3 = ((\boldsymbol{R}^4, F_3),\ g_3^0,\ g_3,\ h_3, h^0),\ \text{where } F_3 = \begin{bmatrix} 0 & 0 & 0 & -0.08 \\ 1 & 0 & 0 & -0.18 \\ 0 & 1 & 0 & -0.22 \\ 0 & 0 & 1 & -1.1 \end{bmatrix},$$

$h_3 = [-6,\ 74,\ -44.3,\ -36.6]$, $g_3 = [0.1,\ 0.08,\ 0.07,\ 0.03]^T, g_3^0 = [1,\ 0,\ 0,\ 0]^T$, $h^0 = -15$.

In the case that $n_1 = 5$, $n_2 = 0$, a 5-dimensional so-called linear system $\sigma = ((\boldsymbol{R}^5, F_4),\ x_4^0,\ g_4,\ h_4)$ obtained by the CLS method is expressed as follows:

$$F_4 = \begin{bmatrix} 0 & 0 & 0 & 0 & 0 \\ 1 & 0 & 0 & 0 & -0.04 \\ 0 & 1 & 0 & 0 & -0.11 \\ 0 & 0 & 1 & 0 & 0.52 \\ 0 & 0 & 0 & 1 & -0.4 \end{bmatrix},\ h_4 = [-6,\ 74,\ -44.2,\ -36.4,\ 38.1],$$

$g_4^0 = [1,\ 0,\ 0,\ 0,\ 0]^T, g_4 = [0.1,\ 0.1,\ 0.09,\ 0.15,\ 0.11]^T,\ h^0 = -15$.

In this example, original signals are considered as modified impulse responses of a 5-dimensional so-called linear system and the desirable modified impulse responses are obtained by the CLS method.

The following table shows that the 4-dimensional so-called linear system reconstructs the original signal with 3.8 and 4 % error to signal ratio and with

0.01 and 0.04 ratio of matrices. Therefore, a good approximate realization was obtained. For reference, a 5-dimensional so-called linear system is also obtained by the CLS method.

Just we expected, the following table and Fig. 4.4 truly indicate that the 4-dimensional so-called linear system is a good approximation for a given 5-dimensional so-called linear system.

dimen-ion	ratio of matrices	mean values of square root for sum of			cosine ① and ②	error ratio
		signal ①	signal by CLS ②	error ③	$\cos\theta$	③/①
$I(0)_-(4,0)$	0.01	4.9	4.99	0.186	0.9995	0.038
$I(1)_-(4,0)$	0.04	5.17	5.26	0.2	0.9994	0.04
$I(0)_-(5,0)$	0	4.9	4.9	0	1	0
$I(1)_-(5,0)$	0	5.17	5.17	0	1	0

For the notations $I(0)_-(n_1, n_2)$ and $I(1)_-(n_1, n_2)$, see Definition (4.11).

Example 4.24. Let the signals be the modified impulse responses of the following 6-dimensional so-called linear system: $\sigma = ((\boldsymbol{R}^6, F), x^0, g, h)$, where

$$F = \begin{bmatrix} 0 & 0 & 0 & 0 & 0 & 0 \\ 1 & 0 & 0 & 0 & 0 & -0.0384 \\ 0 & 1 & 0 & 0 & 0 & -0.0272 \\ 0 & 0 & 1 & 0 & 0 & 0.164 \\ 0 & 0 & 0 & 1 & 0 & 0.48 \\ 0 & 0 & 0 & 0 & 1 & -0.5 \end{bmatrix}, \ h^0 = -15, \ x^0 = [-10, \ 0, \ 1, \ 0, \ 0, 0]^T,$$

$$h = [1, \ 2, \ -5, \ -1, \ 3, \ -2], \ g = [1, \ 0, \ 0, \ 0, \ 0, \ 0]^T.$$

Then the approximate realization problem is solved as follows:

covariance matrix	eigenvalues						
	1	2	3	4	5	6	7
$H^T_{\underline{a}\ (6,50)}(6,0) H_{\underline{a}\ (6,50)}(6,0)$	34538	22808	8983	533	6.7	0.1	
$H^T_{\underline{a}\ (7,50)}(7,0) H_{\underline{a}\ (7,50)}(7,0)$	36180	24840	9320	552	7.8	0.1	0
covariance matrix	square root of eigenvalues						
$H^T_{\underline{a}\ (6,50)}(6,0) H_{\underline{a}\ (6,50)}(6,0)$	186	158	95	23	2.6	0.3	
$H^T_{\underline{a}\ (7,50)}(7,0) H_{\underline{a}\ (7,50)}(7,0)$	190	158	98	23.5	2.8	0.3	0

covariance matrix	eigenvalues						
	1	2	3	4	5	6	7
$H^T_{\underline{a}\ (6,50)}(5,1) H_{\underline{a}\ (6,50)}(5,1)$	46000	22000	10000	564	6.9	0.001	
$H^T_{\underline{a}\ (7,50)}(6,1) H_{\underline{a}\ (7,50)}(6,1)$	46000	27000	11000	577	7	0.1	0
covariance matrix	square root of eigenvalues						
$H^T_{\underline{a}\ (6,50)}(5,1) H_{\underline{a}\ (6,50)}(5,1)$	214	148	100	23.7	2.6	0.03	
$H^T_{\underline{a}\ (7,50)}(6,1) H_{\underline{a}\ (7,50)}(6,1)$	214	164	104	24	2.6	0.3	0

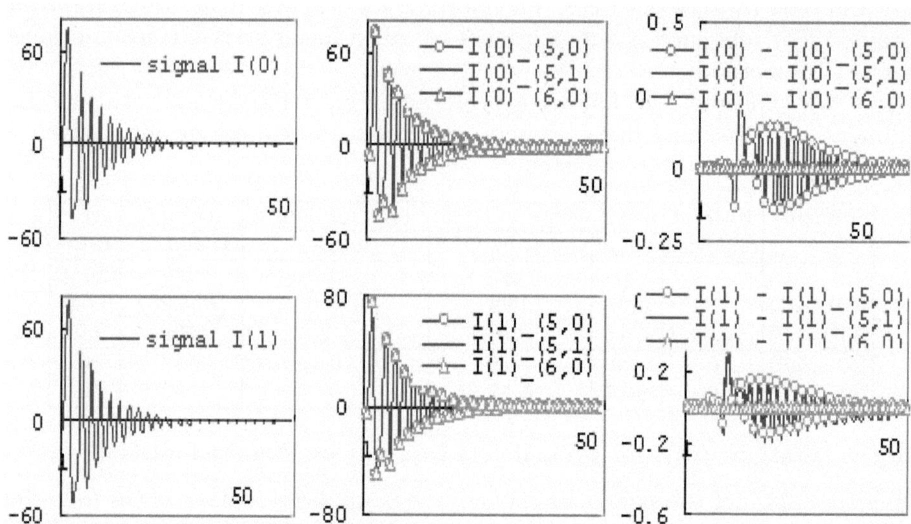

Fig. 4.5. The left are original modified impulse responses $I(0)$ and $I(1)$. The middle are obtained modified impulse responses $I(0)\_(5,0)$, $I(0)\_(5,1)$, $I(0)\_(6,0)$, $I(1)\_(5,0)$, $I(1)\_(5,1)$ and $I(1)\_(6,0)$ by the CLS method, The right is the difference between $I(0)$ and $I(0)\_(5,0)$, $I(0)\_(5,1)$ or $I(0)\_(6,0)$ and the difference between $I(1)$ and $I(1)\_(5,0)$, $I(1)\_(5,1)$ or $I(1)\_(6,0)$ in Example (4.24).

1) Since the ratio $\frac{0.3}{186} = 0.002$ obtained by the square root of $H^T_{\underline{a}\,(6,50)}(6,0) \times H_{\underline{a}\,(6,50)}(6,0)$ is very small and the ratio $\frac{0.03}{214} = 0.0001$ obtained by the square root of $H^T_{\underline{a}\,(6,50)}(5,1)H_{\underline{a}\,(6,50)}(5,1)$ is also very small, the approximate almost system obtained by the CLS method may be good.

2) After determining the numbers n_1 and n_2 of dimensions which are 5 and 0, we execute the approximate realization algorithm by the CLS method.

The approximate almost system obtained by the CLS method is obtained as follows;

$$\sigma_3 = ((\boldsymbol{R}^5,\ F_3),\ g_3^0,\ g_3,\ h_3,\ h^0),\ \text{where } F_3 = \begin{bmatrix} 0 & 0 & 0 & 0 & -0.04 \\ 1 & 0 & 0 & 0 & -0.05 \\ 0 & 1 & 0 & 0 & 0.045 \\ 0 & 0 & 1 & 0 & 0.45 \\ 0 & 0 & 0 & 1 & -0.41 \end{bmatrix},$$

$h_3 = [-6,\ 74,\ -45,\ -35.7,\ 46.2]$, $g_3 = [0.1,\ 0.1,\ 0.1,\ 0.18,\ 0.13]^T$,
$g_3^0 = [1,\ 0,\ 0,\ 0,\ 0]^T$, $h^0 = -15$.

In the case that $n_1 = 6$ and $n_2 = 0$, a 6-dimensional so-called linear system $\sigma_4 = ((\boldsymbol{R}^6,\ F_4),\ x_4^0,\ g_4,\ h_4)$ obtained by the CLS method is expressed as follows:

$$F_4 = \begin{bmatrix} 0\,0\,0\,0\,0 & 0 \\ 1\,0\,0\,0\,0 & -0.04 \\ 0\,1\,0\,0\,0 & -0.03 \\ 0\,0\,1\,0\,0 & 0.16 \\ 0\,0\,0\,1\,0 & 0.48 \\ 0\,0\,0\,0\,1 & -0.5 \end{bmatrix} , \; h_4 = [-6,\ 74,\ -45,\ -35.7,\ 46.2,\ -40.5],$$

$g_4^0 = [1,\ 0,\ 0,\ 0,\ 0,\ 0]^T, g_4 = [0.1,\ 0.1,\ 0.1,\ 0.12,\ 0.18,\ 0.12]^T, \, , \; h^0 = -15.$

The other so-called linear system σ with a 6-dimension obtained by the CLS method is also provided. The modified impulse responses are shown in Fig. 4.5.

In this example, original signals are considered as modified impulse responses of a 6-dimensional so-called linear system and the desirable modified impulse responses are obtained by the CLS method.

The following table shows that the 5-dimensional so-called linear system reconstructs the original signal with 0.5 and 0.5 % error to signal ratio and with 0.002 and 0.007 ratio of matrices. Therefore, a good approximate realization was obtained. For reference, a 6-dimensional so-called linear system obtained by the CLS method is also shown.

Just as we expected, the following table and Fig. 4.5 truly indicate that the 5-dimensional so-called linear system is a good approximation for the given 6-dimensional so-called linear system.

dimen-ion	ratio of matrices	mean values of square root for sum of			cosine ① and ②	error ratio
		signal ①	signal by CLS ②	error ③	$\cos\theta$	③/①
I(0)_(5,0)	0.002	2.58	2.58	0.012	0.99999	0.005
I(1)_(5,0)	0.007	2.71	2.71	0.013	0.99999	0.005
I(0)_(6,0)	0	2.58	2.58	0	1	0
I(1)_(6,0)	0	2.71	2.71	0	1	0

For the notations $I(0)\_(n_1, n_2)$ and $I(1)\_(n_1, n_2)$, see Definition (4.11).

4.6 Noisy Realization of So-called Linear Systems

In this section, we discuss noisy realization of so-called linear systems which are nonlinear systems. The noisy realization of nonlinear is presented here for the first time.

In order to make our discussion simple, we assume that the set Y of outout is the set R of real numbers, namely 1-output.

For noise $\{\bar{\gamma}(t) : t \in N\}$ added to an unknown so-called linear system σ, we will obtain the observed data $\{\hat{\gamma}(|\omega|) + \bar{\gamma}(|\omega|) : \omega \in U^*\}$.

For any given data $\{\hat{\gamma}(|\omega|) + \bar{\gamma}(|\omega|) : \omega \in U^*\}$, σ which satisfies $\{a_\sigma(\omega) = \hat{\gamma}(|\omega|) : \omega \in U^*\}$ is called a noisy realization of an input response map a.

We can propose the following noisy realization problem:

For any given $\{\hat{\gamma}(|\omega|) + \bar{\gamma}(|\omega|) : \omega \in U^*\}$, find a so-called linear system σ which satisfies $a_\sigma(\omega) \approx \hat{\gamma}(|\omega|)$ for any $\omega \in U^*$.

A situation for noisy realization problem 4.25

Let the observed object be a so-called linear system and noise be added to the output. Then we will obtain the data $\{\gamma(t) = \hat{\gamma}(t) + \bar{\gamma}(t) : 0 \leq t \leq \underline{N}\}$ for some integer $\underline{N} \in N$, where $\hat{\gamma}(t)$ is the exact signal which come from the observed so-called linear system and $\bar{\gamma}(t)$ is the noise added at the time of observation.

Problem 4.26. Problem statement of noisy realization for so-called linear systems

Let $H_{\underline{a}\ (p,\bar{p})}$ be the measured finite-sized Input/output matrix. Then find out the cleaned-up Input/output matrix $\hat{H}_{\underline{a}\ (p,\bar{p})}$ such that $H_{\underline{a}\ (p,\bar{p})} = \hat{H}_{\underline{a}\ (p,\bar{p})} + \bar{H}_{\underline{a}\ (p,\bar{p})}$ holds.

Namely, find out a minimal dimensional linear system $\sigma = ((\boldsymbol{R}^n, F), g, h))$ which realizes $\hat{H}_{\underline{a}\ (p,\bar{p})}$.

Theorem 4.27. *Algorithm for noisy realization*

Let a partial input response map $\underline{a}$ be a considered object which is a so-called linear system. Then a noisy realization $\sigma_r = ((\boldsymbol{R}^n, F_r), g_r^0, g_r, h_r, h^0)$ of $\underline{a}$ is given by the following algorithm:

1) Based on the square root of eigenvalues for a matrix $H_{\underline{a}\ (p,\bar{p})}(p,0) H_{\underline{a}\ (p,\bar{p})}(p,0)^T$, determine the value n_1 of rank for the Input/output matrix $H_{\underline{a}\ (p,\bar{p})}(p,0)$, where $n_1 \leq p$.

Namely, determine the value n_1 of rank for the matrix $H_{\underline{a}\ (p,\bar{p})}(p,0)$ such that a set of the square root of eigenvalues for the covariance matrix composed of relatively small and equally-sized numbers may be found, in which the signal part may be divided from the noisy part.

2) The CLS method is used as follows:

① Let a matrix $A_1 \in \boldsymbol{R}^{1 \times (n_1+1)}$ be $A_1 = [\alpha_{11}, \alpha_{12}, \cdots, \alpha_{1n_1}, -1]$.

② Choose the coefficients $\{\alpha_{1i} : 1 \leq i \leq n_1\}$ such that $\sum_{j=1}^{n_1+1} \underline{S}_l^{j-1} \bar{I}_{\underline{a}} \cdot \underline{S}_l^{j-1} \bar{I}_{\underline{a}}$ takes a minimum value, where $\{\underline{S}_l^i \bar{I}_{\underline{a}} \in \boldsymbol{R}^{L \times 1} : 0 \leq i \leq n_1\}$ are given by the equation $[\bar{I}_{\underline{a}}, \underline{S}_l \bar{I}_{\underline{a}}, \cdots, \underline{S}_l^{n_1} \bar{I}_{\underline{a}}]^T := A_1^T [A_1 A_1^T]^{-1} A_0 H_{\underline{a}\ T\ (n_1+1,L)}^T(n_1+1,0)$ and $H_{\underline{a}\ (n_1+1,L)}(n_1+1,0) :=$ $[I_{\underline{a}}(0), \cdots, S_l^{n_1-1} I_{\underline{a}}(0), S_l^{n_1} I_{\underline{a}}(0)]$. And $\cdot$ denotes the inner product of two vectors.

③ Let $h_{1r} \in \boldsymbol{R}^{1 \times n_1}$ be $h_{1r} = [I_{\underline{a}}(1) - \bar{I}_{\underline{a}}(1), I_{\underline{a}}(2) - \bar{I}_{\underline{a}}(2), \cdots, I_{\underline{a}}(n_1) - \bar{I}_{\underline{a}}(n_1)]$.

3) Based on the square root of eigenvalues for a matrix $H_{\underline{a}\ (n_1+p,\bar{p})}(n_1,p) H_{\underline{a}\ (n_1+p,\bar{p})}(n_1,p)^T$, determine the value n_2 of rank for the matrix $H_{\underline{a}\ (n_1+p,\bar{p})}(n_1,p)$, where $n_2 \leq p$.

Namely, determine the value n_2 of rank for the matrix $H_{\underline{a}\ (n_1+p,\bar{p})}(n_1,p)$ such that a set of the square root of eigenvalues for the covariance matrix composed of relatively small and equally-sized numbers is excluded, where the signal part effected by the set may be the noisy part.

4) The CLS method is used as follows:

① Let a matrix $A_2 \in \boldsymbol{R}^{1 \times (n_1+n_2+1)}$ be $A_2 = [\alpha_{21}, \alpha_{22}, \cdots, \alpha_{2n_1+n_2}, -1]$.

② Choose the coefficients $\{\alpha_{2i} : 1 \leq i \leq n_1 + n_2\}$ such that $\sum_{j=1}^{n_1+n_2+1} \underline{S}_l^{j-1} \bar{I}_{\underline{a}} \cdot \underline{S}_l^{j-1} \bar{I}_{\underline{a}}$ takes a minimum value, where $\{\underline{S}_l^i \bar{I}_{\underline{a}} \in \boldsymbol{R}^{L \times 1} :$

$0 \le i \le n_1 + n_2 \}$ *are given by the equation* $[\bar{I}_{\underline{a}}, \underline{S}_l \bar{I}_{\underline{a}}, \cdots, \underline{S}_l^{n_1} \bar{I}_{\underline{a}}]^T :=$
$A_2^T [A_2 A_1^T]^{-1} A_2 H_{\underline{a}\ (n_1+n_2+1,L)}^T (n_1, n_2 + 1)$ *and* $H_{\underline{a}\ (n_1+n_2+1,L)} (n_1, n_2 + 1) :=$
$[I_{\underline{a}}(0), \cdots, S_l^{n_1-1} I_{\underline{a}}(0), I_{\underline{a}}(1), \cdots, S_l^{n_2-1} I_{\underline{a}}(1), S_l^{n_2} I_{\underline{a}}(1)]$. *And* **.** *denotes the inner product of two vectors.*
③ *Let* $F_r \in \mathbf{R}^{(n_1+n_2)\times(n_1+n_2)}$ *be given as the same as in Definition (4.11). Let* g_r^0 *be* $g_r^0 = \mathbf{e}_1$ *and* g_r *be* $g_r = \mathbf{e}_{n_1+1} - \mathbf{e}_1$, *where*

$$\mathbf{e}_i = [0, \cdots, 0, \overset{i}{1}, 0, \cdots, 0]^T \in \mathbf{R}^{n_1+n_2}.$$

④ *Let* h_r *be* $h_s =$
$[h_{1r}, I_{\underline{a}}(1) - \bar{I}_{\underline{a}}(1), I_{\underline{a}}(1^2) - \bar{I}_{\underline{a}}(1^2), \cdots, I_{\underline{a}}(1^{n_2-1}) - \bar{I}_{\underline{a}}(1^{n_2-1})]$.

[proof]. By 1) and 3), the noisy part in the data can be excluded in the sense of the number of dimensions by checking what part is the noisy part. The matrices A_1 in 2) and A_2 in 4) correspond to the matrix A in Proposition (2.14). Hence, if we determine the coefficients $\{\alpha_{ij} : 0 \le i \le 1,\ 1 \le j \le n_{i+1}\}$, we can obtain the noisy part of the finite Input/output matrices $H_{\underline{a}\ (n_1+1,\bar{p})}(n_1 + 1, 0)$ and $H_{\underline{a}\ (n_1+n_2+1,\bar{p})}(n_1, n_2 + 1)$ by using Proposition (2.14).

Therefore, we obtain the cleaned-up Input/output matrices $\hat{H}_{\underline{a}\ (n_1+1,\bar{p})}(n_1 + 1, 0)$ and $\hat{H}_{\underline{a}\ (n_1+n_2+1,\bar{p})}(n_1, n_2 + 1)$. Finally, we apply Proposition (3.15) to the $\hat{H}_{\underline{a}\ (n_1+1,\bar{p})}(n_1 + 1, 0)$ and $\hat{H}_{\underline{a}\ (n_1+n_2+1,\bar{p})}(n_1, n_2 + 1)$.

Remark 1: Let S and N be the norm of a signal and a noise. Then the selected ratio of matrices in the algorithm may be considered as $\frac{N}{S+N}$.

Remark 2: This noisy realization method is very new.

Example 4.28. Let signals be the modified impulse responses of the following 2-dimensional so-called linear system $\sigma = ((R^2, F), x^0, g, h)$,

where $F = \begin{bmatrix} 0 & 0.4 \\ 1 & 0.5 \end{bmatrix}$, $h = [10,\ 2]$, $x^0 = [-1, 1]^T$, $g = [1,\ 0]^T$.

The almost linear system which corresponds to the so-called linear system is given by $\sigma = ((R^2, F), g^0, g, h, h^0)$, where $g^0 = [1.4,\ -1.5]^T$, $h^0 = 11$. Let added noises be given in Fig. 4.6.

Then the noisy realization problem is solved as follows:

covariance matrix	eigenvalues				
	1	2	3	4	5
$H_{\underline{a}\ (3,40)}^T(3,0) H_{\underline{a}\ (3,40)}(3,0)$	165.2	3.7	0.98		
$H_{\underline{a}\ (4,40)}^T(4,0) H_{\underline{a}\ (4,40)}(4,0)$	166	5.1	2.1	0.65	
$H_{\underline{a}\ (5,40)}^T(5,0) H_{\underline{a}\ (5,40)}(5,0)$	167	6	3	1.5	0.4
covariance matrix	square root of eigenvalues				
	1	2	3	4	5
$H_{\underline{a}\ (5,40)}^T(5,0) H_{\underline{a}\ (5,40)}(5,0)$	13	2.4	1.7	1.2	0.6

covariance matrix	eigenvalues				
	1	2	3	4	5
$H^T_{\underline{a}\ (3,40)}(1,2)H_{\underline{a}\ (3,40)}(1,2)$	697	170.7	0.7		
$H^T_{\underline{a}\ (4,40)}(1,3)H_{\underline{a}\ (4,40)}(1,3)$	793	210	3.6	0.8	
$H^T_{\underline{a}\ (5,40)}(1,4)H_{\underline{a}\ (5,40)}(1,4)$	832	262	4	3.2	0.7
covariance matrix	square root of eigenvalues				
	1	2	3	4	5
$H^T_{\underline{a}\ (5,40)}(1,4)H_{\underline{a}\ (5,40)}(1,4)$	29	16.2	2	1.8	0.8

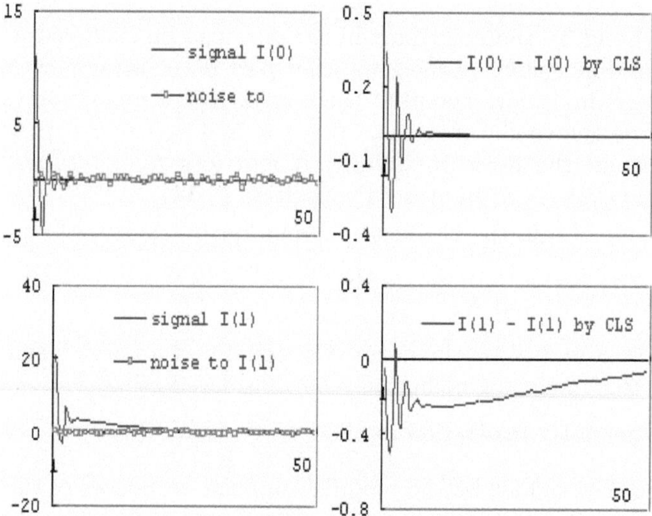

Fig. 4.6. The left are exact modified impulse responses $I(0)$ and $I(1)$ and noises added to $I(0)$ and $I(1)$. The right are the difference between the exact ones and the obtained modified impulse responses by CLS in Example (4.28).

1) Since a set $\{2.4,\ 1.7,\ 1.2,\ 0.6\}$ is composed of relatively small and equally-sized numbers in the square root of $H^T_{\underline{a}\ (5,40)}(5,0)H_{\underline{a}\ (5,40)}(5,0)$, the noisy realization so-called linear system obtained by the CLS method may be good for a 1-dimensional space.

2) After determining the number n_1 of dimensions which is 1, we will continue the noisy realization algorithm by the CLS method.

Therefore, the modified impulse response $I(0)$ of a so-called linear system obtained by the CLS method is obtained by a 1-dimensional so-called linear system.

3) Since a set $\{2,\ 1.8,\ 0.8\}$ is composed of relatively small and equally-sized numbers in the square root of $H^T_{\underline{a}\ (5,40)}(1,4)H_{\underline{a}\ (5,40)}(1,4)$, the almost linear system obtained by the CLS method may be somewhat good by adding a further 1-dimensional space.

4) After determining the number n_2 of dimensions which is 1, we will continue the noisy realization algorithm by the CLS method.

Therefore, the modified impulse response $I(1)$ of a so-called linear system obtained by the CLS method is constructed by adding an additional 1-dimensional space.

The 2- dimensional system is given by $\sigma_o = ((\boldsymbol{R}^2, \ F_o), \ g_o^0, g_o, h_o, \ h^0)$, where

$$F_o = \begin{bmatrix} -0.41 & -2.1 \\ 0 & 0.93 \end{bmatrix}, \ g_o^0 = \mathbf{e}_1, \ g_o = [-1, \ 1]^T, \ h_o = [10.7, \ 21.1] \text{ and } h^0 = 11.$$

The obtained modified impulse responses $I(0)$ and $I(1)$ are shown in Fig. 4.6.

In this example, original signals are considered as the modified impulse responses of a 2-dimensional linear system and the desirable modified impulse responses are obtained by the CLS method. The model obtained by the CLS method is the 2-dimensional so-called linear system which has the same number of dimensions as the number of the original system.

The following table indicates that the 2-dimensional so-called linear system reconstructs the original signal with a 4 and 6 % error to signal ratio and with 0.18 and 0.07 noise to signal ratio, please refer to Remark 1 in Theorem 4.26 for the noise to signal ratio.

Just as we expected, the following table and Fig. 4.6 indicate that the model obtained by the CLS method is a good 2-dimensional system for the original 2-dimensional system.

dimen-ion	ratio of matrices	mean values of square root for sum of			cosine	error ratio
		signal	signal by CLS	error	① and ②	
		①	②	③	$\cos\theta$	③/①
$I(0)$_$(1,1)$	0.18	0.243	0.234	0.01	0.9999	0.04
$I(1)$_$(1,1)$	0.07	0.489	0.497	0.03	0.998	0.06

For the notations $I(0)$_(n_1, n_2) and $I(1)$_(n_1, n_2), see Definition (4.11).

Example 4.29. Let signals be the modified impulse responses of the following 2-dimensional so-called linear system $\sigma = ((R^2, F), x^0, g, h)$,

where $F = \begin{bmatrix} 0 & 0.7 \\ 1 & 0.2 \end{bmatrix}, \ h = [10, \ 3], \ x^0 = [-12.2, -16]^T, \ g = [1, \ 0]^T.$

The almost linear system which corresponds to the so-called linear system is given by $\sigma = ((R^2, F), g^0, g, h, h^0)$, where $g^0 = [1, \ 0.6]^T, \ h^0 = -170.$

Let added noises be given in Fig. 4.7.

Then the noisy realization problem is solved as follows:

covariance matrix	eigenvalues				
	1	2	3	4	5
$H_{\underline{a}\ (3,60)}^T(3,0)H_{\underline{a}\ (3,60)}(3,0)$	2276	22.1	6.1		
$H_{\underline{a}\ (4,60)}^T(4,0)H_{\underline{a}\ (4,60)}(4,0)$	2857	24	8.1	3.7	
$H_{\underline{a}\ (5,60)}^T(5,0)H_{\underline{a}\ (5,60)}(5,0)$	3383	30	10.1	5.4	2.6
covariance matrix	square root of eigenvalues				
	1	2	3	4	5
$H_{\underline{a}\ (5,60)}^T(5,0)H_{\underline{a}\ (5,60)}(5,0)$	58.2	5.5	3.2	2.3	1.9

covariance matrix	eigenvalues				
	1	2	3	4	5
$H^T_{\underline{a}\ (3,60)}(2,1)H_{\underline{a}\ (3,60)}(2,1)$	4058	40.8	5.7		
$H^T_{\underline{a}\ (4,60)}(2,2)H_{\underline{a}\ (4,60)}(2,2)$	5959	133	6.7	2.3	
$H^T_{\underline{a}\ (5,60)}(2,3)H_{\underline{a}\ (5,60)}(2,3)$	7709	150	8.12	6.3	3.3
covariance matrix	square root of eigenvalues				
	1	2	3	4	5
$H^T_{\underline{a}\ (5,60)}(2,3)H_{\underline{a}\ (5,60)}(2,3)$	87.8	12.2	2.8	2.5	1.8

1) Since a set $\{3.2,\ 2.3,\ 1.9\}$ is composed of relatively small and equally-sized numbers in the square root of $H^T_{\underline{a}\ (5,60)}(5,0)H_{\underline{a}\ (5,60)}(5,0)$, the noisy realization of a so-called linear system obtained by the CLS method may be good for a 2-dimensional space.

2) After determining the number n_1 of dimensions which is 2, we will continue the noisy realization algorithm by the CLS method.

Therefore, the modified impulse response $I(0)$ of a so-called linear system obtained by the CLS method is constructed for a 2-dimensional space.

3) Since a set $\{2.8,\ 2.5,\ 1.8\}$ is composed of relatively small and equally-sized numbers in the square root of $H^T_{\underline{a}\ (5,60)}(2,3)H_{\underline{a}\ (5,60)}(2,3)$, the almost linear system obtained by the CLS method may be somewhat good by adding nothing.

4) After determining the number n_2 of dimensions which is 0, we will continue the noisy realization algorithm by the CLS method.

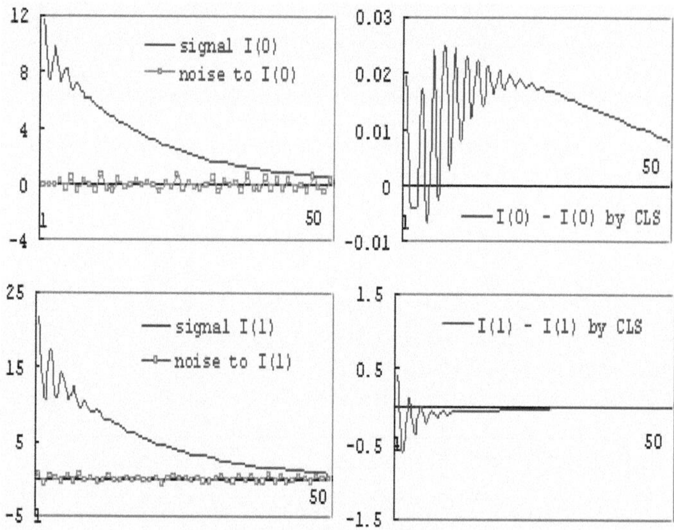

Fig. 4.7. The left are exact modified impulse responses $I(0)$ and $I(1)$ and noises added to $I(0)$ and $I(1)$. The right is the difference between the exact ones and obtained modified impulse responses by CLS in Example (4.29).

Therefore, the modified impulse response $I(1)$ of a so-called linear system obtained by the CLS method is obtained by adding nothing.

Therefore, the modified impulse responses $I(0)$ and $I(1)$ of an almost linear system obtained by the CLS method is obtained by a 2-dimensional so-called linear system.

The system is given by $\sigma_o = ((\boldsymbol{R}^2,\ F_o),\ g_o^0, g_o, h_o,\ h^0)$, where $F_o = \begin{bmatrix} 0 & 0.7 \\ 1 & 0.2 \end{bmatrix}$, $g_o^0 = \mathbf{e}_1$, $g_o = [1.15,\ -0.52]^T$, $h_o = [11.8,\ 7.5]$ and $h^0 = -170$.

The obtained modified impulse responses $I(0)$ and $I(1)$ are illustrated in Fig. 4.7.

In this example, original signals are considered as the modified impulse responses of a 2-dimensional so-called linear system and the desirable modified impulse responses are obtained by the CLS method. The model obtained by the CLS method is a 2-dimensional so-called linear system which has the same number of dimensions as the number of the original system.

The following table indicates that the 2-dimensional so-called linear system reconstructs the original signal with a 0.2 and 2 % error to signal ratio and with 0.05 and 0.03 noise to signal ratio, please refer to Remark 1 in Theorem 4.26 for the noise to signal ratio.

Just as we expected, the following table and Fig. 4.7 indicate that the model obtained by the CLS method is a good 2-dimensional system for the original 2-dimensional system.

dimen-ion	ratio of matrices	mean values of square root for sum of			cosine ① and ②	error ratio
		signal ①	signal by CLS ②	error ③	$\cos\theta$	③/①
$I(0)_-(2,0)$	0.05	0.591	0.590	0.002	1	0.002
$I(1)_-(2,0)$	0.03	0.988	1.01	0.018	0.9999	0.02

For the notations $I(0)_-(n_1, n_2)$ and $I(1)_-(n_1, n_2)$, see Definition (4.11).

Example 4.30. Let signals be the modified impulse responses of the following 3-dimensional so-called linear system $\sigma = ((R^3, F), x^0, g, h)$,

where $F = \begin{bmatrix} 0 & 0 & -0.7 \\ 1 & 0 & 0.6 \\ 0 & 1 & 0.7 \end{bmatrix}$, $h = [2,\ 5,\ -3]$, $x^0 - [0,\ 1.5,\ -2]^T$, $g = [1,\ 0,\ 0]^T$.

The almost linear system which corresponds to the so-called linear system is given by $\sigma = ((R^3, F), g^0, g, h, h^0)$, where $g^0 = [1.4,\ -2.7,\ 2.1]^T$, $h^0 = 13.5$.

Let added noises be given in Fig. 4.8.

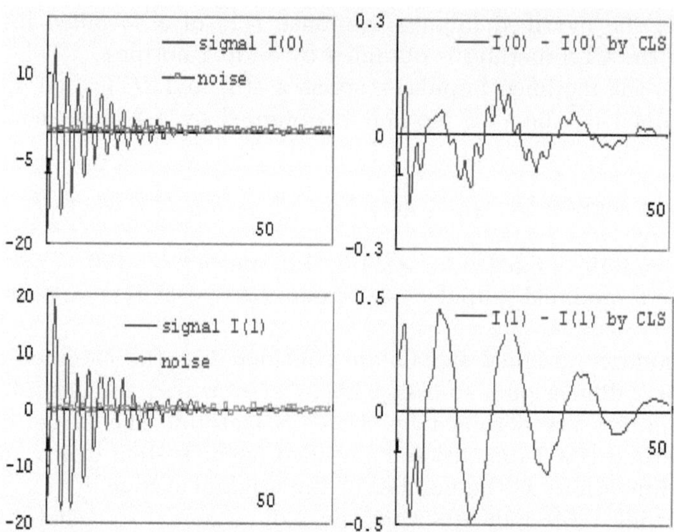

Fig. 4.8. The left are the exact modified impulse responses $I(0)$ and $I(1)$ and noises added to $I(0)$ and $I(1)$. The right is the difference between the exact ones and the obtained modified impulse responses by CLS in Example (4.30).

Then the noisy realization problem is solved as follows:

covariance matrix	eigenvalues					
	1	2	3	4	5	6
$H_{\underline{a}\,(4,20)}^T(4,0)H_{\underline{a}\,(4,20)}(4,0)$	3644	30	5.7	0.81		
$H_{\underline{a}\,(5,20)}^T(5,0)H_{\underline{a}\,(5,20)}(5,0)$	4134	31	11	1.2	0.5	
$H_{\underline{a}\,(6,20)}^T(6,0)H_{\underline{a}\,(6,20)}(6,0)$	4494	31	12.6	1.9	0.6	0.3
covariance matrix	square root of eigenvalues					
	1	2	3	4	5	6
$H_{\underline{a}\,(6,20)}^T(6,0)H_{\underline{a}\,(6,20)}(6,0)$	67	5.6	3.5	1.4	0.8	0.5

covariance matrix	eigenvalues					
	1	2	3	4	5	6
$H_{\underline{a}\,(4,20)}^T(3,1)H_{\underline{a}\,(4,20)}(3,1)$	4762	82	23	0.5		
$H_{\underline{a}\,(5,20)}^T(3,2)H_{\underline{a}\,(5,20)}(3,2)$	6176	225	33.3	0.6	0.4	
$H_{\underline{a}\,(6,20)}^T(3,3)H_{\underline{a}\,(6,20)}(3,3)$	7290	273	80.3	0.6	0.4	0.2
covariance matrix	square root of eigenvalues					
	1	2	3	4	5	6
$H_{\underline{a}\,(6,20)}^T(3,3)H_{\underline{a}\,(6,20)}(3,3)$	85.4	16.5	9	0.8	0.6	0.4

1) Since a set $\{1.4,\ 0.8,\ 0.5\}$ is composed of relatively small and equally-sized numbers in the square root of $H_{\underline{a}\,(6,20)}^T(6,0)H_{\underline{a}\,(6,20)}(6,0)$, the noisy realization of a so-called linear system obtained by the CLS method may be good for a 3-dimensional space.

2) After determining the number n_1 of dimensions which is 3, we will continue the noisy realization algorithm by the CLS method.

Therefore, the modified impulse response $I(0)$ of a so-called linear system obtained by the CLS method is constructed by a 3-dimensional so-called linear system.

3) Since a set $\{0.8,\ 0.6,\ 0.4\}$ is composed of relatively small and equally-sized numbers in the square root of $H^T_{\underline{a}\ (6,20)}(3,3)H_{\underline{a}\ (6,20)}(3,3)$, an almost linear system obtained by the CLS method may be somewhat good by adding nothing.

4) After determining the number n_2 of dimensions which is 0, we will continue the noisy realization algorithm by the CLS method.

Therefore, the modified impulse response $I(1)$ of a so-called linear system obtained by the CLS method is constructed by adding nothing.

Therefore, the modified impulse responses $I(0)$ and $I(1)$ of an almost linear system obtained by the CLS method is constructed by a 3-dimensional almost linear system.

The system is given by $\sigma_o = ((\boldsymbol{R}^3,\ F_o),\ g^0_o, g_o, h_o,\ h^0)$, where $F_o =$

$$\begin{bmatrix} 0 & 0 & -0.7 \\ 1 & 0 & 0.61 \\ 0 & 1 & 0.74 \end{bmatrix},\ g^0_o = \mathbf{e}_1,\ g_o = [3.5,\ 0.35,\ -3.8]^T,\ h_o = [-16.9,\ 13.9,\ -14.5]$$

and $h^0 = 13.5$.

The obtained modified impulse responses $I(0)$ and $I(1)$ are illustrated in Fig. 4.8.

In this example, original signals are considered as the modified impulse responses of a 3-dimensional linear system and the desirable modified impulse responses are obtained by the CLS method. A model obtained by the CLS method is a 3-dimensional so-called linear system which has the same number of dimensions as the number of the original system.

The following table indicates that the 3-dimensional so-called linear system reconstructs the original signal with a 1 and 4 % error to signal ratio and with 0.04 and 0.03 noise to signal ratio, please refer to Remark 1 in Theorem 4.26 for the noise to signal ratio.

Just as we expected, the following table and Fig. 4.8 indicate that the model obtained by the CLS method is a good 3-dimensional system for the original 3-dimensional system.

dimen-ion	ratio of matrices	mean values of square root for sum of			cosine ① and ②	error ratio
		signal ①	signal by CLS ②	error ③	$\cos\theta$	③/①
$I(0)$_$(3,0)$	0.04	0.72	0.718	0.008	0.9999	0.01
$I(1)$_$(3,0)$	0.03	0.85	0.845	0.03	0.999	0.04

For the notations $I(0)$_(n_1, n_2) and $I(1)$_(n_1, n_2), see Definition (4.11).

Example 4.31. Let signals be the modified impulse responses of the following 4-dimensional so-called linear system $\sigma = ((R^4, F), x^0, g, h)$, where $F = \begin{bmatrix} 0 & 0 & 0 & -0.5 \\ 1 & 0 & 0 & -0.6 \\ 0 & 1 & 0 & 1.1 \\ 0 & 0 & 1 & 0.4 \end{bmatrix}$, $h = [2,\ 3,\ 0,\ -4]$, $x^0 = [-1,\ 0,\ 0.1,\ 0]^T$, $g = [1,\ 0,\ 0,\ 0]^T$.

The almost linear system which corresponds to the so-called linear system is given by $\sigma = ((R^4, F), g^0, g, h, h^0)$, where $g^0 = [1,\ -1,\ -0.1,\ 0.1]^T$, $h^0 = -2$.

Let added noises be given in Fig. 4.9.

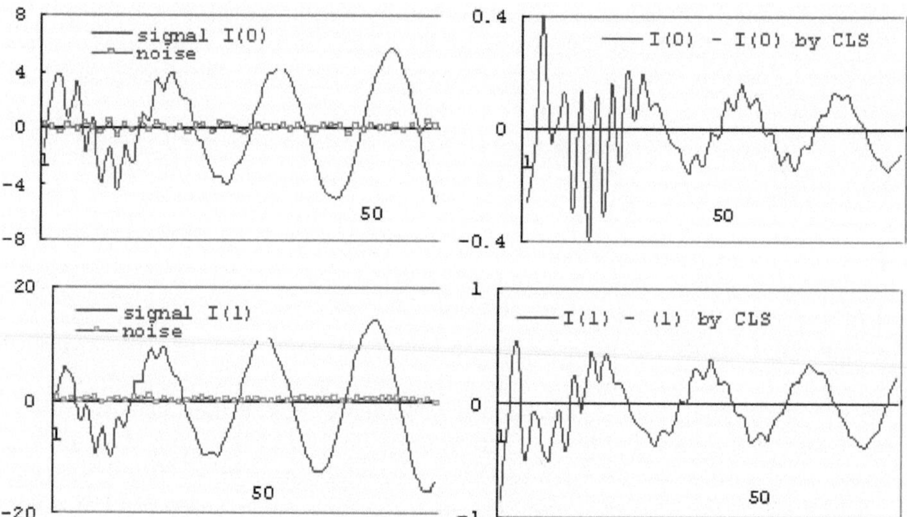

Fig. 4.9. The left are exact modified impulse responses $I(0)$ and $I(1)$ and noises added to $I(0)$ and $I(1)$. The right is the difference between the exact ones and the obtained modified impulse responses by CLS in Example (4.31).

Then the noisy realization problem is solved as follows:

covariance matrix	eigenvalues						
	1	2	3	4	5	6	7
$H^T_{\underline{a}\ (5,80)}(5,0)\,H_{\underline{a}\ (5,80)}(5,0)$	5577	2731	149	19.3	3.8		
$H^T_{\underline{a}\ (6,80)}(6,0)\,H_{\underline{a}\ (6,80)}(6,0)$	5721	4422	164	29.8	4.5	3	
$H^T_{\underline{a}\ (7,80)}(7,0)\,H_{\underline{a}\ (7,80)}(7,0)$	6402	5721	192	31	5.9	3.5	2.3
covariance matrix	square root of eigenvalues						
	1	2	3	4	5	6	7
$H^T_{\underline{a}\ (7,80)}(7,0)\,H_{\underline{a}\ (7,80)}(7,0)$	80	75.6	13.9	5.6	2.4	1.9	1.5

covariance matrix	eigenvalues						
	1	2	3	4	5	6	7
$H^T_{\underline{a}\ (5,80)}(4,1)H_{\underline{a}\ (5,80)}(4,1)$	15313	1853	149	21.9	3.3		
$H^T_{\underline{a}\ (6,80)}(4,2)H_{\underline{a}\ (6,80)}(4,2)$	24108	3811	182	24.3	3.4	1.9	
$H^T_{\underline{a}\ (7,80)}(4,3)H_{\underline{a}\ (7,80)}(4,3)$	31101	7992	320	26	3.4	2.5	1.5
covariance matrix	square root of eigenvalues						
	1	2	3	4	5	6	7
$H^T_{\underline{a}\ (7,80)}(4,3)H_{\underline{a}\ (7,80)}(4,3)$	176	89.4	17.9	5.1	1.8	1.6	1.2

1) Since a set $\{2.4,\ 1.9,\ 1.5\}$ is composed of relatively small and equally-sized numbers in the square root of $H^T_{\underline{a}\ (7,80)}(7,0)H_{\underline{a}\ (7,80)}(7,0)$, the noisy realization of a so-called linear system obtained by the CLS method may be good for a 4-dimensional space.

2) After determining the number n_1 of dimensions which is 4, we will continue the noisy realization algorithm by the CLS method.

Therefore, the modified impulse response $I(0)$ of a so-called linear system obtained by the CLS method is constructed for a 4-dimensional space.

3) Since a set $\{1.8,\ 1.6,\ 1.21\}$ is composed of relatively small and equally-sized numbers in the square root of $H^T_{\underline{a}\ (7,80)}(4,3)H_{\underline{a}\ (7,80)}(4,3)$, an almost linear system obtained by the CLS method may be somewhat good by adding nothing.

4) After determining the number n_2 of dimensions which is 0, we will continue the noisy realization algorithm by the CLS method.

Therefore, the modified impulse response $I(1)$ of a so-called linear system obtained by the CLS method is constructed by adding nothing.

Therefore, the modified impulse responses $I(0)$ and $I(1)$ of the so-called linear system obtained by the CLS method is realized by a 4-dimensional so-called linear system.

The system is given by $\sigma_o = ((\mathbf{R}^4,\ F_o),\ g_o^0, g_o, h_o,\ h^0)$, where $F_o =$

$$\begin{bmatrix} 0 & 0 & 0 & -0.6 \\ 1 & 0 & 0 & -0.53 \\ 0 & 1 & 0 & 1.2 \\ 0 & 0 & 1 & 0.29 \end{bmatrix},\ g_o^0 = \mathbf{e}_1,\ g_o = [-0.11,\ -0.87,\ 1.23,\ 1.7]^T,$$

$h_o = [-1.1,\ 3.1,\ 3.3,\ 0.66]$ and $h^0 = -2$. The obtained modified impulse responses $I(0)$ and $I(1)$ are illustrated in Fig. 4.9.

In this example, original signals are considered as the modified impulse responses of a 4-dimensional linear system and the desirable modified impulse responses are obtained by the CLS method. The model obtained by the CLS method is a 4-dimensional so-called linear system which has the same number of dimensions as the number of the original system.

The following table indicates that the 4-dimensional so-called linear system reconstructs the original signal with a 5 and 3 % error to signal ratio and with 0.03 and 0.01 noise to signal ratio, please refer to Remark 1 in Theorem 4.26 for the noise to signal ratio.

Just as we expected, the following table and Fig. 4.9 indicate that the model obtained by the CLS method is somewhat good 4-dimensional system for the original 4-dimensional system.

dimen-ion	ratio of matrices	mean values of square root for sum of			cosine ① and ②	error ratio
		signal ①	signal by CLS ②	error ③	$\cos\theta$	③/①
$I(0)_-(4,0)$	0.03	0.442	0.435	0.02	0.999	0.05
$I(1)_-(4,0)$	0.01	1.176	1.159	0.04	0.9995	0.03

For the notations $I(0)_-(n_1, n_2)$ and $I(1)_-(n_1, n_2)$, see Definition (4.11).

Example 4.32. Let signals be the modified impulse responses of the following 4-dimensional so-called linear system $\sigma = ((R^4, F), x^0, g, h)$,

$$\text{where } F = \begin{bmatrix} 0\ 0\ 0\ -0.6 \\ 1\ 0\ 0\ -0.8 \\ 0\ 1\ 0\ \ 0.4 \\ 0\ 0\ 1\ -0.3 \end{bmatrix}, \ h = [10,\ 2,\ 0,\ -4], \ h^0 = 6.37,$$

$$x^0 = [0.16087,\ 1.5087,\ -0.665217,\ -0.434783]^T, \ g = [1,\ 0,\ 0,\ 0]^T.$$

The almost linear system which corresponds to the so-called linear system is given by $\sigma = ((R^4, F), g^0, g, h, h^0)$, where $g^0 = [0.1,\ -1,\ 2,\ -0.1]^T$, $h^0 = 6.37$.
Let added noises be given in Fig. 4.10.

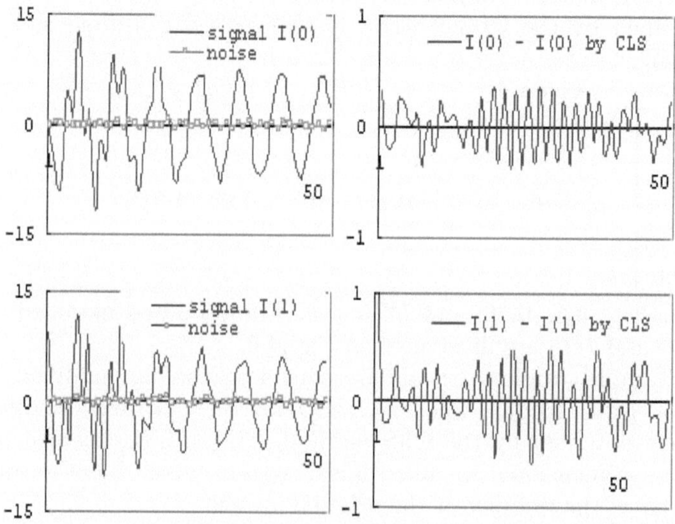

Fig. 4.10. The left are exact modified impulse responses $I(0)$ and $I(1)$ and noises added to $I(0)$ and $I(1)$. The right is the difference between the exact ones and the obtained modified impulse responses by CLS in Example (4.32).

Then the noisy realization problem is solved as follows:

covariance matrix	eigenvalues						
	1	2	3	4	5	6	7
$H_{\underline{a}\ (5,50)}^T(5,0)H_{\underline{a}\ (5,50)}(5,0)$	4288	2679	1128	57.7	4.1		
$H_{\underline{a}\ (6,50)}^T(6,0)H_{\underline{a}\ (6,50)}(6,0)$	4895	3418	1388	70.8	7.5	1.7	
$H_{\underline{a}\ (7,50)}^T(7,0)H_{\underline{a}\ (7,50)}(7,0)$	4913	4777	1495	96.5	9.6	4.5	0.4
covariance matrix	square root of eigenvalues						
	1	2	3	4	5	6	7
$H_{\underline{a}\ (7,50)}^T(7,0)H_{\underline{a}\ (7,50)}(7,0)$	70.1	69.1	38.7	9.8	3.1	2.1	0.6

covariance matrix	eigenvalues						
	1	2	3	4	5	6	7
$H_{\underline{a}\ (5,50)}^T(4,1)H_{\underline{a}\ (5,50)}(4,1)$	4471	2554	926	112	5.1		
$H_{\underline{a}\ (6,50)}^T(4,2)H_{\underline{a}\ (6,50)}(4,2)$	4806	3058	1508	116	5.4	2.5	
$H_{\underline{a}\ (7,50)}^T(4,3)H_{\underline{a}\ (7,50)}(4,3)$	4999	3718	2061	120	5.4	3.8	1.7
covariance matrix	square root of eigenvalues						
	1	2	3	4	5	6	7
$H_{\underline{a}\ (7,50)}^T(4,3)H_{\underline{a}\ (7,50)}(4,3)$	70.7	61	45.4	11	2.3	1.9	1.3

1) Since a set $\{3.1,\ 2.1,\ 0.6\}$ is composed of relatively small and equally-sized numbers in the square root of $H_{\underline{a}\ (7,50)}^T(7,0)H_{\underline{a}\ (7,50)}(7,0)$, a noisy realization of a so-called linear system obtained by the CLS method may be good for a 4-dimensional space.

2) After determining the number n_1 of dimensions which is 4, we will continue the noisy realization algorithm by the CLS method.

Therefore, modified impulse response $I(0)$ of a so-called linear system obtained by the CLS method is constructed for a 4-dimensional space.

3) Since a set $\{2.3,\ 1.9,\ 1.3\}$ is composed of relatively small and equally-sized numbers in the square root of $H_{\underline{a}\ (7,50)}^T(4,3)H_{\underline{a}\ (7,50)}(4,3)$, an almost linear system obtained by the CLS method may be somewhat good by adding nothing.

4) After determining the number n_2 of dimensions which is 0, we will continue the noisy realization algorithm by the CLS method.

Therefore, the modified impulse response $I(1)$ of a so-called linear system obtained by the CLS method is constructed by adding nothing.

Therefore, the modified impulse responses $I(0)$ and $I(1)$ of an almost linear system obtained by the CLS method is realized by a 4-dimensional almost linear system.

The system is given by $\sigma_o = ((\boldsymbol{R}^4,\ F_o),\ g_o^0, g_o, h_o,\ h^0)$, where

$$F_o = \begin{bmatrix} 0 & 0 & 0 & -0.66 \\ 1 & 0 & 0 & -0.8 \\ 0 & 1 & 0 & 0.41 \\ 0 & 0 & 1 & -0.36 \end{bmatrix},\ g_o^0 = \mathbf{e}_1,\ g_o = [0.44,\ -0.67,\ -0.34,\ 0.54]^T,$$

$h_o = [-0.92,\ -7,\ -8.6,\ 3.9]$ and $h^0 = 6.37$.

The obtained modified impulse responses $I(0)$ and $I(1)$ are illustrated in Fig. 4.10.

In this example, original signals are considered as the modified impulse responses of a 4-dimensional linear system and the desirable modified impulse responses are obtained by the CLS method. The model obtained by the CLS method is a 4-dimensional so-called linear system which has the same number of dimensions as the number of the original system.

The following table indicates that the 4-dimensional so-called linear system reconstructs the original signal with a 4 and 6 % error to signal ratio and with 0.04 and 0.03 noise to signal ratio, please refer to Remark 1 in Theorem 4.26 for noise to signal ratio.

Just as we expected, the following table and Fig. 4.10 indicate that the model obtained by the CLS method is a good 4-dimensional system for the original 4-dimensional system.

dimen-ion	ratio of matrices	mean values of square root for sum of			cosine ① and ②	error ratio
		signal ①	signal by CLS ②	error ③	$\cos\theta$	③/①
$I(0)\_(4,0)$	0.04	0.811	0.807	0.03	0.9970	0.04
$I(1)\_(4,0)$	0.03	0.766	0.771	0.06	0.9979	0.06

For the notations $I(0)\_(n_1, n_2)$ and $I(1)\_(n_1, n_2)$, see Definition (4.11).

Example 4.33. Let signals be the modified impulse responses of the following 5-dimensional so-called linear system $\sigma = ((R^5, F), x^0, g, h)$, where

$$F = \begin{bmatrix} 0 & 0 & 0 & 0 & 0 \\ 1 & 0 & 0 & 0 & -0.04 \\ 0 & 1 & 0 & 0 & -0.1 \\ 0 & 0 & 1 & 0 & 0.5 \\ 0 & 0 & 0 & 1 & -0.4 \end{bmatrix}, \ h = [1,\ 2,\ -5,\ -1,\ 3], \ x^0 = [-10,\ 0,\ 1,\ 0,\ 0]^T,$$

$g = [1,\ 0,\ 0,\ 0,\ 0]^T$.

The almost linear system which corresponds to the so-called linear system is given by $\sigma = ((R^4, F), g^0, g, h, h^0)$, where $g^0 = [10,\ -10,\ -1,\ 1,\ 0]^T$, $h^0 = -15$.

Let added noises be given in Fig. 4.11.

Then the noisy realization problem is solved as follows:

covariance matrix	eigenvalues						
	1	2	3	4	5	6	7
$H^T_{\underline{a}\ (5,40)}(5,0) H_{\underline{a}\ (5,40)}(5,0)$	2.6×10^5	17000	8795	849	19		
$H^T_{\underline{a}\ (6,40)}(6,0) H_{\underline{a}\ (6,40)}(6,0)$	2.6×10^5	20000	8957	852	20	8.9	
$H^T_{\underline{a}\ (7,40)}(7,0) H_{\underline{a}\ (7,40)}(7,0)$	3.6×10^5	22172	9137	882	24	12.5	4.7
covariance matrix	square root of eigenvalues						
	1	2	3	4	5	6	7
$H^T_{\underline{a}\ (7,40)}(7,0) H_{\underline{a}\ (7,40)}(7,0)$	600	149	95.6	29.7	4.9	3.5	2.28

covariance matrix	eigenvalues						
	1	2	3	4	5	6	7
$H^T_{\underline{a}\ (6,40)}(5,1)H_{\underline{a}\ (6,40)}(5,1)$	321000	23800	10470	917	21.3	4.5	
$H^T_{\underline{a}\ (7,40)}(5,2)H_{\underline{a}\ (7,40)}(5,2)$	384000	24500	15000	1125	22	5.7	4.4

covariance matrix	square root of eigenvalues						
	1	2	3	4	5	6	7
$H^T_{\underline{a}\ (7,40)}(5,2)H_{\underline{a}\ (7,40)}(5,2)$	620	495	122	33.5	4.7	2.4	2.1

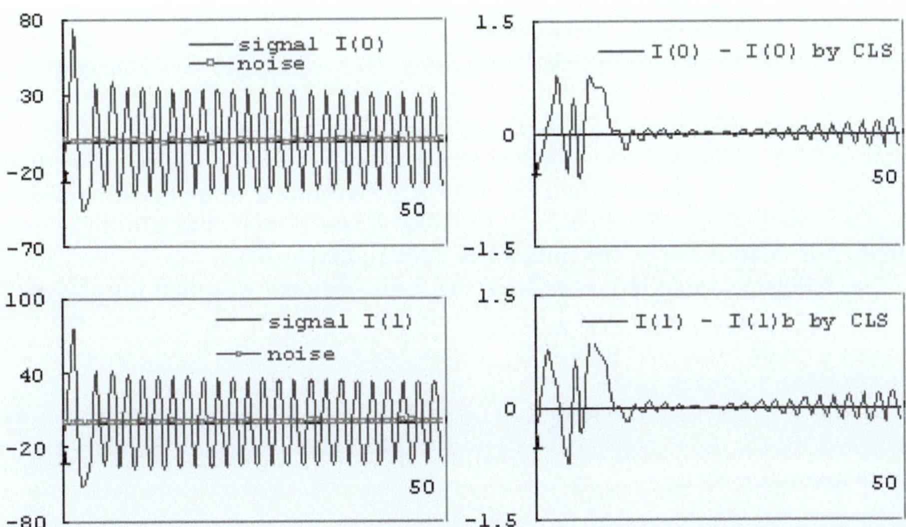

Fig. 4.11. The left are exact modified impulse responses $I(0)$ and $I(1)$ and noises added to $I(0)$ and $I(1)$. The right is the difference between the exact ones and the obtained modified impulse responses by CLS in Example (4.33).

1) Since a set $\{3.52,\ 2.28\}$ is composed of relatively small and equally-sized numbers in the square root of $H^T_{\underline{a}\ (7,40)}(7,0)H_{\underline{a}\ (7,40)}(7,0)$, a noisy realization of a so-called linear system obtained by the CLS method may be good for a 5-dimensional space.

2) After determining the number n_1 of dimensions which is 5, we will continue the noisy realization algorithm by the CLS method.

Therefore, the modified impulse response $I(0)$ of a so-called linear system obtained by the CLS method is constructed for a 5-dimensional space.

3) Since a set $\{2.4,\ 2.1\}$ is composed of relatively small and equally-sized numbers in the square root of $H^T_{\underline{a}\ (7,40)}(5,2)H_{\underline{a}\ (7,40)}(5,2)$, an almost linear system obtained by the CLS method may be somewhat good by adding nothing.

4) After determining the number n_2 of dimensions which is 0, we will continue the noisy realization algorithm by the CLS method.

Therefore, the modified impulse response $I(1)$ of a so-called linear system obtained by the CLS method is constructed by adding nothing.

Therefore, the modified impulse responses $I(0)$ and $I(1)$ of an almost linear system obtained by the CLS method is realized by a 5-dimensional almost linear system.

The system is given by $\sigma_o = ((\boldsymbol{R}^5,\ F_o),\ g_o^0, g_o, h_o,\ h^0)$, where

$$F_o = \begin{bmatrix} 0\ 0\ 0\ 0 & -0.02 \\ 1\ 0\ 0\ 0 & -0.09 \\ 0\ 1\ 0\ 0 & -0.17 \\ 0\ 0\ 1\ 0 & 0.23 \\ 0\ 0\ 0\ 1 & -0.66 \end{bmatrix},\ g_o^0 = \mathbf{e}_1,\ g_o = [0.11,\ 0.14,\ 0.15,\ 0.33,\ 0.26]^T,\ h_o =$$

$[-5.4,\ 74,\ -44.5,\ -37.2,\ 38.6]$ and $h^0 = -15$.

The obtained modified impulse responses $I(0)$ and $I(1)$ are illustrated in Fig. 4.11.

In this example, original signals are considered as the modified impulse responses of a 5-dimensional so-called linear system and the desirable modified impulse responses are obtained by the CLS method. The model obtained by the CLS method is a 5-dimensional so-called linear system which has the same number of dimensions as the number of the original system.

The following table indicates that the 5-dimensional so-called linear system reconstructs the original signal with a 1 and 1 % error to signal ratio and with 0.01 and 0.003 noise to signal ratio, please refer to Remark 1 in Theorem 4.26 for the noise to signal ratio.

Just as we expected, the following table and Fig. 4.11 indicate that the model obtained by the CLS method is a good 5-dimensional system for the original 5-dimensional system.

dimen-sion	ratio of matrices	mean values of square root for sum of			cosine	error ratio
		signal	signal by CLS	error	① and ②	
		①	②	③	$\cos\theta$	③/①
$I(0)_-(5,0)$	0.01	4.90	4.91	0.03	0.99997	0.01
$I(1)_-(5,0)$	0.003	5.167	5.165	0.04	0.99997	0.01

For the notations $I(0)_-(n_1, n_2)$ and $I(1)_-(n_1, n_2)$, see Definition (4.11).

Example 4.34. Let signals be the modified impulse responses of the following 6-dimensional so-called linear system $\sigma = ((R^6, F), x^0, g, h)$, where

$$F = \begin{bmatrix} 0\ 0\ 0\ 0\ 0 & 0 \\ 1\ 0\ 0\ 0\ 0 & -0.04 \\ 0\ 1\ 0\ 0\ 0 & -0.03 \\ 0\ 0\ 1\ 0\ 0 & 0.2 \\ 0\ 0\ 0\ 1\ 0 & 0.5 \\ 0\ 0\ 0\ 0\ 1 & -0.5 \end{bmatrix},\ h = [1,\ 2,\ -5,\ -1,\ 3,\ -2],\ x^0 = [-10,\ 0,\ 1,\ 0,\ 0,\ 0]^T,$$

$g = [1,\ 0,\ 0,\ 0,\ 0,\ 0]^T$.

The almost linear system which corresponds to the so-called linear system is given by $\sigma = ((R^6, F), g^0, g, h, h^0)$, where $g^0 = [10,\ -10,\ -1,\ 1,\ 0,\ 0]^T$, $h^0 = -15$.

Let added noises be given in Fig. 4.12.

Then the noisy realization problem is solved as follows:

covariance matrix	eigenvalues							
	1	2	3	4	5	6	7	8
$H^T_{\underline{a}\,(7,35)}(7,0)H_{\underline{a}\,(7,35)}(7,0)$	36700	25000	9300	540	9.1	6.7	4.2	
$H^T_{\underline{a}\,(8,35)}(8,0)H_{\underline{a}\,(8,35)}(8,0)$	38000	26100	9460	544	9.1	7.3	6.4	3
covariance matrix	square root of eigenvalues							
	1	2	3	4	5	6	7	8
$H^T_{\underline{a}\,(9,35)}(9,0)H_{\underline{a}\,(9,35)}(9,0)$	195	162	96	23.3	3	2.7	2.5	1.7

covariance matrix	eigenvalues							
	1	2	3	4	5	6	7	8
$H^T_{\underline{a}\,(7,35)}(5,1)H_{\underline{a}\,(7,35)}(5,1)$	46000	23000	10000	540	9.4	3		
$H^T_{\underline{a}\,(8,35)}(5,2)H_{\underline{a}\,(8,35)}(5,2)$	61000	23000	13000	680	10	3.9	2.4	
$H^T_{\underline{a}\,(9,35)}(5,3)H_{\underline{a}\,(9,35)}(5,3)$	67000	29000	13000	1010	11.6	3.7	3.2	1.9
covariance matrix	square root of eigenvalues							
$H^T_{\underline{a}\,(9,35)}(5,3)H_{\underline{a}\,(9,35)}(5,3)$	259	170	114	31.8	3.4	1.9	1.8	1.4

1) Since a set $\{2.7,\ 2.5,\ 1.7\}$ is composed of relatively small and equally-sized numbers in the square root of $H^T_{\underline{a}\,(8,35)}(8,0)H_{\underline{a}\,(8,35)}(8,0)$, a noisy realization of a so-called linear system obtained by the CLS method may be good for 5-dimensional space.

2) After determining the number n_1 of dimensions which is 5, we will continue the noisy realization algorithm by the CLS method.

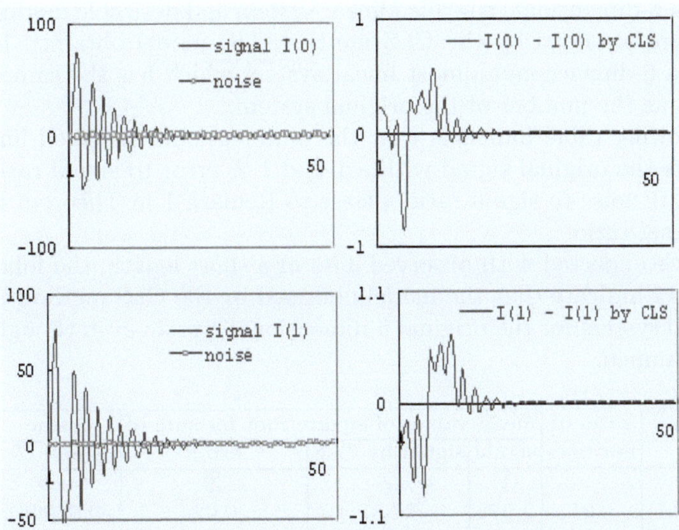

Fig. 4.12. The left are exact modified impulse responses $I(0)$ and $I(1)$ and noises added to $I(0)$ and $I(1)$. The right is the difference between the exact ones and the obtained modified impulse responses by CLS in Example (4.34).

Therefore, the modified impulse response $I(0)$ of a so-called linear system obtained by the CLS method is constructed for a 5-dimensional space.

3) Since a set $\{1.9,\ 1.8,\ 1.4\}$ is composed of relatively small and equally-sized numbers in the square root of $H^T_{\underline{a}\ (9,35)}(5,3)H_{\underline{a}\ (9,35)}(8,3)$, an almost linear system obtained by the CLS method may be somewhat good by adding a 1-dimensional space.

4) After determining the number n_2 of dimensions which is 1, we will continue the noisy realization algorithm by the CLS method.

Therefore, the modified impulse response $I(1)$ of a so-called linear system obtained by the CLS method is constructed by adding 1-dimensional space.

Therefore, the modified impulse responses $I(0)$ and $I(1)$ of an almost linear system obtained by the CLS method is realized by a 6-dimensional almost linear system.

In paticular, $I(0)$ has a 5-dimensional space and $I(1)$ has a 1-dimensional space.

The system is given by $\sigma_o = ((\boldsymbol{R}^6,\ F_o),\ g^0_o, g_o, h_o,\ h^0)$, where

$$F_o = \begin{bmatrix} 0\ 0\ 0\ 0 & -0.04 \\ 1\ 0\ 0\ 0 & -0.044 \\ 0\ 1\ 0\ 0 & -0.013 \\ 0\ 0\ 1\ 0 & 0.024 \\ 0\ 0\ 0\ 1 & -0.6 \end{bmatrix},\ g^0_o = e_1,\ g_o = [0.09,\ 0.08,\ 0.11,\ 0.3,\ 0.27]^T,\ h_o =$$

$[-6.2,\ 73.8,\ -44.9,\ -35.7,\ 47.1]$, and $h^0 = -15$.

The obtained modified impulse responses $I(0)$ and $I(1)$ are illustrated in Fig. 4.12.

In this example, original signals are considered as the modified impulse responses of a 6-dimensional so-called linear system and desirable modified impulse responses are obtained by the CLS method. The model obtained by the CLS method is a 6-dimensional almost linear system which has the same number of dimensions as the number of the original system.

The following table indicates that the 6-dimensional so-called linear system reconstructs the original signal with a 1 and 1 % error to signal ratio and with 0.01 and 0.01 noise to signal ratio, please to Remark 1 in Theorem 4.26 for the noise to signal ratio.

Just as we expected with observed data of a short length, the following table and Fig. 4.12 indicate that the model obtained by the CLS method is a good 6-dimensional system for the original 6-dimensional system even though the signal has been damped.

dimen-ion	ratio of matrices	mean values of square root for sum of			cosine ① and ②	error ratio
		signal	signal by CLS	error		
		①	②	③	$\cos\theta$	③/①
$I(0)_-(5,1)$	0.01	2.578	2.587	0.02	0.99996	0.01
$I(1)_-(5,1)$	0.01	2.714	2.721	0.03	0.99992	0.01

For the notations $I(0)_-(n_1, n_2)$ and $I(1)_-(n_1, n_2)$, see Definition (4.11).

4.7 Historical Notes and Concluding Remarks

Approximate realization and noisy realization problems of so-called linear systems were studied from the viewpoint of Input/output matrix norm and the CLS method. The matrix norm is used for determining the dimension number of state space and the CLS method is used for determining the parameters of so-called linear systems, which are a sort of non-linear systems.

For our treatement of the approximate and noisy realization problems, as we said, there may be a time for using singular value decomposition and the Constrained Least Square (CLS) in Kalman [1997]. In the reference, Kalman also pointed out that the identification problem from noisy data should be treated without any prejudice, hence, should be view in a statistical sense, not a probabilistic sense. Here, we only insist that the signal and the noise are not correlated. Then we discussed noisy realization problems of non-linear systems, that could not been treated.

In order to ascertain that our method for approximate and noisy realization is effective for non-linear cases, we provided several examples. Based on the result of the examples, we have shown that the ratio of the square root of singular values implies the degree of approximation. For our noisy realization problems, we have shown that we can determine the dimension number of so-called linear systems when a set of relatively small and equally-sized numbers of the square root of singular values can be found.

However, for noisy realization problems, we could not fully apply the noisy realization algorithm to modified impulse responses which has rapid damping or values near zero, which is shown in Example 4.30. For this case, we have to apply our method with observed data of a short length.

As stated in the Historical notes and concluding remarks of chapter 3, our methods can be roughly summarized as follows:

Intuitively, our several examples for approximate realization problems demonstrate that the smaller the ratio of matrices is, the smaller the error to signal ratio is. The ratio 0.01 Input/output matrix ratio implies a range of 1 to 6 % error to signal ratio.

The several examples suggest that our two features can be expressed as follows:

(1) The ratio of matrices determines a degree of the crossed angle between directions of the approximate signal and the original signal.
(2) The CLS method determines the coefficients of linearly dependent vectors such that the error between the approximate signal and original signal has a minimum value in the sense of the square norm while conserving the crossed angle.

Intuitively, our several examples for noisy realization problems show that the smaller the ratio of matrices is, the smaller the error to signal ratio is. The ratio within a 0.05 input/output matrix ratio implies an error to signal ratio within 6 %.

The several examples suggest that our two features can be expressed as follows:

(1) The ratio of matrices determines a degree of the crossed angle between directions of the obtained signal and the original signal.
(2) The CLS method determines the coefficients of linearly dependent vectors such that the error between the obtained signal and original signal has a minimum value in the sense of the square norm while conserving the crossed angle.

5 Approximate and Noisy Realization of Almost Linear Systems

Let the set of output's values Y be a linear space over the field $\boldsymbol{R}$. In the reference [Matsuo and Hasegawa, 2003], we introduced almost linear systems that are in a subclass of pseudo linear systems, which are very close to linear systems.

At first, their realization theory was stated. Namely, it was shown that any almost linear systems can be characterized by time-invariant, affine input response maps and any time-invariant, affine input response maps, that is, any input/output maps with causality, time-invariance and affinity can be completely characterized by two modified impulse responses, where the modified impulse response may be a slightly revised version of an impulse response in linear systems. The existence theorem and uniqueness theorem were also proved.

Secondly, details of finite dimensional almost linear systems were investigated. A criterion for the canonical finite dimensional almost linear systems and representation theorems of isomorphic classes for canonical almost linear systems were given. Moreover, a criterion for the behavior of finite dimensional almost linear systems and a procedure to obtain the canonical almost linear systems were given. The criterion is the finite rank condition of an Input/output matrix, which is a natural extension of a finite rank of a Hankel matrix in linear systems.

Thirdly, their partial realization was discussed according to the above results. An algorithm to obtain an almost linear system from the given partial input response map was given.

We can easily understand that the above results of our systems are the same as ones obtained in linear system theory.

In chapter 4, we stated fundamental facts about so-called linear systems for preparation of their approximate and noisy realization problems. Since so-called linear systems are in a subclass of almost linear systems, the problems were discussed in the sense of almost linear systems in order to make a unified treatment as much as we could. Therefore, the fundamental facts about almost linear systems were stated. Hence, please refer to the facts needed for our discussion about approximate and noisy realization problems in chapter 4.

Y. Hasegawa: Approxi. & Noisy Reali. of Discrete-Time Dyn. Sys., LNCIS 376, pp. 95–121, 2008.
springerlink.com   © Springer-Verlag Berlin Heidelberg 2008

5.1 Basic Facts of Almost Linear Systems

Definition 5.1. *Almost Linear System*
1) A system given by the following system equations is written as a collection
$\sigma = ((X, F), g^0, g, h, h^0)$ *and is said to be an almost linear system.*

$$\begin{cases} x(t+1) = Fx(t) + g^0 + g\omega(t+1) \\ x(0) \quad = 0 \\ \hat{\gamma}(t) \quad = h^0 + hx(t) \end{cases}$$

*for any $t \in N$, $x(t) \in X$, $\gamma(t) \in Y$, and X is a linear space over the
field $\mathbf{R}$, F is a linear operator on X, g^0, $g \in X$ and $h : X \to Y$ is a linear
operator.*
*2) The input response map $a_\sigma : U^* \to Y; \omega \mapsto h^0 + h(\sum_{j=1}^{|\omega|} F^{|\omega|-j}(g^0 + g\omega(j)))$
is said to be the behavior of σ.*
3) For the almost linear system σ and any $i \geq 1$,
$I_\sigma(1)(i) := a_\sigma(0^{i-1}|1^1) - a_\sigma(0^{i-1}) = hF^{i-1}(g^0 + g)$ and
*$I_\sigma(0)(i) := a_\sigma(0^i) - a_\sigma(0^{i-1}) = hF^{i-1}g^0$ are said to be modified impulse
responses of σ, where $0^0 := 1$.*
*Note that there is a one-to-one correspondense between the behavior of σ
and the modified impulse responses $I_\sigma(0)$ and $I_\sigma(1) \in F(N, Y)$ of σ by
the relations $a_\sigma(\omega) = (\sum_{j=1}^{|\omega|}(I_\sigma(0)(|\omega| - j + 1) + I_\sigma(1)(|\omega| - j + 1) \times \omega(j))$.*
*4) An almost linear system σ is said to be quasi-reachable if the linear hull
of the reachable set $\{\sum_{j=1}^{|\omega|} F^{|\omega|-j}(g^0 + g\omega(j)); \omega \in U^*\}$ is equal to X and
an almost linear system σ is called to be observable if $hF^i x_1 = hF^i x_2$ for
any $i \in N$ implies $x_1 = x_2$.*
*Especially, an almost linear σ is called reachable if the reachable set
$\{\sum_{j=1}^{|\omega|} F^{|\omega|-j}(g^0 + g\omega(j)); \omega \in U^*\}$ is equal to X.*
*5) An almost linear system σ is called canonical if σ is reachable and
observable.*
*6) A system σ is said to be intrinsically canonical if σ is reachable and
observable.*

Example 5.2. $A(N \times \{0, 1\}, K) := \{\lambda = \sum_{n,u} \lambda(n, u)\mathbf{e}_{(n,u)}$ (finite sum) $; n \in
N, u \in \{0, 1\}\}$, where $\mathbf{e}_{(n,u)}$ is given by the following equations for $n, n' \in N$
and $u, u' \in \{0, 1\}$. If $n = n'$ and $u = u'$ imply $\mathbf{e}_{(n,u)}(n', u') = 1$. If $n \neq n'$
or $u \neq u'$ imply $\mathbf{e}_{(n,u)}(n', u') = 0$. Then $A(N \times \{0, 1\}, K)$ is clearly a linear
space. Let S_r be $S_r \mathbf{e}_{(n,u)} = \mathbf{e}_{(n+1,u)}$. Then $S_r \in L(A(N \times \{0, 1\}, K))$ and S_r
is irrelevant to the input value's set $\{0, 1\}$. S_r is a right shift operator. Let
$\bar{\eta} := e_{(0,1)} - e_{(0,0)}$ and let a linear map $\bar{a} : A(N \times \{0, 1\}, K) \to Y$ be $\bar{a}(\mathbf{e}_{(n,u)}) =
a(u^{n+1}) - a(u^n)$ for any time-invariant, affine input response map $a \in F(U^*, Y)$.
Then a collection $((A(N \times \{0, 1\}, K), S_r), \mathbf{e}_{(0,0)}, \bar{\eta}, \bar{a}, a(1))$ is a quasi-reachable
almost linear system that realizes a.

Let $F(N, Y) := \{$ any function $f : N \to Y\}$ and let $S_l \gamma(t) = \gamma(t+1)$ for any
$\gamma \in F(N, Y)$ and $t \in N$. Then $S_l \in L(F(N, Y))$. Let a map $\chi^0 \in F(N, Y)$ be

$(\chi^0)(t) := a(\omega|0) - a(\omega)$ and $\bar{\chi} \in F(N, Y)$ be $(\bar{\chi})(t) := a(\omega|1) - a(\omega|0)$ for any $t \in N$, a time-invariant, affine input response map $a \in F(U^*, Y)$ and $\omega \in U^*$ such that $|\omega| = t$. Moreover, let a linear map 0 be $F(N, Y) \rightarrow Y; \gamma \mapsto \gamma(0)$. Then a collection $((F(N, Y), S_l), \chi^0, \bar{\chi}, 0, a(1))$ is an observable almost linear system that realizes a.

Theorem 5.3. *The following two almost linear systems are canonical realizations of any time-invariant, affine input response map $a \in F(U^*, Y)$.*
1) $(A(N \times \{0,1\}, K)/_{=a}, \hat{S}_r), [e_{(0,0)}], \hat{\bar{\eta}}, \hat{\bar{a}}, a(1))$,
   where $A(N \times \{0,1\}, K)/_{=a}$ is a quotient space obtained by equivalence
   relation $\sum_{(n,u)} \lambda_1(n, u)\mathbf{e_{(n,u)}} = \sum_{(n',u')} \lambda_2(n', u')\mathbf{e_{(n',u')}} \Longleftrightarrow$
   $\sum_{(n,u)} \lambda(n, u)(a(u^{n+1}) - a(u^n)) = \sum_{(n,u)} \lambda(n, u)(a(u^{n+1}) - a(u^n))$.
   Moreover, $\hat{S}_r \in L(A(N \times \{0,1\}, K)/_{=a})$ is given by $\hat{S}_r[e(n, u)] = [e_{(n+1,u)}]$
   for $[e_{(n,u)}] \in A(N \times \{0,1\}, K)/_{=a}$, and $\hat{\bar{\eta}} = [e_{(0,1)}] - [e_{(0,0)}]$, $\hat{\bar{a}}$ is given by
   $\hat{\bar{a}} : A(N \times \{0,1\}, K)/_{=a} \rightarrow Y; [e(n, u)] \mapsto a(u^{n+1}) - a(u^n)$.
2) $((\ll S_l^N(\chi(U)) \gg, S_l), \chi^0, \bar{\chi}, 0, a(1))$,
   where $\ll S_l^N(\chi(U)) \gg$ is the smallest linear space that contains
   $S_l^N(\chi(U)) := \{S_l^i(\chi^0 + \bar{\chi}u); u \in K, i \in N, S_l^i(\chi^0 + \bar{\chi}u)(t) = (\chi(u)(i+t+1) =$
   $a(\omega|u) - a(\omega)$ for $\omega \in U^*$, $|\omega| = i + t\}$.

Theorem 5.4. *Realization Theorem of almost linear systems*

Eistence: For any time-invariant, affine input response map $a \in F(U^*, Y)$, there
   exist at least two canonical almost linear systems that realize a.
Uniqueness: Let σ_1 and σ_2 be any two canonical almost linear systems which
   realize a time-invariant, affine input response map $a \in F(U^*, Y)$. Then there
   exists an isomorphism $T : \sigma_1 \rightarrow \sigma_2$.

For the isomorphism of almost linear systems, see Definition (4.7).

5.2 Finite Dimensional Almost Linear Systems

Based on the realization theory (5.4), we want to review the fundamental facts about almost linear systems in this section. The facts are as follows:
1) when almost linear system is finite dimensional.
2) when finite dimensional almost linear system is canonical.
3) how we find a standard almost linear system.
4) a criterion for an Input/output relation to be the behavior of finite
   dimensional almost linear systems.
5) a procedure to obtain the standard system which realizes a given input
   response map.
6) how to find a partial realization σ from a given partial input/output data.
7) how to find a partial realization σ from a given partial inpur/output data
   in real time.
In chapter 4, since the above facts were stated when discussing our discussing approximate and noisy realization problems in the sense of almost linear systems, they are omitted.

There is a fact about finite dimensional linear spaces that an n-dimensional linear space over the field $\boldsymbol{R}$ is isomorphic to $\boldsymbol{R}^n$ and $L(\boldsymbol{R}^n, \boldsymbol{R}^m)$ is isomorphic to $\boldsymbol{R}^{m \times n}$ (See Halmos [1958]). Therefore, without loss of generality, we can consider n-dimensional almost linear system as $\sigma = ((\boldsymbol{R}^n, F), g^0, g, h, h^0)$, where $F \in \boldsymbol{R}^{n \times n}$, g, $g^0 \in \boldsymbol{R}^n$ and $h \in \boldsymbol{R}^{p \times n}$.

Proposition 5.5. *An almost linear system* $\sigma = ((\boldsymbol{R}^n, F), g^0, g, h, h^0)$ *is intrinsically canonical if and only if the following two conditions hold.*
rank $[g, Fg, F^2g, \cdots, F^{n-1}g] = n$
rank $[h^T, (hF)^T, \cdots, (hF^{n-1})^T] = n$.

Definition 5.6. *For any time-invariant, affine input response map* $a \in F$ (U^*, Y), *the corresponding linear input/output map* $A : (A(N \times \{0, 1\}, \boldsymbol{R}), S_r) \to$ $(F(N, Y), S_l)$ *satisfies* $A(\mathbf{e}_{(\mathbf{s}, \mathbf{u})})(t) = a(u^{s+t+1}) - a(u^{s+t})$ *for any* $u \in \{0, 1\}$.

Therefore, the map A *can be represented by the infinite matrix* $(I/O)_a$. *This* $(I/O)_a$ *is said to be an Input/output matrix of a. For the Input/output matrix* $(I/O)_a$, *see Definition* (4.9).

For a partial time-invariant, affine input response map $\underline{a} \in F(U_{\underline{N}}^*, Y)$, *the matrix* $(I/O)_{\underline{a} \ (p, \underline{N}-p)}$ *is said to be a finite-sized Input/output matrix of $\underline{a}$, where* $0 \le s \le p$, $0 \le t \le \underline{N} - p$ *and* $u \in \{0, 1\}$. *For* $(I/O)_{\underline{a} \ (p, \underline{N}-p)}$, *see section* (4.4). *Since* $I_{\underline{a}}(u)(i + j) = \underline{a}(u^{i+j+1}) - \underline{a}(u^{i+j})$ *holds for* $u \in \{0, 1\}$, *column vectors of* $(I/O)_{\underline{a}}$ *are denoted by* $S_l^i I_{\underline{a}}(u)$.

Let a matrix $(I/O)_{\underline{a} \ (p, \underline{N}-p)}(v, w)$ *denote* $(I/O)_{\underline{a} \ (p, \underline{N}-p)}(v, w)$
$:= [I_{\underline{a}}(0), S_l I_{\underline{a}}(0), \cdots, S_l^{v-1} I_{\underline{a}}(0), I_{\underline{a}}(1), S_l I_{\underline{a}}(1), \cdots, S_l^{w-1} I_{\underline{a}}(1)]$.
When we treat actually approximate and noisy realization problems, we will use a notation $H_{\underline{a} \ (n_1+n_2, \underline{N}-n_1-n_2)}(n_1, n_2)$ *expressed as follows:*
$H_{\underline{a} \ (n_1+n_2, \underline{N}-n_1-n_2)}(n_1, n_2) = [I_{\underline{a}}(0), \cdots, S_l^{n_1-1} I_{\underline{a}}(0), I_{\underline{a}}(1), \cdots, S_l^{n_2-1} I_{\underline{a}}(1)]$.

5.3　Approximate Realization of Almost Linear Systems

In this section, we discuss approximate realization problems of almost linear systems. Here, we will discuss the approximate realization problem of almost linear systems, which is stated as follows:
　<For any given finite-length modified impulse response of an almost linear system, find an almost linear system which approximates it.>
　The approximate realization of almost linear system is presented here for the first time.

　In order to make our discussion simple, we assume that the set Y of outout is the set $\boldsymbol{R}$ of real numbers, namely 1-output.

Theorem 5.7. *Algorithm for approximate realization*
　Let a partial input response map $\underline{a}$ *be a considered object which is an almost linear system. Then an approximate realization* $\sigma_r = ((\boldsymbol{R}^n, F_r), g_r^0, g_r, h_r, h^0)$ *of* $\underline{a}$ *is given by the following algorithm:*
1) Based on the ratio of the square root of eigenvalues for a matrix
　　$H_{\underline{a} \ (p, \bar{p})}(p, 0) H_{\underline{a} \ (p, \bar{p})}(p, 0)^T$, *determine the value n_1 of rank for the matrix*
　　$H_{\underline{a} \ (p, \bar{p})}(p, 0)$, *where* $n_1 \le p$.

Namely, determine the value n_1 of rank for the matrix $H_{\underline{a}\ (p,\bar{p})}(p,0)$ such that the ratio of the squqre root of eigenvalues for the covariance matrix becomes very small. The small ratio means the nearness of approximation degree.

2) We use the CLS method as follows:

① Let a matrix $A_1 \in \mathbf{R}^{1\times(n_1+1)}$ be $A_1 = [\alpha_{11}, \alpha_{12}, \cdots, \alpha_{1n_1}, -1]$.

② Choose the coefficients $\{\alpha_{1i} : 1 \leq i \leq n_1\}$ such that $\sum_{j=1}^{n_1+1} \underline{S}_l^{j-1}\bar{I}_{\underline{a}} \cdot \underline{S}_l^{j-1}\bar{I}_{\underline{a}}$ takes a minimum value, where $\{\underline{S}_l^i\bar{I}_{\underline{a}} \in \mathbf{R}^{L\times1} : 0 \leq i \leq n_1\}$ are given by the equation $[\bar{I}_{\underline{a}}(0), \underline{S}_l\bar{I}_{\underline{a}}(0), \cdots, \underline{S}_l^{n_1}\bar{I}_{\underline{a}}(0)]^T := A_1^T[A_1A_0^T]^{-1}A_1H_{\underline{a}\ (n_1+1,L)}^T(n_1+1,0)$ and $H_{\underline{a}\ (n_1,L)}^T(n_1,0):= [I_{\underline{a}}(0), \cdots, S_l^{n_1-1}I_{\underline{a}}(0), S_l^{n_1}I_{\underline{a}}(0)]$. And $\cdot$ denotes the inner product of two vectors.

③ Let $h_{1r} \in \mathbf{R}^{1\times n_1}$ be $h_{1r} = [(I_{\underline{a}}(0))(0) - (\bar{I}_{\underline{a}}(0))(0), (S_lI_{\underline{a}}(0))(0) - (S_l\bar{I}_{\underline{a}}(0))(0), \cdots, (S_l^{n_1-1}I_{\underline{a}}(0))(0) - (S_l^{n_1-1}\bar{I}_{\underline{a}}(0))(0)]$.

3) Based on the ratio of the square root of eigenvalues for a matrix $H_{\underline{a}\ (n_1+p,\bar{p})}(n_1,p)H_{\underline{a}\ (n_1+p,\bar{p})}(n_1,p)^T$, determine the value n_2 of rank for the matrix $H_{\underline{a}\ (n_1+p,\bar{p})}(n_1,p)$, where $n_2 \leq p$.

Namely, determine the value n_2 of rank for the matrix $H_{\underline{a}\ (p,\bar{p})}(p,0)$ such that the ratio of the squqre root of eigenvalues for the covariance matrix becomes very small. The small ratio means the nearness of approximation degree.

4) The CLS method is used as follows:

① Let a matrix $A_2 \in \mathbf{R}^{1\times(n_1+n_2+1)}$ be $A_2 = [\alpha_{21}, \alpha_{22}, \cdots, \alpha_{2n_1+n_2}, -1]$.

② Choose the coefficients $\{\alpha_{2i} : 1 \leq i \leq n_1 + n_2\}$ such that $\sum_{j=1}^{n_1+n_2+1} \underline{S}_l^{j-1}\bar{I}_{\underline{a}} \cdot \underline{S}_l^{j-1}\bar{I}_{\underline{a}}$ takes a minimum value, where $\{\underline{S}_l^i\bar{I}_{\underline{a}} \in \mathbf{R}^{L\times1} : 0 \leq i \leq n_1 + n_2\}$ are given by the equation $[\bar{I}_{\underline{a}}, \underline{S}_l\bar{I}_{\underline{a}}, \cdots, \underline{S}_l^{n_1}\bar{I}_{\underline{a}}]^T := A_2^T[A_2A_2^T]^{-1}A_2H_{\underline{a}\ (n_1+n_2,L)}^T(n_1,n_2+1)$ and $H_{\underline{a}\ (n_1,L)}^T(n_1,n_2+1):= [I_{\underline{a}}(0), \cdots, S_l^{n_1-1}I_{\underline{a}}(0), I_{\underline{a}}(1), \cdots, S_l^{n_2-1}I_{\underline{a}}(1), S_l^{n_2}I_{\underline{a}}(1)]$. And $\cdot$ denotes the inner product of two vectors.

③ Let $F_r \in \mathbf{R}^{(n_1+n_2)\times(n_1+n_2)}$ be given as the same as in Definition (4.11). Let g_r^0 be $g_r^0 = \mathbf{e}_1$ and g_r be $g_r = \mathbf{e}_{n_1+1} - \mathbf{e}_1$, where $\mathbf{e}_i = [0, \cdots, 0, \overset{i}{1}, 0, \cdots, 0]^T \in \mathbf{R}^{n_1+n_2}$.

④ Let h_r be $h_r = [h_{1r}, (I_{\underline{a}}(1))(0) - (\bar{I}_{\underline{a}}(1))(0), (S_lI_{\underline{a}}(1))(0) - (S_l\bar{I}_{\underline{a}}(1))(0), \cdots, (S_l^{n_2-1}I_{\underline{a}}(1))(0) - (S_l^{n_2-1}\bar{I}_{\underline{a}}(1))(0)]$.

For the real time standard system $\sigma_r = ((\mathbf{R}^n, F_r), g_r^0, g_r, h_r, h^0)$, its modified impulse responses $I(0)(i) := h_rF_r^ig_r^0$ and $I(1)(i) := h_rF_r^i(g_r^0 + g_r)$ are written by $I(0)\_(n_1, n_2)$ and $I(1)\_(n_1, n_2)$ respectively, where $n := n_1 + n_2$

[proof]. This is the same as the Algorithm for approximate realization (4.13).

Example 5.8. Let the signals be the modified impulse responses of the following 3-dimensional almost linear system: $\sigma = ((\boldsymbol{R}^3, F), g, g^0, h, h^0)$, where $F =$
$$\begin{bmatrix} 0 & 0.4 & 0 \\ 1 & 0.5 & 0 \\ 0 & 0 & 0.7 \end{bmatrix}, \ g^0 = [2, \ 1, \ 0]^T, \ h = [10, \ 2, \ 5], \ g = [0, \ 0, \ 1]^T, \ h^0 = 1.$$

Then the approximate realization problem is solved as follows:

covariance matrix	eigenvalues			
	1	2	3	4
$H^T_{\underline{a}\ (3,50)}(3,0)\,H_{\underline{a}\ (3,50)}(3,0)$	3697	79	0	
covariance matrix	square root of eigenvalues			
$H^T_{\underline{a}\ (3,50)}(3,0)\,H_{\underline{a}\ (3,50)}(3,0)$	61	8.9	0	
covariance matrix	eigenvalues			
	1	2	3	4
$H^T_{\underline{a}\ (3,50)}(2,1)\,H_{\underline{a}\ (3,50)}(2,1)$	4745	111	6.3	
$H^T_{\underline{a}\ (4,50)}(2,2)\,H_{\underline{a}\ (4,50)}(2,2)$	6080	161	10	0
covariance matrix	square root of eigenvalues			
$H^T_{\underline{a}\ (4,50)}(2,2)\,H_{\underline{a}\ (4,50)}(2,2)$	78	12.7	3.2	0

1) Since the ratio $\frac{3.2}{78} = 0.04$ obtained by the squqre root of
$H^T_{\underline{a}\ (4,50)}(2,2)\,H_{\underline{a}\ (4,50)}(2,2)$ is not so small, the approximate almost linear system obtained by the CLS method may not be so good.
2) After determining the numbers n_1 and n_2 of dimensions which are 2 and 0, we execute the approximate realization algorithm by the CLS method.

The almost linear system $\sigma_1 = ((\boldsymbol{R}^2, \ F_1), \ x_1^0, \ g_1, \ h_1, h^0)$ obtained by the CLS method is expressed as follows:

$$F_1 = \begin{bmatrix} 0 & 0.4 \\ 1 & 0.5 \end{bmatrix}, \ h_1 = [22, \ 9], \ g_1 = [0.26, \ -0.15]^T, g_1^0 = [1, \ 0]^T, \ h^0 = 1.$$

For reference, a 3-dimensional almost linear system $\sigma_2 = ((\boldsymbol{R}^3, \ F_2), \ x_2^0, \ g_2, \ h_2, h^0)$ obtained by the CLS method is expressed as follows:

$$F_2 = \begin{bmatrix} 0 & 0.4 & -0.7 \\ 1 & 0.5 & 1 \\ 0 & 0 & 0.7 \end{bmatrix}, \ h_2 = [22, \ 9, \ 27], \ g_2 = [-1, \ 0, \ 1]^T, g_2^0 = [1, \ 0, \ 0]^T,$$
$h^0 = 1.$

In this example, the original signals are considered as the modified impulse responses of a 3-dimensional almost linear system and the desirable modified impulse responses are obtained by the CLS method.

The following table indicates that the 2-dimensional almost linear system reconstructs the original signal with 0 and 9 % error to signal ratio and with 0 and 0.04 ratio of matrices, and the 3-dimensional almost linear system completely reconstructs the original system.

Therefore, an approximate realization could be obtained.

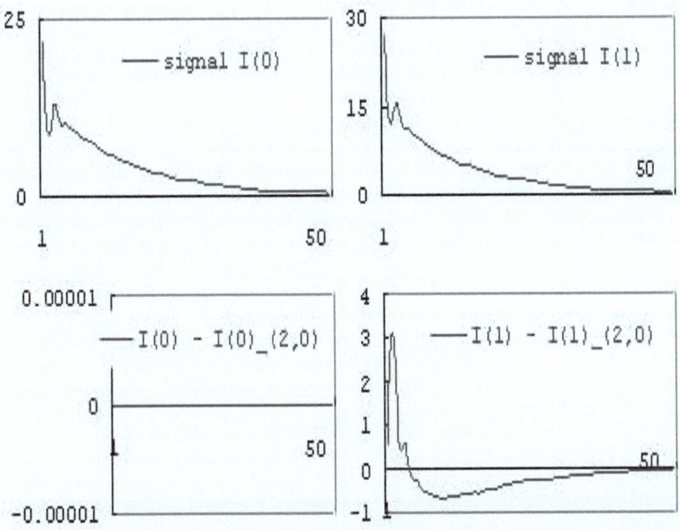

Fig. 5.1. The left are the original modified impulse response $I(0)$ of a 3-dimensional almost linear system and the difference between the original signal and the obtained one by a 2-dimensional almost linear system. The right are the original modified impulse response $I(1)$ of a 3-dimensional almost linear system and the difference between the original signal and the obtained one by a 2-dimensional almost linear system in Example (5.8).

Just as we thought, the following table and Fig. 5.1 truly indicate that the 2-dimensional almost linear system given by the CLS method is not so such a good approximation.

dimen-ion	ratio of matrices	mean values of square root for sum of			cosine ① and ②	error ratio
		signal	signal by CLS	error		
		①	②	③	$\cos\theta$	③/①
$I(0)$_$(2,0)$	0	0.8	0.8	0	1	0
$I(1)$_$(2,0)$	0.04	0.92	0.92	0.081	0.996	0.09
$I(0)$_$(2,1)$	0	0.8	0.8	0	1	0
$I(1)$_$(2,1)$	0	0.92	0.92	0	1	0

For the notations $I(0)$_(n_1, n_2) and $I(1)$_(n_1, n_2), see Algorithm (5.7).

Example 5.9. Let the signals be the modified impulse responses of the following 4-dimensional almost linear system: $\sigma = ((\mathbf{R}^4, F), g, g^0, h, h^0)$, where $F =$

$$\begin{bmatrix} 0 & 0 & 0.7 & 0 \\ 1 & 0 & 0.4 & 0 \\ 0 & 1 & -0.2 & 0 \\ 0 & 0 & 0.1 & 0.1 \end{bmatrix}, g^0 = [1, 0, 0, 0]^T, h = [9, 15, -5, 10],$$

$g = [0, 0, 0, 1]^T, h^0 = 1.$

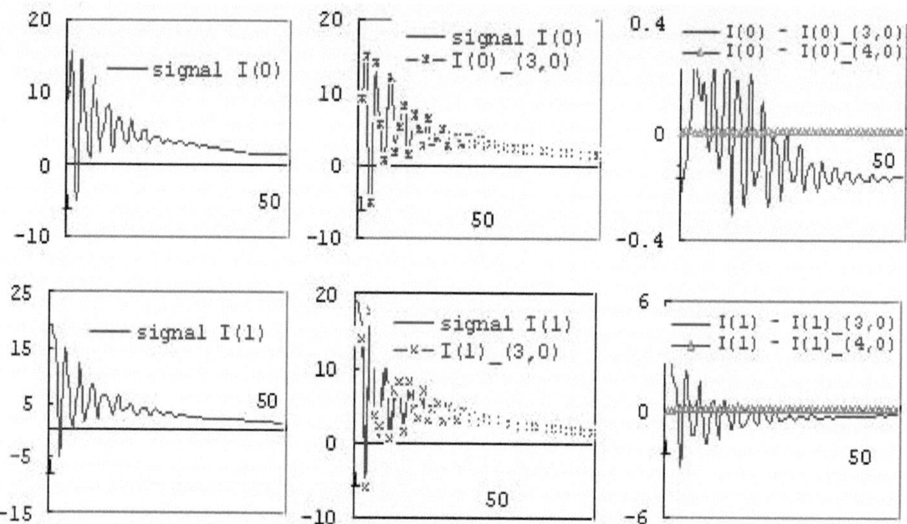

Fig. 5.2. The left are the original modified impulse response $I(0)$ and $I(1)$ of an original 4-dimensional almost linear system. The middle are the original modified impulse responses and the obtained ones by a 3-dimensional almost linear system. The right are the difference between the original signals and the obtained ones by a 3 or 4-dimensional almost linear system in Example (5.9).

Then the approximate realization problem is solved by the following algorithm:

covariance matrix	eigenvalues				
	1	2	3	4	5
$H^T_{\underline{a}\ (4,50)}(4,0)H_{\underline{a}\ (3,50)}(4,0)$	2734	716	393	0.32	
$H^T_{\underline{a}\ (5,50)}(5,0)H_{\underline{a}\ (5,50)}(5,0)$	3174	800	484	0.32	0
covariance matrix	square root of eigenvalues				
$H^T_{\underline{a}\ (4,50)}(4,0)H_{\underline{a}\ (4,50)}(4,0)$	52.3	26.8	19.9	0.6	
$H^T_{\underline{a}\ (5,50)}(5,0)H_{\underline{a}\ (5,50)}(5,0)$	56	28	22	0.6	0

covariance matrix	eigenvalues				
	1	2	3	4	5
$H^T_{\underline{a}\ (4,50)}(3,1)H_{\underline{a}\ (4,50)}(3,1)$	3408	735	470	22.3	
$H^T_{\underline{a}\ (5,50)}(4,1)H_{\underline{a}\ (5,50)}(4,1)$	4197	761	501	26	0
covariance matrix	square root of eigenvalues				
$H^T_{\underline{a}\ (4,50)}(3,1)H_{\underline{a}\ (4,50)}(3,1)$	58.4	27.1	21.7	4.7	
$H^T_{\underline{a}\ (5,50)}(4,1)H_{\underline{a}\ (5,50)}(4,1)$	65	27.6	22.3	5	0

1) Since the ratio $\frac{0.6}{52.3} = 0.01$ obtained by the squqre root of $H^T_{\underline{a}\ (4,50)}(4,0)H_{\underline{a}\ (4,50)}(4,0)$ is small and the ratio $\frac{5}{65} = 0.08$ obtained by the squqre root of $H^T_{\underline{a}\ (5,50)}(4,1)H_{\underline{a}\ (5,50)}(4,1)$ is a little large, the approximate almost linear system obtained by the CLS method may not be good.

2) After determining the number n_1 and n_2 of dimensions which are 3 and 0, we execute the approximate realization algorithm by the CLS method.

The almost linear system $\sigma_1 = ((\boldsymbol{R}^3,\ F_1),\ g_1^0,\ g_1,\ h_1)$ obtained by the CLS method is expressed as follows:

$$F_1 = \begin{bmatrix} 0 & 0 & 0.72 \\ 1 & 0 & 0.41 \\ 0 & 1 & -0.22 \end{bmatrix},\ h_1 = [9.2,\ 15.1,\ -5.1],\ g_1 = [0.21,\ 0.13,\ -0.28]^T,$$

$g_1^0 = [1,\ 0,\ 0]^T,\ h^0 = 1.$

For reference, a 4-dimensional almost linear system
$\sigma_2 = ((\boldsymbol{R}^4,\ F_2),\ g_2^0,\ g_2,\ h_2, h^0)$ obtained by the CLS method is expressed as follows:

$$F_2 = \begin{bmatrix} 0 & 0 & 0 & -0.07 \\ 1 & 0 & 0 & 0.66 \\ 0 & 1 & 0 & 0.42 \\ 0 & 0 & 1 & -0.1 \end{bmatrix},\ h_2 = [9,\ 15,\ -5,\ 14.3],\ g_2 = [-7\ -4,\ 2,\ 10]^T,$$

$g_2^0 = [1,\ 0,\ 0,\ 0]^T,\ h^0 = 1.$

In this example, the original signals are considered as the modified impulse responses of a 4-dimensional almost linear system and the desirable modified impulse responses are obtained by the CLS method.

The following table indicates that the 3-dimensional almost linear system reconstructs the original signal with 4 and 26 % error to signal ratio and with 0.01 and 0.08 ratio of matrices, and the 4-dimensional almost linear system completely reconstructs the original system.

Therefore, a somewhat good approximate realization could be obtained.

The following table and Fig. 5.2 truly indicate that the 3-dimensional almost linear system obtained by the CLS method is a somewhat good approximation within our expectations. Hence, there exists a somewhat good approximation for the given system except the peak values in the modified impulse response $I(1)$.

dimen- ion	ratio of matrices	mean values of square root for sum of			cosine ① and ②	error ratio
		signal ①	signal by CLS ②	error ③	$\cos\theta$	③/①
$I(0)$_$(3,0)$	0.01	0.686	0.694	0.025	0.999	0.04
$I(1)$_$(3,0)$	0.08	0.77	0.783	0.15	0.98	0.2
$I(0)$_$(4,0)$	0	0.686	0.686	0	1	0
$I(1)$_$(4,0)$	0	0.771	0.772	0	1	0

For the notations $I(0)$_(n_1, n_2) and $I(1)$_(n_1, n_2), see Algorithm (5.7).

Example 5.10. Let the signals be the modified impulse responses of the following 5-dimensional almost linear system: $\sigma = ((\boldsymbol{R}^5, F), g, g^0, h, h^0)$, where

$$F = \begin{bmatrix} 0 & 0 & -0.112 & 0 & 0 \\ 1 & 0 & 0.038 & 0 & 0 \\ 0 & 1 & 0 & 0 & 0 \\ 0 & 0 & 1 & 0 & 0.52 \\ 0 & 0 & 0 & 0.5 & -0.4 \end{bmatrix},\ g^0 = [1,\ 0,\ 0,\ 0,\ 0]^T,\ h = [10,\ 2,\ -5,\ -1,\ 3],$$

$g = [0,\ 0,\ 0,\ 1,\ 0]^T,\ h^0 = 1.$

Then the approximate realization problem is solved as follows:

covariance matrix	eigenvalues					
	1	2	3	4	5	6
$H^T_{\underline{a}\ (5,50)}(5,0)H_{\underline{a}\ (5,50)}(5,0)$	171	37	9.8	4.2	0.01	
$H^T_{\underline{a}\ (6,50)}(6,0)H_{\underline{a}\ (6,50)}(6,0)$	171	37.3	11	4.4	0.01	0
covariance matrix	square root of eigenvalues					
$H^T_{\underline{a}\ (5,50)}(5,0)H_{\underline{a}\ (5,50)}(5,0)$	13	6.1	3.1	2	0.1	
$H^T_{\underline{a}\ (6,50)}(6,0)H_{\underline{a}\ (6,50)}(6,0)$	13	6.1	3.3	2.1	0.1	0

covariance matrix	eigenvalues					
	1	2	3	4	5	6
$H^T_{\underline{a}\ (5,50)}(4,1)H_{\underline{a}\ (5,50)}(4,1)$	298	39	9	4.4	0.19	
$H^T_{\underline{a}\ (6,50)}(5,1)H_{\underline{a}\ (6,50)}(5,1)$	298	39	11	6	0.02	0.01
covariance matrix	square root of eigenvalues					
$H^T_{\underline{a}\ (5,50)}(4,1)H_{\underline{a}\ (5,50)}(4,1)$	17.3	6.2	3	2.1	0.43	
$H^T_{\underline{a}\ (6,50)}(5,1)H_{\underline{a}\ (6,50)}(5,1)$	17.3	6.2	3.3	2.4	0.14	0.1

1) Since the ratio $\frac{0.1}{13} = 0.008$ obtained by the squqre root of
$H^T_{\underline{a}\ (5,50)}(5,0)H_{\underline{a}\ (5,50)}(5,0)$ is small and the ratio $\frac{0.43}{17.3} = 0.02$ obtained by the squqre root of $H^T_{\underline{a}\ (5,50)}(4,1)H_{\underline{a}\ (5,50)}(4,1)$ is also small, the approximate almost linear system obtained by the CLS method may be good.

2) After determining the number n_1 and n_2 of dimensions which are 4 and 0, we execute the approximate realization algorithm by the CLS method.

The almost linear system $\sigma_1 = ((\boldsymbol{R}^4,\ F_1),\ g^0_1,\ g_1,\ h_1)$ obtained by the CLS method is expressed as follows:

$$F_1 = \begin{bmatrix} 0 & 0 & 0 & -0.12 \\ 1 & 0 & 0 & -0.18 \\ 0 & 1 & 0 & -0.2 \\ 0 & 0 & 1 & -0.91 \end{bmatrix},\ h_1 = [10,\ 2,\ -5,\ -2.2],\ g_1 = [0.14,\ 0.05,\ 0.04,\ 1.1]^T,$$

$g^0_1 = [1,\ 0,\ 0,\ 0]^T,\ h^0 = 1.$

For reference, a 5-dimensional almost linear system $\sigma_2 = ((\boldsymbol{R}^5,\ F_2),\ g^0_2,\ g_2,\ h_2, h^0)$ obtained by the CLS method is expressed as follows:

$$F_2 = \begin{bmatrix} 0 & 0 & 0 & 0 & 0.043 \\ 1 & 0 & 0 & 0 & -0.014 \\ 0 & 1 & 0 & 0 & -0.1 \\ 0 & 0 & 1 & 0 & 0.33 \\ 0 & 0 & 0 & 1 & -0.28 \end{bmatrix},\ h_2 = [10,\ 2,\ -5,\ -2.2,\ 1.46],$$

$g_2 = [0.21,\ 0.18,\ 0.16,\ 1.7,\ 0.78]^T, g^0_2 = [1,\ 0,\ 0,\ 0,\ 0]^T,\ h^0 = 1.$

In this example, the original signals are considered as the modified impulse responses of a 5-dimensional almost linear system and the desirable modified impulse responses are obtained by the CLS method.

The following table indicates that the 4-dimensional almost linear system reconstructs the original signal with 1.7 and 2.1 % error to signal ratio and with 0.008 and 0.02 ratio of matrices, and the 5-dimensional almost linear system completely reconstructs the original system.

The following table and Fig. 5.3 truly indicate that the 4-dimensional almost linear system obtained by the CLS method is a good approximation within our expectations. For reference, the modified impulse respomses of the same dimensional almost linear system as the original system are shown. Hence, there exists a good approximation for the given system except for the peak values in the modified impulse response $I(1)$.

dimen-ion	ratio of matrices	mean values of square root for sum of			cosine ① and ②	error ratio
		signal ①	signal by CLS ②	error ③	$\cos\theta$	③/①
$I(0)_-(4,0)$	0.008	0.234	0.234	0.004	0.9999	0.017
$I(1)_-(4,0)$	0.02	0.229	0.230	0.005	0.9998	0.021
$I(0)_-(5,0)$	0	0.234	0.234	0	1	0
$I(1)_-(5,0)$	0	0.229	0.229	0	1	0

For the notations $I(0)_-(n_1, n_2)$ and $I(1)_-(n_1, n_2)$, see Algorithm (5.7).

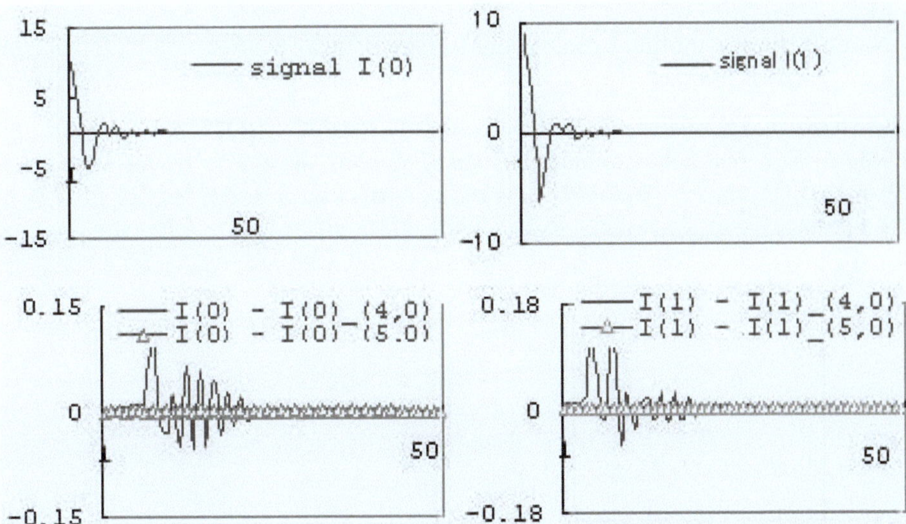

Fig. 5.3. The left are the original modified impulse response $I(0)$ of a 5-dimensional almost linear system and the difference between the original signal and the obtained one by a 4 or 5-dimensional almost linear system. The right are the original modified impulse response $I(1)$ of a 5-dimensional almost linear system and the difference between the original signal and the obtained one by a 4 or 5-dimensional almost linear system in Example (5.10).

5.4 Noisy Realization of Almost Linear systems

In this section, we discuss noisy realization of almost linear systems.

We will obtain the observed data $\{\hat{\gamma}(|\omega|) + \bar{\gamma}(|\omega|) : \omega \in U^*\}$ for noise $\{\bar{\gamma}(t) : t \in N\}$ added to the unknown almost linear system σ with the behavior a.

For any given $\{\hat{\gamma}(|\omega|) + \bar{\gamma}(|\omega|) : \omega \in U^*\}$, a system σ which satisfies $a_\sigma(\omega) \approx \hat{\gamma}(|\omega|)$ for any $\omega \in U^*$ is called a noisy realization of a.

In order to make our discussion simple, we assume that the set Y of outout is the set $\boldsymbol{R}$ of real numbers, namely 1-output.

We can propose the following noisy realization problem:

For any given $\{\hat{\gamma}(|\omega|) + \bar{\gamma}(|\omega|) : \omega \in U^*\}$, find an almost linear system σ which satisfies $a_\sigma(\omega) = \hat{\gamma}(|\omega|)$ for any $\omega \in U^*$.

A situation for noisy realization problem 5.11

Let the observed object be an almost linear system and noise be added to its output. Then we will obtain the data $\{\gamma(t) = \hat{\gamma}(t) + \bar{\gamma}(t) : 0 \leq t \leq \underline{N}\}$ for some integer $\underline{N} \in N$, where $\hat{\gamma}(t)$ is the exact signal which comes from the observed almost linear system and $\bar{\gamma}(t)$ is the noise added at observation.

Problem 5.12. Problem statement of noisy realization for almost linear systems

Let $H_{\underline{a}\ (p,\bar{p})}$ be the measured finite-sized Input/output matrix. Then find the cleaned-up Input/output matrix $\hat{H}_{\underline{a}\ (p,\bar{p})}$ such that $H_{\underline{a}\ (p,\bar{p})} = \hat{H}_{\underline{a}\ (p,\bar{p})} + \bar{H}_{\underline{a}\ (p,\bar{p})}$ holds.

Namely, find a minimal dimensional almost linear system $\sigma = ((\boldsymbol{R}^n, F), g^0, g, h, h^0))$ which realizes $\hat{H}_{\underline{a}\ (p,\bar{p})}$.

Definition 5.13. *A canonical almost linear system* $\sigma_r = ((\boldsymbol{R}^n, F_r), g_r^0, g_r, h_r, h^0)$ *is said to be a real time standard system if* $g_r^0 = \mathbf{e}_1$, $\mathbf{e}_i = F_r^{i-1}\mathbf{e}_1$ *for* $i \leq n_1$ *and* $F_r^{n_1}\mathbf{e}_1 = \sum_{i=1}^{n_1} \alpha_{0i} F_r^{i-1}\mathbf{e}_1$ *hold.* $g_r = \mathbf{e}_{n_1+1} - \mathbf{e}_1$, $\mathbf{e}_{n_1+i} = F_r^{i-1}\mathbf{e}_{n_1+1}$ *for* $i \leq n_2$ *and* $F_r^{n_2}\mathbf{e}_{n_1+1} = \sum_{i=1}^{n_2} \alpha_{1i} F_r^{i-1}\mathbf{e}_{n_1+1}$ *hold.* F_r *is given by the following.*

$$
F_r = \begin{bmatrix}
0 & \cdots & 0 & \alpha_{11} & 0 & \cdots & \cdots & 0 & \alpha_{21} \\
1 & \ddots & & \alpha_{12} & 0 & \cdots & & 0 & \alpha_{22} \\
\vdots & \ddots & 0 & \vdots & \vdots & & & \vdots & \vdots \\
0 & & 1 & \alpha_{1n_1} & \vdots & & & \vdots & \alpha_{2n_1} \\
0 & 0 & \cdots & 0 & 0 & \cdots & \cdots & 0 & \alpha_{2n_1+1} \\
0 & 0 & \cdots & \vdots & 1 & \ddots & & \vdots & \alpha_{2n_1+2} \\
0 & 0 & \cdots & \vdots & 0 & \ddots & \ddots & \vdots & \vdots \\
0 & \ddots & \cdots & \vdots & \vdots & \ddots & 1 & 0 & \alpha_{2n-1} \\
0 & 0 & \cdots & 0 & 0 & \cdots & 0 & 1 & \alpha_{2n}
\end{bmatrix}.
$$

Theorem 5.14. *Algorithm for noisy realization*

Let $\underline{a}$ *be a considered object which is an almost linear system. Then a noisy realization* $\sigma_r = ((\boldsymbol{R}^n, F_r), g_r^0, g_r, h_r, h^0)$ *of* $\underline{a}$ *is given by the following algorithm:*

1) *Based on the square root of eigenvalues for a matrix $H_{\underline{a}\ (p,\bar{p})}(p,0)H_{\underline{a}\ (p,\bar{p})}(p,0)^T$, determine the value n_1 of rank for the matrix $H_{\underline{a}\ (p,\bar{p})}(p,0)$, where $n_1 \leq p$. Namely, determine the value n_1 of rank for the matrix $H_{\underline{a}\ (p,\bar{p})}(p,0)$ such that a set of the squqre root of eigenvalues for the covariance matrix composed of relatively small and equally-sized numbers is excluded, where the signal part effected by the set may be a noisy part.*

2) *The CLS method is used as follows:*

   ① *Let a matrix $A_1 \in \boldsymbol{R}^{1\times(n_1+1)}$ be $A_1 = [\alpha_{11}, \alpha_{12}, \cdots, \alpha_{1n_1}, -1]$.*

   ② *Choose the coefficients $\{\alpha_{0i} : 1 \leq i \leq n_1\}$ such that $\sum_{j=1}^{n_1+1} \underline{S}_l^{j-1}\bar{I}_{\underline{a}} \cdot \underline{S}_l^{j-1}\bar{I}_{\underline{a}}$ takes a minimum value, where $\{\underline{S}_l^i \bar{I}_{\underline{a}} \in \boldsymbol{R}^{L\times 1} : 0 \leq i \leq n_1\}$ are given by the equation $[\bar{I}_{\underline{a}}, \underline{S}_l\bar{I}_{\underline{a}}, \cdots, \underline{S}_l^{n_1}\bar{I}_{\underline{a}}]^T :=$ $A_1^T[A_1A_1^T]^{-1}A_1H_{\underline{a}\ T\ (n_1+1,L)}^T(n_1+1,0)$ and $H_{\underline{a}\ (n_1+1,L)}(n_1+1,0):=$ $[I_{\underline{a}}(0), \cdots, S_l^{n_1-1}I_{\underline{a}}(0), S_l^{n_1}I_{\underline{a}}(0)]$. And $\cdot$ denotes the inner product of two vectors.*

   ③ *Let $h_{1r} \in \boldsymbol{R}^{1\times n_1}$ be $h_{1r} = [I_{\underline{a}}(1) - \bar{I}_{\underline{a}}(1), I_{\underline{a}}(2) - \bar{I}_{\underline{a}}(2), \cdots, I_{\underline{a}}(n_1) - \bar{I}_{\underline{a}}(n_1)]$.*

3) *Based on the square root of eigenvalues for a matrix $H_{\underline{a}\ (n_1+p,\bar{p})}(n_1,p)H_{\underline{a}\ (n_1+p,\bar{p})}(n_1,p)^T$, determine the value n_2 of rank for the matrix $H_{\underline{a}\ (n_1+p,\bar{p})}(n_1,p)$, where $n_2 \leq p$.*

Namely, determine the value n_2 of rank for the matrix $H_{\underline{a}\ (n_1+p,\bar{p})}(n_1,p)$ such that a set of the squqre root of eigenvalues for the covariance matrix composed of relatively small and equally-sized numbers is excluded, where the signal part effected by the set may be a noisy part.

4) *The CLS method is used as follows:*

   ① *Let a matrix $A_2 \in \boldsymbol{R}^{1\times(n_1+n_2+1)}$ be $A_2 = [\alpha_{21}, \alpha_{22}, \cdots, \alpha_{2n_1+n_2}, -1]$.*

   ② *Choose the coefficients $\{\alpha_{2i} : 1 \leq i \leq n_1 + n_2\}$ such that $\sum_{j=1}^{n_1+n_2+1} \underline{S}_l^{j-1}\bar{I}_{\underline{a}} \cdot \underline{S}_l^{j-1}\bar{I}_{\underline{a}}$ takes a minimum value, where $\{\underline{S}_l^i \bar{I}_{\underline{a}} \in \boldsymbol{R}^{L\times 1} : 0 \leq i \leq n_1 + n_2\}$ are given by the equation $[\bar{I}_{\underline{a}}, \underline{S}_l\bar{I}_{\underline{a}}, \cdots, \underline{S}_l^{n_1}\bar{I}_{\underline{a}}]^T :=$ $A_2^T[A_2A_2^T]^{-1}A_2H_{\underline{a}\ (n_1+n_2+1,L)}^T(n_1,n_2+1)$ and $H_{\underline{a}\ (n_1+n_2+1,L)}(n_1,n_2+1):=$ $[I_{\underline{a}}(0), \cdots, S_l^{n_1-1}I_{\underline{a}}(0), I_{\underline{a}}(1), \cdots, S_l^{n_2-1}I_{\underline{a}}(1), S_l^{n_2}I_{\underline{a}}(1)]$. And $\cdot$ denotes the inner product of two vectors.*

   ③ *Let $F_r \in \boldsymbol{R}^{(n_1+n_2)\times(n_1+n_2)}$ be given as the same as in Definition (4.11). Let g_r^0 be $g_r^0 = \mathbf{e}_1$ and g_r be $g_r = \mathbf{e}_{n_1+1} - \mathbf{e}_1$, where*

$$\mathbf{e}_i = [0, \cdots, 0, \overset{i}{1}, 0, \cdots, 0]^T \in \boldsymbol{R}^{n_1+n_2}.$$

   ④ *Let h_r be*

$$h_r = [h_{1r}, I_{\underline{a}}(1) - \bar{I}_{\underline{a}}(1), I_{\underline{a}}(1^2) - \bar{I}_{\underline{a}}(1^2), \cdots, I_{\underline{a}}(1^{n_2-1}) - \bar{I}_{\underline{a}}(1^{n_2-1})].$$

For the real time standard system $\sigma_r = ((\boldsymbol{R}^n, F_r), g_r^0, g_r, h_r, h_r^0)$, its modified impulse responses $I(0)(i) := h_rF_r^ig_r^0$ and $I(1)(i) := h_rF_r^i(g_r^0 + g_r)$ may be written by $I(0)\_(n_1,n_2)$ and $I(1)\_(n_1,n_2)$ repectively.

[proof]. This algorithm is the same as the Algorithm for noisy realization (4.27) except 1).

Remark 1: A determination method of the degree n in the almost linear system $\sigma = ((\boldsymbol{R}^n, F_s), g, h_s)$ can be found in the Principal Component Method. This method is popular.

Remark 2: Let S and N be the norm of a signal and a noise. Then the selected ratio of matrices in the algorithm may be considered as $\frac{N}{S+N}$.

Remark 3: This noisy realization method is very new.

Remark 4: For a noisy case, the AIC is famous for determing only linear systems including dimensions of the state space.

Example 5.15. Let signals be the modified impulse responses of the following 2-dimensional almost linear system $\sigma = ((R^2, F), g^0, g, h, h^0)$,

where $F = \begin{bmatrix} 0.9 & 0 \\ 0 & -0.8 \end{bmatrix}$, $h = [15, \ -6]$, $g^0 = [1, \ 0]^T$, $g = [0, \ 1]^T$, $h^0 = 1$.

Let added noises be given in Fig. 5.4.

Then the noisy realization problem is solved as follows:

covariance matrix	eigenvalues				
	1	2	3	4	5
$H_{\underline{a}\ (3,85)}(3,0)H^T_{\underline{a}\ (3,85)}(3,0)$	2920	5	4.4		
$H_{\underline{a}\ (4,85)}(4,0)H^T_{\underline{a}\ (4,85)}(4,0)$	3553	5.5	5.3	2.7	
$H_{\underline{a}\ (5,85)}(5,0)H^T_{\underline{a}\ (5,85)}(5,0)$	4058	6.4	5.3	3.5	2.7
covariance matrix	square root of eigenvalues				
$H_{\underline{a}\ (5,85)}(5,0)H^T_{\underline{a}\ (5,85)}(5,0)$	63.7	2.5	2.3	1.9	1.6

covariance matrix	eigenvalues				
	1	2	3	4	5
$H_{\underline{a}\ (3,85)}(1,2)H^T_{\underline{a}\ (3,85)}(1,2)$	3335	163	3.6		
$H_{\underline{a}\ (4,85)}(1,3)H^T_{\underline{a}\ (4,85)}(1,3)$	4089	190	4.6	2.7	
$H_{\underline{a}\ (5,85)}(1,4)H^T_{\underline{a}\ (5,85)}(1,4)$	4759	230	4.6	3.2	2.1
covariance matrix	square root of eigenvalues				
$H_{\underline{a}\ (5,85)}(1,4)H^T_{\underline{a}\ (5,85)}(1,4)$	69	15.1	2.1	1.8	1.4

1) Since a set {2.5, 2.3, 1.9, 1.6} is composed of relatively small and equally-sized numbers in the squqre root of $H_{\underline{a}\ (5,85)}(5,0)H^T_{\underline{a}\ (5,85)}(5,0)$, an almost linear system obtained by the CLS method may be realized for a one-dimensional space.
2) After determining the number n_1 of dimensions which is 1, we will continue the noisy realization algorithm by the CLS method.

Therefore, the modified impulse response $I(0)$ of an almost linear system obtained by the CLS method is constructed by a one-dimensional space.
3) Since a set {2.1, 1.8, 1.4} is composed of relatively small and equally-sized numbers in the squqre root of $H_{\underline{a}\ (5,85)}(1,5)H^T_{\underline{a}\ (5,85)}(1,5)$, the almost linear system obtained by the CLS method may be realized by adding another one-dimensional space.

4) After determining the number n_2 of dimensions which is 1, we execute the noisy realization algorithm by the CLS method.

Therefore, the modified impulse responses $I(0)$ and $I(1)$ of an almost linear system obtained by the CLS method is obtained by a (1,2)-dimensional almost linear system.

The system is given by $\sigma_o = ((\boldsymbol{R}^2, \ F_o), \ g_o^0, g_o, h_o, \ h^0)$, where

$$F_o = \begin{bmatrix} 0.9 & 1.7 \\ 0 & -0.8 \end{bmatrix}, \ g_o^0 = \mathbf{e}_1, \ g_o = [-1, \ 1]^T, \ h_o = [15, \ 8.9] \text{ and } h^0 = 1.$$

The obtained modified impulse responses $I(0)$ and $I(1)$ are illustrated in Fig. 5.4.

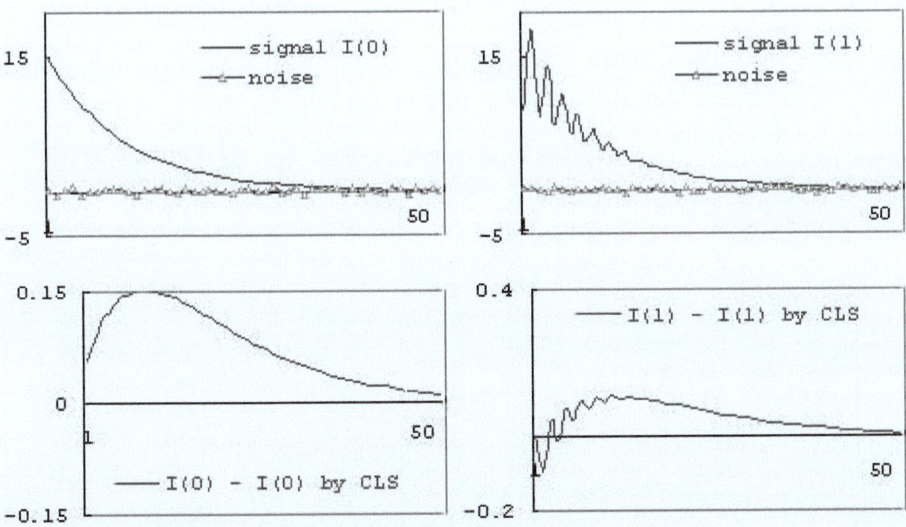

Fig. 5.4. The left are the exact modified impulse response $I(0)$ of a 2-dimensional almost linear system with noise and the difference between the original signal and the obtained one $I(0)_{(1,1)}$ by the CLS method. The right are the original modified impulse response $I(1)$ of a 2-dimensional almost linear system with noise and the difference between the original signal and the obtained one $I(1)_{(1,1)}$ by the CLS method in Example (5.15).

In this example, the original signals are considered as the modified impulse responses of a 2-dimensional almost linear system and the desirable modified impulse responses are obtained by the CLS method.

The following table indicates that the 2-dimensional almost linear system reconstructs the original signal with a 2 and 1 % error to signal ratio and with 0.04 and 0.03 noise to signal ratio, please refer to Remark 2 in Theorem 5.14 for the noise to signal ratio.

The following table and Fig. 5.4 truly indicate that the 3-dimensional almost linear system obtained by the CLS method is a good approximation within our expectations. Hence, there exists a good approximation for the given 2-dimensional almost linear system.

dimen-ion	ratio of matrices	mean values of square root for sum of			cosine	error ratio
		signal ①	signal by CLS ②	error ③	① and ② $\cos\theta$	③/①
$I(0)_-(1,1)$	0.04	0.688	0.678	0.01	0.99993	0.02
$I(1)_-(1,1)$	0.03	0.687	0.683	0.01	0.99992	0.01

For the notations $I(0)_-(n_1, n_2)$ and $I(1)_-(n_1, n_2)$, see Algorithm (5.14).

Example 5.16. Let signals be the modified impulse responses of the following 2-dimensional almost linear system $\sigma = ((R^2, F), g^0, g, h)$,

where $F = \begin{bmatrix} 0.9 & 0.6 \\ 0 & -0.1 \end{bmatrix}$, $h = [17, \ -8]$, $g^0 = [1, \ 0]^T$, $g = [0, \ 1]^T$, $h^0 = 1$.

Let added noises be given in Fig. 5.5.

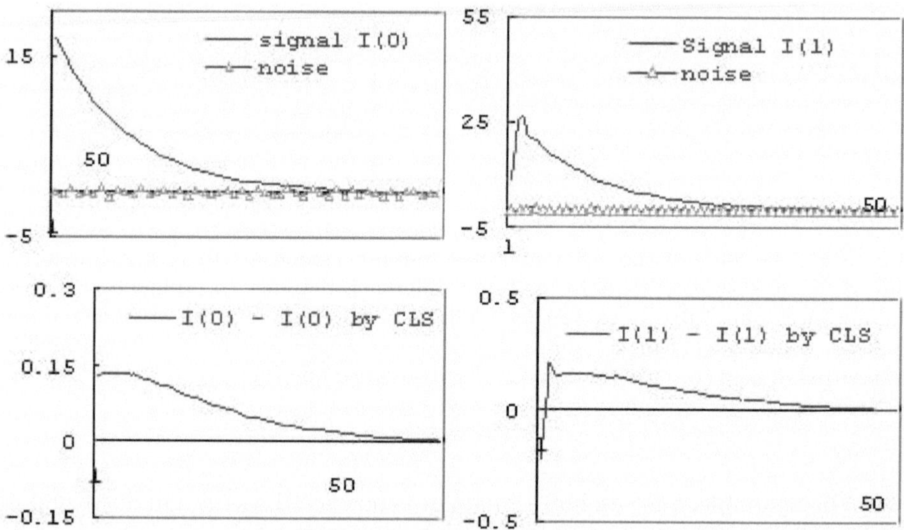

Fig. 5.5. The left are the exact modified impulse response $I(0)$ of a 2-dimensional almost linear system with noise and the difference between the original signal and the obtained one $I(0)_{(1,2)}$ by the CLS method. The right are the original modified impulse response $I(1)$ of a 2-dimensional almost linear system with noise and the difference between the original signal and the obtained one $I(1)_{(1,2)}$ by the CLS method in Example (5.16).

Then the noisy realization problem is solved by the following algorithm:

covariance matrix	eigenvalues				
	1	2	3	4	5
$H_{\underline{a}\ (3,85)}(3,0)H_{\underline{a}\ (3,85)}^{T}(3,0)$	3754	12.3	5.6		
$H_{\underline{a}\ (4,85)}(4,0)H_{\underline{a}\ (4,85)}^{T}(4,0)$	4557	14.3	7.7	3.1	
$H_{\underline{a}\ (5,85)}(5,0)H_{\underline{a}\ (5,85)}^{T}(5,0)$	5212	15	10	5	1.9
covariance matrix	square root of eigenvalues				
$H_{\underline{a}\ (5,85)}(5,0)H_{\underline{a}\ (5,85)}^{T}(5,0)$	72.2	3.9	3.2	2.2	1.4

covariance matrix	eigenvalues				
	1	2	3	4	5
$H_{\underline{a}\ (3,85)}(1,2)H_{\underline{a}\ (3,85)}^{T}(1,2)$	7917	198	9.2		
$H_{\underline{a}\ (4,85)}(1,3)H_{\underline{a}\ (4,85)}^{T}(1,3)$	10465	211	9.9	3.7	
$H_{\underline{a}\ (5,85)}(1,4)H_{\underline{a}\ (5,85)}^{T}(1,4)$	12534	224	10	5.6	3.1
covariance matrix	square root of eigenvalues				
$H_{\underline{a}\ (5,85)}(1,4)H_{\underline{a}\ (5,85)}^{T}(1,4)$	112	15	3.2	2.4	1.8

1) Since a set $\{3.9,\ 3.2,\ 2.2,\ 1.4\}$ is composed of relatively small and equally-sized numbers in the squqre root of $H_{\underline{a}\ (5,85)}(5,0)H_{\underline{a}\ (5,85)}^{T}(5,0)$, the almost linear system obtained by the CLS method may be realized for a one-dimensional space.
2) After determining the number n_1 of dimensions which is 1, we will continue the noisy realization algorithm by the CLS method.

Therefore, the modified impulse response $I(0)$ of an almost linear system obtained by the CLS method is obtained by a one-dimensional space.
3) Since a set $\{3.2,\ 2.4,\ 1.8\}$ is composed of relatively small and equally-sized numbers in the squqre root of $H_{\underline{a}\ (5,85)}(1,4)H_{\underline{a}\ (5,85)}^{T}(1,4)$, the almost linear system obtained by the CLS method may be realized by adding another one-dimensional space.
4) After determining the number n_2 of dimensions which is 1, we execute the noisy realization algorithm by the CLS method.

Therefore, the modified impulse responses $I(0)$ and $I(1)$ of an almost linear system obtained by the CLS method is one of a (1,1)-dimensional almost linear system.

The system is given by the following $\sigma_o = ((\boldsymbol{R}^2,\ F_o),\ g_o^0, g_o, h_o,\ h^0)$, where

$$F_o = \begin{bmatrix} 0.9 & 1.6 \\ 0 & -0.1 \end{bmatrix},\ g_o^0 = \mathbf{e}_1,\ g_o = [-1,\ 1]^T,\ h_o = [16.9,\ 9.2] \text{ and } h^0 = 1.$$

The obtained modified impulse responses $I(0)$ and $I(1)$ are illustrated in Fig. 5.5.

In this example, the original signals are considered as the modified impulse responses of a 2-dimensional almost linear system and the desirable modified impulse responses are obtained by the CLS method. The model obtained by the CLS method is a 2-dimensional almost linear system.

The following table and Fig. 5.5 truly indicate that the 3-dimensional almost linear system obtained by the CLS method is a good approximation within our expectations. Hence, there exists a good approximation for the given 2-dimensional almost linear system.

dimen-ion	ratio of matrices	mean values of square root for sum of		error	cosine ① and ②	error ratio
		signal	signal by CLS	error		
		①	②	③	$\cos\theta$	③/①
$I(0)_-(1,1)$	0.05	0.78	0.77	0.01	0.99999	0.01
$I(1)_-(1,1)$	0.03	1.152	1.14	0.01	0.99997	0.01

For the notations $I(0)_-(n_1, n_2)$ and $I(1)_-(n_1, n_2)$, see Algorithm (5.14).

Example 5.17. Let signals be the modified impulse responses of the following 3-dimensional almost linear system $\sigma = ((R^3, F), g^0, g, h, h^0)$, where

$$F = \begin{bmatrix} 0 & 0 & 0 \\ 1 & 0.7 & 0 \\ 0 & 0 & -0.8 \end{bmatrix}, \ h = [12, \ -4, \ 3], \ g^0 - [1, \ 0, \ 0]^T, \ g = [2, \ 0, \ 1]^T, \ h^0 = 1.$$

Let added noises be given in Fig. 5.6.

Then the noisy realization problem is solved by the following algorithm:

covariance matrix	eigenvalues					
	1	2	3	4	5	6
$H_{\underline{a}\ (4,40)}(4,0)H^T_{a\ (4,40)}(4,0)$	190	48	5.3	3.1		
$H_{\underline{a}\ (5,40)}(5,0)H^T_{a\ (5,40)}(5,0)$	191	51	5.6	4	2.1	
$H_{\underline{a}\ (6,40)}(6,0)H^T_{a\ (6,40)}(6,0)$	191	52	5.8	4.7	3.1	1.5
covariance matrix	square root of eigenvalues					
$H_{\underline{a}\ (6,40)}(6,0)H^T_{a\ (6,40)}(6,0)$	13.8	7.2	2.4	2.2	1.8	1.2

covariance matrix	eigenvalues					
	1	2	3	4	5	6
$H_{\underline{a}\ (4,40)}(2,2)H^T_{a\ (4,40)}(2,2)$	2109	271	6.4	1.5		
$H_{\underline{a}\ (5,40)}(2,3)H^T_{a\ (5,40)}(2,3)$	2118	376	15.2	4.5	1.3	
$H_{\underline{a}\ (6,40)}(2,4)H^T_{a\ (6,40)}(2,4)$	2149	426	16.6	6.2	2.6	1.3
covariance matrix	square root of eigenvalues					
$H_{\underline{a}\ (6,40)}(2,4)H^T_{a\ (6,40)}(2,4)$	46.4	20.6	4.1	2.5	1.6	1.1

1) Since a set $\{2.4, 2.2, 1.8, 1.2\}$ is composed of relatively small and equally-sized numbers in the squqre root of $H_{\underline{a}\ (6,40)}(6,0)H^T_{\underline{a}\ (6,40)}(6,0)$, the almost linear system obtained by the CLS method may be realized for a two-dimensional space.

2) After determining the number n_1 of dimensions which is 2, we will continue the noisy realization algorithm by the CLS method.

Therefore, the modified impulse response $I(0)$ of an almost linear system obtained by the CLS method is obtained for a 2-dimensional space.

3) Since a set $\{2.5, 1.6, 1.1\}$ is composed of relatively small and equally-sized numbers in the squqre root of $H_{\underline{a}\ (6,40)}(2,4)H^T_{\underline{a}\ (6,40)}(2,4)$, the almost linear system obtained by the CLS method may be realized by adding another one-dimensional space.

4) After determining the number n_2 of dimensions which is 1, we execute the
noisy realization algorithm by the CLS method.

Therefore, the modified impulse responses $I(0)$ and $I(1)$ of an almost linear
system obtained by the CLS method is one of a (2,1)-dimensional almost linear
system.

Therefore, the modified impulse responses $I(0)$ and $I(1)$ of an almost linear
system obtained by the CLS method can be constructed by a 3-dimensional
almost linear system.

The system is given by $\sigma = ((\boldsymbol{R}^3, \ F_o), \ g_o^0, g_o, h_o, \ h^0)$, where

$$F_o = \begin{bmatrix} 0 & 0.01 & 2.5 \\ 1 & 0.69 & 2.9 \\ 0 & 0 & -0.85 \end{bmatrix}, \ g_o^0 = \mathbf{e}_1, \ g_o = [-1, \ 0, \ 1]^T, \ h_o = [12.1, \ -3.9, \ 38.9] \text{ and}$$

$h^0 = 1$.

The obtained modified impulse responses $I(0)$ and $I(1)$ are illustrated in
Fig. 5.6.

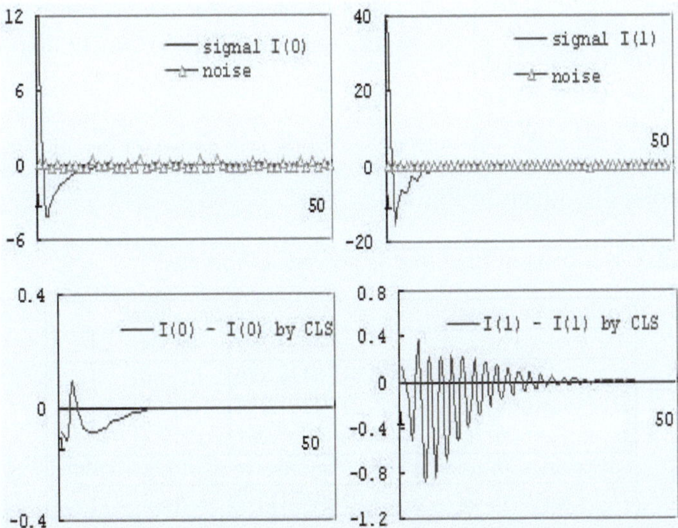

Fig. 5.6. The left are the exact modified impulse response $I(0)$ of a 3-dimensional
almost linear system with noise and the difference between the original signal and
the obtained one $I(0)_{(2,1)}$ by the CLS method. The right are the original modified
impulse response $I(1)$ of a 3-dimensional almost linear system with noise and the
difference between the original signal and the obtained one $I(1)_{(2,1)}$ by the CLS method
in Example (5.17).

In this example, the original signal $I(0)$ is considered as the modified impulse
response of a 2-dimensional linear space and the original signal $I(1)$ is considered
as the modified impulse response of an added 1-dimensional linear space. The
desirable modified impulse responses are attempted to be obtained by the CLS
method. The model obtained by the CLS method is a (2,1)-dimensional almost

linear system which has the same number of dimensions as the number of the original system.

Just as we expected, the following table and Fig. 5.6 indicate that the model obtained by the CLS method is a good (2,1)-dimensional system for the original (2,1)-dimensional system.

dimen-sion	ratio of matrices	mean values of square root for sum of			cosine	error ratio
		signal	signal by CLS	error	① and ②	
		①	②	③	$\cos\theta$	③/①
$I(0)_-(2,1)$	0.17	0.265	0.266	0.01	0.9998	0.02
$I(1)_-(2,1)$	0.05	0.862	0.858	0.04	0.9991	0.04

For the notations $I(0)_-(n_1, n_2)$ and $I(1)_-(n_1, n_2)$, see Algorithm (5.14).

Example 5.18. Let signals be the modified impulse responses of the following 4-dimensional almost linear system $\sigma = ((R^4, F), g^0, g, h, h^0)$,

$$\text{where } F = \begin{bmatrix} 0 & -0.5 & 0 & -0.8 \\ 1 & 0.9 & 0 & 0 \\ 0 & 0 & 0 & -0.9 \\ 0 & 0 & 1 & 0.4 \end{bmatrix}, \ h = [12, \ -4, \ 0, \ -0.1], \ g^0 = [1, \ 0, \ 0, \ 0]^T,$$

$$g = [-1, \ 0, \ -1, \ 1]^T, \ h^0 = 1.$$

Let added noises be given in Fig. 5.7.

Then the noisy realization problem is solved as follows:

covariance matrix	eigenvalues					
	1	2	3	4	5	6
$H_{\underline{a}\ (4,50)}(4, 0)H_{\underline{a}\ (4,50)}^T(4, 0)$	411	276	11	6.6		
$H_{\underline{a}\ (5,50)}(5, 0)H_{\underline{a}\ (5,50)}^T(5, 0)$	418	284	11.5	8.7	3.8	
$H_{\underline{a}\ (6,50)}(6, 0)H_{\underline{a}\ (6,50)}^T(6, 0)$	420	297	12	9.4	7.1	1.3
covariance matrix	square root of eigenvalues					
$H_{\underline{a}\ (6,50)}(6, 0)H_{\underline{a}\ (6,50)}^T(6, 0)$	20.5	17.2	3.5	3.1	2.7	1.1

covariance matrix	eigenvalues						
	1	2	3	4	5	6	7
$H_{\underline{a}\ (5,50)}(2, 3)H_{\underline{a}\ (5,50)}^T(2, 3)$	4953	2959	357	187	2.7		
$H_{\underline{a}\ (6,50)}(2, 4)H_{\underline{a}\ (6,50)}^T(2, 4)$	6212	4144	408	188	6.2	2.1	
$H_{\underline{a}\ (7,50)}(2, 5)H_{\underline{a}\ (7,50)}^T(2, 5)$	6737	5783	415	235	6.2	5.5	1.7
covariance matrix	square root of eigenvalues						
$H_{\underline{a}\ (7,50)}(2, 5)H_{\underline{a}\ (7,50)}^T(2, 5)$	82	76	20.4	15.3	2.5	2.3	1.3

1) Since a set {3.5, 3.1, 2.7, 1.1} is composed of relatively small and equally-sized numbers in the square root of $H_{\underline{a}\ (6,50)}(6, 0)H_{\underline{a}\ (6,50)}^T(6, 0)$, the almost linear system obtained by the CLS method may be realized for a 2-dimensional space.

2) After determining the number n_1 of dimensions which is 2, we will continue the noisy realization algorithm by the CLS method.

Therefore, the modified impulse response $I(0)$ of an almost linear system obtained by the CLS method is characterized by a 2-dimensional almost linear system.

3) Since a set $\{2.5,\ 2.3,\ 1.3\}$ is composed of relatively small and equally-sized numbers in the squqre root of $H_{\underline{a}\ (7,50)}(2,5)\,H_{\underline{a}\ (7,50)}^{T}(2,5)$, the almost linear system obtained by the CLS method may be realized by adding another two-dimensional space.

4) After determining the number n_2 of dimensions which is 2, we will continue the noisy realization algorithm by the CLS method.

Therefore, the modified impulse response $I(1)$ of an almost linear system obtained by the CLS method is realized by adding another 2-dimensional space.

Therefore, the modified impulse responses $I(0)$ and $I(1)$ of an almost linear system obtained by the CLS method can be realized by a (2,2)-dimensional almost linear system.

The system is given as follows: $\sigma_o = ((\boldsymbol{R}^4,\ F_o),\ g_o^0, g_o, h_o,\ h^0)$,

where $F_o = \begin{bmatrix} 0 & -0.46 & 0 & 0.8 \\ 1 & 0.88 & 0 & -0.78 \\ 0 & 0 & 0 & -0.9 \\ 0 & 0 & 1 & 0.4 \end{bmatrix}$, $g_o^0 = \mathbf{e}_1$, $g_o = [-1,\ 0,\ 1,\ 0]^T$,

$h_o = [12.2,\ -4.4,\ -0.18,\ -10]$ and $h^0 = 1$.

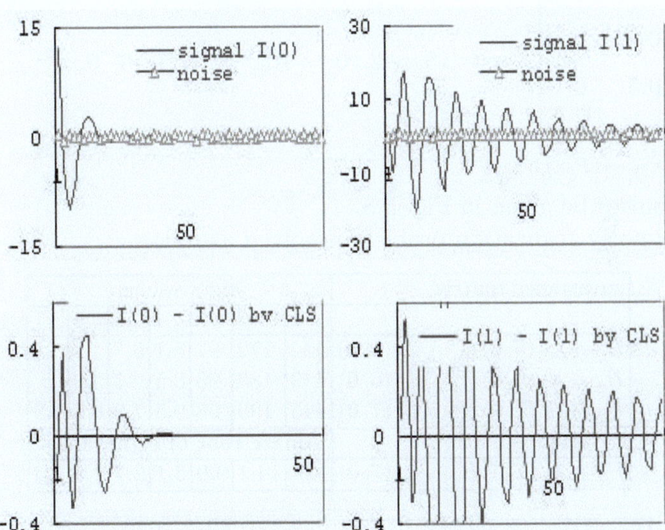

Fig. 5.7. The left are the exact modified impulse response $I(0)$ of a 4-dimensional almost linear system with noise and the difference between the original signal and the obtained one $I(0)_{(2,2)}$ by the CLS method. The right are the original modified impulse response $I(1)$ of a 4-dimensional almost linear system with noise and the difference between the original signal and the obtained one $I(1)_{(2,2)}$ by the CLS method in Example (5.18).

The obtained modified impulse responses $I(0)$ and $I(1)$ are illustrated in Fig. 5.7.

In this example, the original signal $I(0)$ is considered as the modified impulse responses of a 2-dimensional linear space and the original signal $I(1)$ is considered as the modified impulse response of an added 2-dimensional linear space. The desirable modified impulse responses are attempted to be obtained by the CLS method. The model obtained by the CLS method is a (2,2)-dimensional almost linear system which has the same number of dimensions as the number of the original system.

Just as we expected, the following table and Fig. 5.7 indicate that the model obtained by the CLS method is a good (2,2)-dimensional almost linear system for the original (2,2)-dimensional system.

dimen-ion	ratio of matrices	mean values of square root for sum of			cosine ① and ②	error ratio
		signal ①	signal by CLS ②	error ③	$\cos\theta$	③/①
$I(0)$-(2,2)	0.17	0.353	0.351	0.02	0.9987	0.05
$I(1)$-(2,2)	0.03	1.03	1.02	0.04	0.9991	0.04

For the notations $I(0)$-(n_1, n_2) and $I(1)$-(n_1, n_2), see Algorithm (5.14).

Example 5.19. Let signals be the modified impulse responses of the following 4-dimensional almost linear system $\sigma = ((R^4, F), x^0, g, h)$, where

$$F = \begin{bmatrix} 0 & 0 & -0.5 & 0 \\ 1 & 0 & 0.2 & 0 \\ 0 & 1 & 0.5 & 0 \\ 0 & 0 & 0 & -0.9 \end{bmatrix}, \quad h = [15,\ -2,\ 0,\ -10],\ g^0 = [1,\ 0,\ 0,\ 0]^T,$$

$g = [-1,\ 0,\ 0,\ 1]^T,\ h^0 = 1.$

Let added noises be given in Fig. 5.8.

Then the noisy realization problem is solved as follows:

covariance matrix	eigenvalues						
	1	2	3	4	5	6	7
$H_{\underline{a}\ (5,50)}(5,0)H^T_{\underline{a}\ (5,50)}(5,0)$	424	172	97	8.1	5.7		
$H_{\underline{a}\ (6,50)}(6,0)H^T_{\underline{a}\ (6,50)}(6,0)$	443	183	98	8.5	7	3.8	
$H_{\underline{a}\ (7,50)}(7,0)H^T_{\underline{a}\ (7,50)}(7,0)$	445	199	98	9.5	7.2	6.6	1.9
covariance matrix	square root of eigenvalues						
$H_{\underline{a}\ (7,50)}(7,0)H^T_{\underline{a}\ (7,50)}(7,0)$	21	14.1	9.9	3.1	2.7	2.6	1.4

covariance matrix	eigenvalues						
	1	2	3	4	5	6	7
$H_{\underline{a}\ (5,50)}(3,2)H^T_{\underline{a}\ (5,50)}(3,2)$	1113	230	134	35	5.3		
$H_{\underline{a}\ (6,50)}(3,3)H^T_{\underline{a}\ (6,50)}(3,3)$	1450	233	134	37.8	5.3	5.1	
$H_{\underline{a}\ (7,50)}(3,4)H^T_{\underline{a}\ (7,50)}(3,4)$	1719	237	134	38	6	5.3	5.1
covariance matrix	square root of eigenvalues						
$H_{\underline{a}\ (7,50)}(3,4)H^T_{\underline{a}\ (7,50)}(3,4)$	41.5	15.4	11.6	6.2	2.4	2.3	2.3

1) Since a set $\{3.1,\ 2.7,\ 2.6,\ 1.4\}$ is composed of relatively small and equally-sized numbers in the squqre root of $H_{\underline{a}\ (7,50)}(7,0)H_{\underline{a}\ (7,50)}^{T}(7,0)$, the almost linear system obtained by the CLS method may be good for a 3-dimensional space.

2) After determining the number n_1 of dimensions which is 3, we will continue the noisy realization algorithm by the CLS method.

Therefore, the modified impulse response $I(0)$ of an almost linear system obtained by the CLS method is characterized by 3-dimensional almost linear system.

3) Since a set $\{2.4,\ 2.3,\ 2.3\}$ is composed of relatively small and equally-sized numbers in the squqre root of $H_{\underline{a}\ (7,50)}(3,4)H_{\underline{a}\ (7,50)}^{T}(3,4)$, the almost linear system obtained by the CLS method may be somewhat good by adding another 1-dimensional space.

4) After determining the number n_2 of dimensions which is 1, we will continue the noisy realization algorithm by the CLS method.

Therefore, the modified impulse response $I(1)$ of an almost linear system obtained by the CLS method is realized by adding another 1-dimensional space.

The system is given by $\sigma_o = ((\boldsymbol{R}^4,\ F_o),\ g_o^0, g_o, h_o,\ h^0)$,

$$\text{where } F_o = \begin{bmatrix} 0 & 0 & -0.5 & -0.03 \\ 1 & 0 & 0.2 & -0.1 \\ 0 & 1 & 0.48 & 0.02 \\ 0 & 0 & 0 & -0.92 \end{bmatrix},\ g_o^0 = \mathbf{e}_1,\ g_o = [-1,\ 0,\ 0,\ 1]^T,$$

$h_o = [14.8,\ -1.8,\ 0.31,\ -10.3]$ and $h^0 = 1$.

The obtained modified impulse responses $I(0)$ and $I(1)$ are illustrated in Fig. 5.8.

In this example, the original signal $I(0)$ is characterized by the modified impulse response of a 3-dimensional linear space and the original signal $I(1)$ is characterized as the modified impulse responses of an added 1-dimensional linear space. The desirable modified impulse responses are attempted to be obtained by the CLS method. The model obtained by the CLS method is a (3,1)-dimensional almost linear system which has the same number of dimensions as the number of the original system.

Just as we expected, the following table and Fig. 5.8 indicate that the model obtained by the CLS method is a good (3,1)-dimensional system for the original (3,1)-dimensional system.

dimen-ion	ratio of matrices	mean values of square root for sum of signal	signal by CLS	error	cosine $①$ and $②$	error ratio
		$①$	$②$	$③$	$\cos\theta$	$③/①$
I(0)-(3,1)	0.15	0.362	0.361	0.01	0.9994	0.04
I(1)-(3,1)	0.06	0.459	0.463	0.03	0.998	0.07

For the notations $I(0)$-(n_1, n_2) and $I(1)$-(n_1, n_2), see Algorithm (5.14).

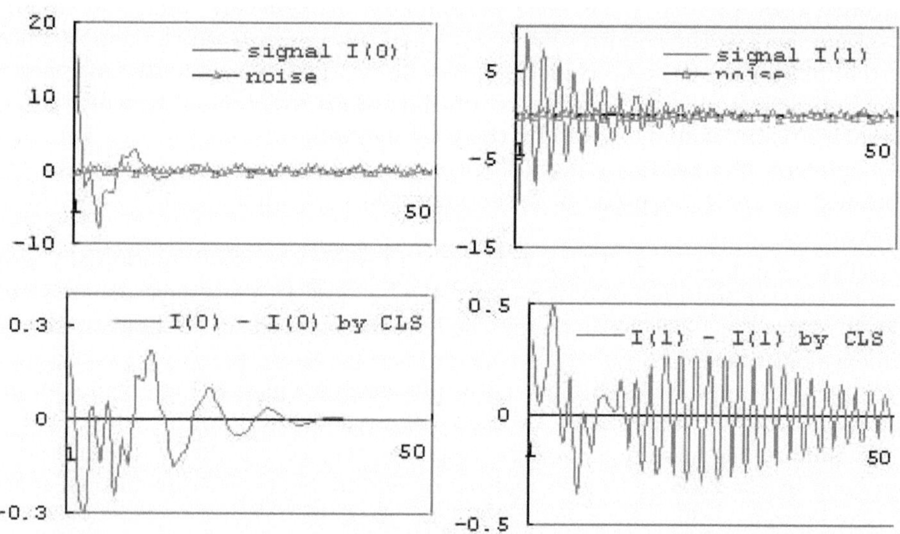

Fig. 5.8. The left are the exact modified impulse response $I(0)$ of a 4-dimensional almost linear system with noise and the difference between the original signal and the obtained one $I(0)_{(3,1)}$ by the CLS method. The right are the original modified impulse response $I(1)$ of a 4-dimensional almost linear system with noise and the difference between the original signal and the obtained one $I(1)_{(3,1)}$ by the CLS method in Example (5.19).

Example 5.20. Let signals be the modified impulse responses of the following 5-dimensional almost linear system $\sigma = ((\boldsymbol{R}^5, F), g^0, g, h, h^0)$,

where $F = \begin{bmatrix} 0 & 0 & 0.4 & 0 & -0.5 \\ 1 & 0 & -0.3 & 0 & 1 \\ 0 & 1 & 0.4 & 0 & -0.9 \\ 0 & 0 & 0 & 0 & 0.4 \\ 0 & 0 & 0 & 1 & 0.6 \end{bmatrix}$, $h = [12, \; -4, \; 2, \; -3, \; 10]$,

$g^0 = [1, \; 0, \; 0, \; 0, \; 0]^T$, $g = [-1, \; 0, \; 0, \; 1, \; 0]^T$, $h^0 = 1$.

Let added noises be given in Fig. 5.9.

Then the noisy realization problem is solved as follows:

covariance matrix	eigenvalues						
	1	2	3	4	5	6	7
$H_{\underline{a}\,(5,50)}(5,0)\,H^T_{\underline{a}\,(5,50)}(5,0)$	283	74	68	7.7	2.1		
$H_{\underline{a}\,(6,50)}(6,0)\,H^T_{\underline{a}\,(6,50)}(6,0)$	286	76	71	8.4	3.7	1.4	
$H_{\underline{a}\,(7,50)}(7,0)\,H^T_{\underline{a}\,(7,50)}(7,0)$	288	77	76	8.8	4.8	2.6	1
covariance matrix	square root of eigenvalues						
$H_{\underline{a}\,(7,50)}(7,0)\,H^T_{\underline{a}\,(7,50)}(7,0)$	16.9	8.8	8.7	2.9	2.2	1.6	1

covariance matrix	eigenvalues							
	1	2	3	4	5	6	7	8
$H_{\underline{a}\ (6,50)}(3,3)H_{\underline{a}\ (6,50)}^{T}(3,3)$	2320	464	192	69	13.8	7.6		
$H_{\underline{a}\ (7,50)}(3,4)H_{\underline{a}\ (7,50)}^{T}(3,4)$	3105	470	193	69	15.7	7.7	4.4	
$H_{\underline{a}\ (8,50)}(3,5)H_{\underline{a}\ (8,50)}^{T}(3,5)$	3892	478	193	72	16	7.6	4.4	3.1
covariance matrix	square root of eigenvalues							
$H_{\underline{a}\ (8,50)}(3,5)H_{\underline{a}\ (8,50)}^{T}(3,5)$	62	22	14	8.5	4	2.8	2.1	1.8

1) Since a set $\{2.9,\ 2.2,\ 1.6,\ 1\}$ is composed of relatively small and equally-sized numbers in the squqre root of $H_{\underline{a}\ (7,50)}(7,0)H_{\underline{a}\ (7,50)}^{T}(7,0)$, the almost linear system obtained by the CLS method may be good for a 3-dimensional space.

2) After determining the number n_1 of dimensions which is 3, we will continue the noisy realization algorithm by the CLS method.

Therefore, the modified impulse response $I(0)$ of an almost linear system obtained by the CLS method is characterized by a 3-dimensional almost linear system.

3) Since a set $\{2.8,\ 2.1,\ 1.8\}$ is composed of relatively small and equally-sized numbers in the squqre root of $H_{\underline{a}\ (8,50)}(3,5)H_{\underline{a}\ (8,50)}^{T}(3,5)$, the almost linear system obtained by the CLS method may be somewhat good by adding another 2-dimensional space.

4) After determining the number n_2 of dimensions which is 2, we will continue the noisy realization algorithm by the CLS method.

Therefore, the modified impulse response $I(1)$ of an almost linear system obtained by the CLS method is constructed by adding another 2-dimensional space.

The system is given by $\sigma_o = ((\boldsymbol{R}^5,\ F_o),\ g_o^0, g_o, h_o,\ h^0)$,

$$\text{where } F_o = \begin{bmatrix} 0 & 0 & 0.4 & 0 & -0.45 \\ 1 & 0 & -0.3 & 0 & 1 \\ 0 & 1 & 0.36 & 0 & -0.9 \\ 0 & 0 & 0 & 0 & 0.45 \\ 0 & 0 & 0 & 1 & 0.55 \end{bmatrix},\ g_o^0 = \mathbf{e}_1,\ g_o = [-1\ 0,\ 0,\ 1,\ 0]^T,$$

$h_o = [12.2,\ -4,\ 2,\ -3,\ 9.6]$ and $h^0 = 1$.

The obtained modified impulse responses $I(0)$ and $I(1)$ are illustrated in Fig. 5.9.

In this example, the original signal $I(0)$ is characterized as the modified impulse responses of a 3-dimensional linear space and the original signal $I(1)$ is characterized as the modified impulse response of an added 2-dimensional linear space. The desirable modified impulse responses are attempted to be obtained by the CLS method. The model obtained by the CLS method is a (3,2)-dimensional almost linear system which have the same number of dimensions as the number of the original system.

Just as we expected, the following table and Fig. 5.9 indicate that the model obtained by the CLS method is a good (3,2)-dimensional system for the original (3,2)-dimensional system.

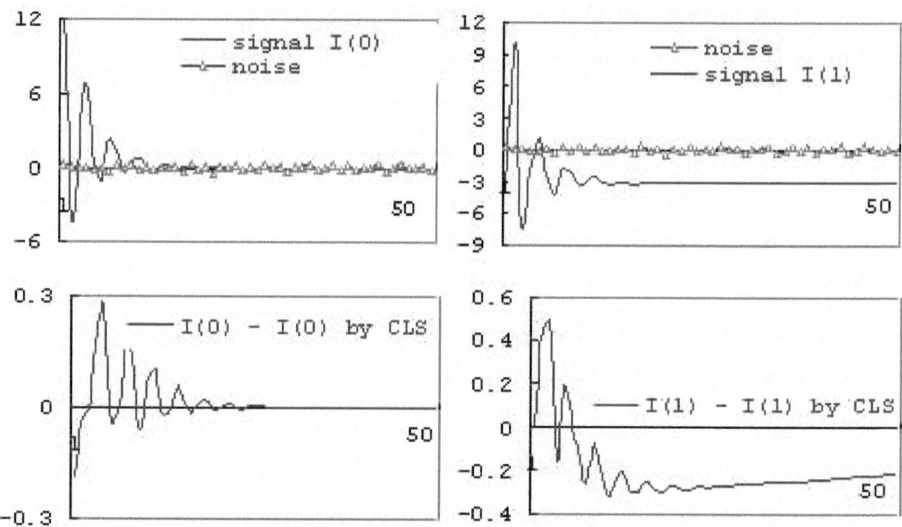

Fig. 5.9. The left are the exact modified impulse response $I(0)$ of a 5-dimensional almost linear system with noise and the difference between the original signal and the obtained one $I(0)_{(3,2)}$ by the CLS method. The right are the original modified impulse response $I(1)$ of a 5-dimensional almost linear system with noise and the difference between the original signal and the obtained one $I(1)_{(3,2)}$ by the CLS method in Example (5.20).

dimen-ion	ratio of matrices	mean values of square root for sum of			cosine	error ratio
		signal	signal by CLS	error	① and ②	
		①	②	③	$\cos\theta$	③/①
$I(0)_-(3,2)$	0.17	0.296	0.296	0.01	0.9994	0.03
$I(1)_-(3,2)$	0.05	0.482	0.454	0.03	0.9988	0.08

For the notations $I(0)_-(n_1, n_2)$ and $I(1)_-(n_1, n_2)$, see Algorithm (5.14).

5.5 Historical Notes and Concluding Remarks

Approximate and noisy realization problems of almost linear systems have been studied with the notion of Input/output matrix norm and the CLS method. The matrix norm is used for determining the dimensions of the state space and the CLS method is used for determining the parameters of almost linear systems, which are a sort of non-linear systems.

For the approximate and noisy realization problems, like I said, there may be a sign of using singular value decomposition and the Constrained Least Square (CLS) method in Kalman [1997]. In this reference, Kalman also pointed out that the identification problem from noisy data should be treated without any prejudice, hence, should be described in a statistical sense, not a probabilistic sense. Here, we only insist that the signal and the noise are not correlated. Then we discussed approximate and noisy realization problems of non-linear systems that could not be treated using existing methods.

In order to insist that our method for approximate and noisy realization are also effective for non-linear cases, we gave several examples. As shown, the numerical results of the examples have demonstrated that the ratio of the squqre root of singular values implies a degree of approximation in the sense of the squqre norm. For our noisy realization problems, we showed that we can determine the dimensions of almost linear systems when a set of equally-sized numbers of the square root of singular values can be found.

In a similar manner in Chapters 3 and 4, our several examples of both approximate and noisy realization problems in almost linear systems suggest that our two features can also be expressed as follows:

(1) The ratio of matrices determines a degree of the crossed angle between directions of the obtained signal and the original signal.
(2) The CLS method determines the coefficients of linearly dependent vectors such that the error between the obtained signal and original signal has a minimum value in the sense of the squqre norm while conseving the crossed angle.

In particular, our several examples of approximate realization have shown that the changing relations among the ratio of matrices and the error to signal ratio are proportional relations and the ratio 0.01 of Input/output matrix ratio ranges from 0.005 to 0.02 for the error to signal ratio.

In addition, our several examples of noisy realization have shown that the changing relations among the ratio of matrices and the error to signal ratio are proportial relations and the ratio 0.01 of Input/output matrix ratio ranges from 0.001 to 0.015 for the error to signal ratio.

6 Approximate and Noisy Realization of Pseudo Linear Systems

Let the set Y of output's values be a linear space over the field $\boldsymbol{R}$. In the reference [Matsuo and Hasegawa, 2003], pseudo linear systems were presented with a main theorem, which says that for any time-invariant input response map, there exist at least two canonical (quasi-reachable and distinguishable) pseudo linear systems which realize, that is, faithfully describe it, and any two canonical pseudo linear systems with the same behavior are isomorphic.

As previously described, the fundamental facts about pseudo linear systems are stated for preparation of their approximate and noisy realization problems.

Firstly, their realization theory is stated.

Secondly, the main facts about finite dimensional pseudo linear systems are stated. A criterion for the canonical finite dimensional pseudo linear systems, representation theorems of isomorphic classes for canonical pseudo linear systems and a procedure to obtain a canonical one are stated.

Thirdly, their partial realization is discussed according to the above results. The main are the following:

An algorithm to obtain a natural partial realization from a given partial time-invariant input response map is given.

We can easily understand that the above results of our systems are the same as ones obtained in linear system theory.

Moreover, for the time-invariant input response map, we can discuss a real time partial realization problem. Namely, by a single experiment, we find a mathematical model from on-line data. An algorithm to obtain a partial realization from the data is given if a physical object is finite dimensional.

6.1 Basic Facts about Pseudo Linear Systems

Definition 6.1. *Pseudo Linear System*
1) A system given by the following equations is written as a collection
$\sigma = ((X, F), g, h, h^0)$ *and it is said to be a pseudo linear system.*

Y. Hasegawa: Approxi. & Noisy Reali. of Discrete-Time Dyn. Sys., LNCIS 376, pp. 123–163, 2008.
springerlink.com

$$\begin{cases} x(t+1) = Fx(t) + g(\omega(t+1)) \\ x(0) \quad = 0 \\ \gamma(t) \quad = h^0 + hx(t) \end{cases}$$

*where X is a linear space over the field $\boldsymbol{R}$, F is a linear operator on X
and $\omega(t) \in U$ for any $t \in N$. And g is a function $: U \mapsto X$, and h is a linear
operator $: X \to Y$ and $h^0 \in Y$.*

*2) The input response map $a_\sigma : U^* \to Y; \omega \mapsto h^0 + h(\sum_{j=1}^{|\omega|}\{((F^{|\omega|-j})g(\omega(j)))$
is said to be a behavior of σ.*

*For a time-invariant input response map $a \in F(U^*, Y)$, σ that satisfies
$a_\sigma = a$ is called a realization of a.*

*3) For the pseudo linear system σ and any $u \in U$, $i \in N$,
$I_\sigma(u)(i) := hF^{i-1}g(u)$ is said to be a modified impulse response of σ,
where $u^0 := 1$. The relation $I_\sigma(u)(i) = a_\sigma(u^i) - a_\sigma(u^{i-1})$ holds.
Note that there is a one-to-one correspondence between the behavior of σ
and the modified impulse responses $I_\sigma(u) \in F(N, Y)$ of σ by
the relations $a_\sigma(\omega) = a_\sigma(1) + \sum_{j=1}^{|\omega|} I_\sigma(\omega(i))(|\omega| - j + 1)$.*

*4) A pseudo linear system σ is said to be quasi-reachable if the linear hull
of the reachable set $\{\sum_{j=1}^{|\omega|}\{((F^{|\omega|-j})g(\omega(j))); \omega \in U^*\}$ is equal to X.
A pseudo linear system σ is called observable if $hF^m x_1 = hF^m x_2$ for any
$m \in N$ implies $x_1 = x_2$.*

*5) A pseudo linear system σ is said to be canonical if σ is quasi-reachable
and observable.*

Example 6.2. $A(N \times U, \boldsymbol{R}) := \{\lambda = \sum_{n,u} \lambda(n, u)e_{(n,u)}(\text{finite sum}); n \in N, u \in U\}$, where $e_{(n,u)}$ is given by the following equations for $n, n' \in N$ and $u, u' \in U$.
If $n = n'$ and $u = u'$, it implies $e_{(n,u)}(n', u') = 1$. If $n \neq n'$ or $u \neq u'$, it
implies $e_{(n,u)}(n', u') = 0$. Then $A(N \times U, \boldsymbol{R})$ is clearly a linear space. Let S_r
be $S_r(e_{(n,u)}) = S_r(e_{(n+1,u)})$, then $S_r \in L(A(N \times U, \boldsymbol{R}))$ and S_r is irrelevant
to the input value's set U. S_r is a right shift operator. Let a map $\eta : U \to
A(N \times U, \boldsymbol{R}); u \mapsto e_{(0,u)}$ and let a linear map $\bar{a} : A(N \times U, \boldsymbol{R}) \to Y$ be $\bar{a}(e_{(n,u)}) =
a(u^{n+1}) - a(u^n)$ for any time-invariant input response map $a \in F(U^*, Y)$. Then a
collection $((A(N \times U, \boldsymbol{R}), S_r), \eta, \bar{a}, a(1))$ is a quasi-reachable pseudo linear system
that realizes a.

Let $F(N, Y) := \{$ any function $f : N \to Y\}$. Let S_l $\gamma(t) = \gamma(t+1)$ for any
$\gamma \in F(N, Y)$ and $t \in N$, then $S_l \in L(F(N, Y))$. Let a map $\chi : U \to F(N, Y)$
be $(\chi(u))(t) := a(\omega|u) - a(\omega)$ for any $u \in U$, $t \in N$, a time-invariant input
response map $a \in F(U^*, Y)$ and ω such that $|\omega| = t$. Moreover, let a linear map
0 be $F(N, Y) \to Y; \gamma \mapsto \gamma(0)$. Then a collection $((F(N, Y), S_l), \chi, 0, a(1))$ is a
distinguishable pseudo linear system that realizes a.

Theorem 6.3. *The following two pseudo linear systems are canonical realizations of any time-invariant input response map $a \in F(U^*, Y)$.*
1) $((A(N \times U, \boldsymbol{R})/_{=a}, \tilde{S}_r), \tilde{\eta}, \tilde{\bar{a}}, a(1))$,

where $A(N \times U, \mathbf{R})/_{=a}$ is a quotient space obtained by equivalence relation $\sum_{n,u} \lambda_1(n,u)e_{(n,u)} = \sum_{\bar{n},\bar{u}} \lambda_2(\bar{n},\bar{u})e_{(\bar{n},\bar{u})} \iff \sum_{n,u}(a(u^{n+1} - a(u^n)) = \sum_{\bar{n},\bar{u}} (a(\bar{u}^{\bar{n}+1} - a(\bar{u}^{\bar{n}})).
And $\tilde{S}_r \in L(A(N \times U, \mathbf{R})/_{=a})$ is given by $\tilde{S}_r[e_{(n,u)}] = [e_{(n+1,u)}]$ for $[e_{(n,u)}] \in A(N \times U, \mathbf{R})/_{=a}$, and $\tilde{\eta}$ is a map : $U \to A(N \times U, \mathbf{R})/_{=a}; u \mapsto [e_{(0,u)}]$, and $\tilde{\bar{a}}$ is given by : $\tilde{\bar{a}} \to Y; [e_{(n,u)}] \mapsto a(u^{n+1}) - a(u^n)$.
2) $((\ll S_l^N(\chi(U)) \gg, S_l), \chi, 0, a(1))$,
where $\ll S_l^N(\chi(U)) \gg$ is the smallest linear space which contains $S_l^N(\chi(U)) := \{S_l^i(\chi(u)); u \in U, i \in N, S_l^i(\chi(u))(t) = (\chi(u))(t+i) = a(\omega|u) - a(\omega), \omega \in U^*, |\omega| = t + i\}$.

Definition 6.4. *Let $\sigma_1 = ((X_1, F_1, g_1, h_1, h^0)$ and $\sigma_2 = ((X_2, F_2, g_2, h_2, h^0)$ be pseudo linear systems, then a linear operator $T : X_1 \to X_2$ is said to be a pseudo linear system morphism $T : \sigma_1 \to \sigma_2$ if T satisfies $TF_1 = F_2T$, $Tg_1 = g_2$ and $h_1 = h_2T$.*
If $T : X_1 \to X_2$ is bijective, then $T : \sigma_1 \to \sigma_2$ is said to be an isomorphism.

Theorem 6.5. *Realization Theorem of Pseudo Linear Systems*
Existence : For any time-invariant input response map $a \in F(U^*, Y)$,
  there exist at least two canonical pseudo linear systems which realize a.
Uniqueness : Let σ_1 and σ_2 be any two canonical pseudo linear systems that
  realize a time-invariant input response map $a \in F(U^*, Y)$.
  Then there exists an isomorphism $T : \sigma_1 \to \sigma_2$.

6.2 Finite Dimensional Pseudo Linear Systems

Based on the realization theory (6.5), we will state facts about finite dimensional pseudo linear systems as previously described.

To state clear facts, we assume that the set U of input values is finite , i.e., $U := \{u_i; 1 \le i \le m\}$ for some $m \in N\}$. This assumption will imply that the g of a pseudo linear system $\sigma = ((X, F), g, h, h^0)$ is completely determined by the finite vectors $\{g(u_i); 1 \le i \le m, m \in N\}$, and it was presented that the assumption is not so special in the reference [Matsuo and Hasegawa, 2003].

We only state the following four facts needed for this chapter.

① : The condition for the finite dimensional pseudo linear system to be canonical.

② : the representation theorem for finite dimensional canonical pseudo linear systems, i.e., we show the real time standard system as a representative.

③ : The criterion for the behavior of finite dimensional pseudo linear systems to be given by the rank condition of an Input/output matrix.

④ : The procedure to obtain the quasi-reachable standard system that realizes a given time-invariant input response map.

Corollary 6.6. *Let T be a pseudo linear system morphism $T : \sigma_1 \to \sigma_2$, then $a_{\sigma_1} = a_{\sigma_2}$ holds.*

The following is a fact about finite dimensional linear spaces:

FACT : < An n-dimensional linear space over the field $\boldsymbol{R}$ is isomorphic to $\boldsymbol{R}^n$ and $L(\boldsymbol{R}^n, \boldsymbol{R}^m)$ is isomorphic to $\boldsymbol{R}^{m \times n}$. (See Halmos [1958]).>
Therefore, without loss of generality, we can consider a n-dimensional pseudo linear system as $\sigma = ((\boldsymbol{R}^n, F), g, h, h^0)$, where $F \in \boldsymbol{R}^{n \times n}$, $g(u) \in \boldsymbol{R}^n$ and $h \in \boldsymbol{R}^{p \times n}$.

Theorem 6.7. *A pseudo linear system $\sigma = ((\boldsymbol{R}^n, F), g, h, h^0)$ is canonical if and only if the following conditions 1) and 2) hold:*
1) rank $[g(u_1), Fg(u_1), \cdots, F^{n-1}g(u_1), g(u_2), Fg(u_2), \cdots, F^{n-1}g(u_2), \cdots,$
$g(u_m), Fg(u_m), \cdots, F^{n-1}g(u_m)] = n$
2) rank $[h^T, (hF)^T, \cdots, \cdots, (hF^{n-1})^T] = n$.

Definition 6.8. *A canonical pseudo linear system $\sigma_s = ((\boldsymbol{R}^n, F_s), g_s, h_s, h^0)$ is said to be a real time standard system if a set $\{(i, u_j) \in N \times U, 1 \leq j \leq m\}$ given by $\mathbf{e}_{m_1 + \cdots + m_{j-1} + i} = F_s^{i-1} g_s(u_j)$ satisfies the following conditions:*
1) $g_s(u_j) = \mathbf{e}_{m_1 + \cdots + m_{j-1} + 1}$ *and* $\mathbf{e}_{m_1 + \cdots + m_{j-1} + i} = F_s^{i-1} g_s(u_j)$ *hold for any*
  i $(1 \leq i \leq m_j,\ j\ (1 \leq j \leq m)$.
2) $F_s^{m_p} g_s(u_p) = \sum_{i=1}^{m_1 + \cdots + m_p} \alpha_{p,i} \mathbf{e}_i$ *holds for any $1 \leq p \leq m$, where $\alpha_{p,i} \in \boldsymbol{R}$*
  and $\mathbf{e}_i = [0, 0, \cdots, 0, \overset{i}{1}, 0, \cdots, 0]^T$.
3) $n = \sum_{i=1}^m m_i$ *holds.*
4) F_s *is given as follows:*

$$
F_s =
\begin{bmatrix}
0\cdots & 0\ \alpha_{11} & 0 \cdots\cdots 0 & \alpha_{21} & 0 \cdots 0 & 0\ \alpha_{m1} \\
1 \ddots & \vdots\ \alpha_{12} & 0 \cdots\quad 0 & \alpha_{22} & \vdots & \alpha_{m2} \\
0 \ddots & \vdots\ \vdots & \vdots & \vdots & \vdots & \vdots \\
\vdots \ddots\ 1\ 0 & \vdots & \vdots & \vdots & \vdots & \vdots \\
0 \ddots 0\ 1\alpha_{1m_1} & \vdots & \vdots\quad \alpha_{2m_1} & \vdots & & \vdots \\
0\cdots 0\ 0\quad 0 & 0 \cdots\cdots 0 & \alpha_{2m_1+1} & 0 \cdots 0 & & \\
0 \cdots & \vdots\quad 1 \ddots & \vdots & \vdots & \vdots & \\
0 \cdots & \vdots\quad 0 \ddots\ddots\vdots & \vdots & \vdots & & \\
0 \cdots & \vdots\quad \vdots \ddots\ 1\ 0 & \vdots & \vdots\ \cdots & & \\
0 \cdots & 0 \cdots 0\ 1\alpha_{2m_1+m_2}\ 0 \ddots & & & & \\
0 \cdots & \cdots\cdots 0\quad 0\quad 0 \ddots & & & & \\
0 \cdots & \cdots\cdots & \vdots & \ddots\ddots\vdots & \vdots & \\
0 \cdots & \cdots\cdots & \vdots & \ddots\ \cdot 0\cdots\cdots 0 & \vdots & \\
0 \cdots & \cdots\cdots & \vdots & \ddots\ 0\ 1 & \vdots & \vdots \\
0 \cdots & \cdots\cdots & \vdots & \ddots\ddots 0 \ddots\ 0 & \vdots & \vdots \\
0 \cdots & \cdots\cdots & \vdots & \ddots\ddots\ \ddots\ 1\ 0\alpha_{mn-1} & & \\
0 \cdots & \cdots\cdots & \cdots & \cdots\cdots 0\cdots\ 0\ 1 & \alpha_{mn} &
\end{bmatrix} .
$$

Theorem 6.9. *Representation Theorem for equivalence classes*
For any finite dimensional canonical pseudo linear system, there exists a uniquely determined isomorphic real time standard system.

[proof] Note that F_s in the real time standard system is the quasi-reachable standard form.

Let $\sigma = ((\mathbf{R}^n, F), g, h, h^0)$ be any finite dimensional canonical pseudo linear system. For the real time standard form $((\mathbf{R}^n, F_s), g_s, h_s, h^0)$ and a linear operator $T : \mathbf{R}^n \to \mathbf{R}^n$ such that $TF = F_s T$ and $Tg = g_s$ hold , let $h_s := h \cdot T^{-1}$. Then T is a pseudo linear system morphism : $\sigma = ((\mathbf{R}^n, F), g, h, h^0) \to \sigma_s = ((\mathbf{R}^n, F_s), g_s, h_s, h^0)$. T is bijective and σ_s is the only real time standard system. By Corollary (6.6), the behaviors of σ and σ_s are the same.

Definition 6.10. *For any time-invariant input response map $a \in F(U^*, Y)$, the corresponding linear input/output map $A : ((A(N \times U, \mathbf{R}), S_r) \to (F(N, Y), S_l)$ satisfies $A(e_{(s,u)})(t) = a(u^{s+t+1}) - a(u^{s+t})$.*

Therefore, the A can be represented by the next infinite matrix $(I/O)_a$. This $(I/O)_a$ is said to be an Input/output matrix of a.

$$
\begin{array}{c}
(s, u) \\[6pt]
(I/O)_a = \\
t
\end{array}
\left(
\begin{array}{ccc}
 & \vdots & \\
 & \vdots & \\
 & \vdots & \\
 & \vdots & \\
\cdots \quad \cdots & a(u^{s+t+1}) - a(u^{s+t}) &
\end{array}
\right)
$$

Since $S_l^s(\chi(u))(t) = (\chi(u))(t+s) = a(\omega|u) - a(\omega), \omega \in U^*, |\omega| = t + s$ holds, the column vectors of Input/output matrix of $(I/O)_a$ may be expressed by $S_l^s(\chi(u)) = S_l^s I(u)$.

Theorem 6.11. *Theorem for existence criterion*
For a time-invariant input response map $a \in F(U^*, Y)$, the following conditions are equivalent:
1) The time-invariant input response map $a \in F(U^*, Y)$ has the behavior of a n-dimensional canonical pseudo linear system.
2) There exist n linearly independent vectors and no more than n linearly independent vectors in a set $\{S_l^i(\chi(u)); u \in U, i \in N, 1 \leq i \leq n\}$.
3) The rank of the Input/output matrix $(I/O)_a$ of a is n.

Theorem 6.12. *Theorem for a realization procedure*
Let a time-invariant input response map $a \in F(U^*, Y)$ satisfy the condition of Theorem (6.11). Then the real time standard system $\sigma_s = ((\mathbf{R}^n, F_s), g_s, h_s, h^0)$ which realizes a can be obtained by the following procedure:
1) Select the linearly independent vectors $\{S_l^j \chi(u_i)); 1 \leq j \leq m_i, 1 \leq i \leq m\}$ in order of the set $\{\chi(u_1), S_l \chi(u_1), \cdots, S_l^{m_1-1} chi(u_1), \chi(u_2), S_l \chi(u_2), \cdots, S_l^{m_2-1} \chi(u_2), \cdots, \chi(u_m), S_l \chi(u_m), \cdots, S_l^{m_m-1} \chi(u_m)\}$. Let $n := \mathrm{rank}\ I/O_a = m_1 + m_2 + \cdots + m_m$.

2) Let the state space be $\mathbf{R}^n$. Let the map $g_s : U \to \mathbf{R}^n$ be $g_s(u_i) := \mathbf{e}_{m_1+\cdots+m_{i-1}+1}$ for $u_i \in U$ and $1 \le i \le m$ and $F_s^j g_s(u_i) := \mathbf{e}_{m_1+\cdots+m_{i-1}+1+j}$ for $1 \le j \le m_i - 1$. And let $F_s^{m_i} g_s(u_i) := \sum_{j=1}^{m_1+\cdots+m_i} \alpha_{i,j} \mathbf{e}_j$ for $u_i \in U$ and $S_l^{m_i} \chi(u_i) := \sum_{j=1}^{m_1+\cdots+m_i} \alpha_{i,j} \chi(u_j)$.

3) Let the output map $h_s = [a(u_1) - a(1), a(u_1^2) - a(u_1), \cdots, a(u_1^{m_1}) - a(u_1^{m_1-1}), \cdots, a(u_m) - a(1), a(u_m^2) - a(u_m), \cdots, a(u_m^{m_m}) - a(u_m^{m_m-1})]$

4) Let $F_s \in \mathbf{R}^{n \times n}$ be the F_s in Definition (6.8).

[proof] Let $R(\chi) = \{S_l^i(\chi(u)); u \in U, i \in N\}$. By Theorem (6.3), $((\ll S_l^N(\chi(U)) \gg, S_l), \chi, 0, a(1))$ is a canonical pseudo linear system that realizes a time-invariant input response map $a \in F(U^*, Y)$. The linearly independent vectors $\{S_l^j(\chi(u_i)) \in \gg = \ll R(\chi) \gg; u_i \in U, 1 \le i \le m, \ 0 \le j \le m_i - 1\}$ are in order of the numerical value. Let a linear map $T :\ll R(\chi) \gg \to \mathbf{R}^n$ be $T(\chi(u_i)) = \mathbf{e}_{m_1+\cdots+m_{i-1}+1}$ for any $i(1 \le i \le n)$ and $T(S_l^j \chi(u_i)) := \mathbf{e}_{m_1+\cdots+m_{i-1}+1+j}$ for $1 \le j \le m_i - 1$. Then, by step 2), $T\chi = g_s$ holds and by step 3), $h_s \cdot T = 0$ holds. And by step 4), $F_s \cdot T = T \cdot F_s$ holds. Consequently, T is bijective and a pseudo linear system morphism : $((\ll S_l^N(\chi(U)) \gg, S_l), \chi, 0, a(1)) \to \sigma_s = ((\mathbf{R}^n, F_s), g_s, h_s, a(1))$.

By Corollary (6.6), the behavior of σ_s is a. It follows from the choice of $\{S_l^i(\chi(u_i)); u_i \in U, i \in N\})$ for $i(1 \le i \le m\}$ are in order of the numerical value and the determination of map T implies that σ_s is the real time standard system.

6.3 Partial Realization of Pseudo Linear Systems

Here we consider a partial realization problem by multi-experiment. Let $\underline{a}$ be an $\underline{N}$ sized time-invariant input response map $(\in F(U_{\underline{N}}^*, Y)$, where $\underline{N} \in N$ and $U_{\underline{N}}^* := \{\omega \in U^*; |\omega| \le \underline{N}\}$. The $\underline{a}$ is said to be a partial time-invariant input response map.

A finite dimensional pseudo linear system $\sigma = ((X, F), g, h.x^0)$ is said to be a partial realization of $\underline{a}$ if $h^0 + h(\sum_{j=1}^{|\omega|} F^{|\omega|-j} g(\omega(j))) = \underline{a}(\omega)$ holds for any $\omega \in U_{\underline{N}}^*$.

A partial realization problem of pseudo linear systems can be stated as follows:

< For any given partial time-invariant input response $\underline{a} \in F(U_{\underline{N}}^*, Y)$, find a partial realization σ of $\underline{a}$ such that the dimensions of state space X of σ is minimum, where the σ is said to be a minimal partial realization of $\underline{a}$.

In section 6.1, we stated a representation theorem for the time-invariant input response maps. The theorem says that any time-invariant input response map can be characterized by the modified impulse response. Note that the modified impulse response $I : U \to F(N, Y)$ can be represented by $(I(u)(t)) = a(u^t) - a(u^{t-1})$ for $u \in U, t \in N$ and the time-invariant input response map $a \in F(U^*, Y)$.

For any given partial time-invariant input response $\underline{a} \in F(U_{\underline{N}}^*, Y)$, this correspondence can determine a partial modified impulse response $\underline{I} : U \to F(N_{\underline{N}-1}, Y)$, where $N_{\underline{N}-1} := \{1, 2, , \underline{N} - 1; \text{ for some } \underline{N} \in N\}$.

$$(I/O)_{\underline{a}\ (p,\underline{N}-p)\atop t} = \begin{pmatrix} & & \vdots \\ & & \vdots \\ & & \vdots \\ & & \vdots \\ \cdots & \cdots & a(u^{s+t+1}) - a(u^{s+t}) \end{pmatrix},$$

$$(s,u)$$

where $0 \le s \le p, 0 \le t \le \underline{N} - p$ and $u \in U$.

When we actually treat approximate and noisy realization problems, we will use a notation $H_{\underline{a}\ (n_1+n_2,\underline{N}-n_1-n_2)}(n_1,n_2)$ expressed as follows:
$$H_{\underline{a}\ (n_1+n_2,\underline{N}-n_1-n_2)}(n_1,n_2) = [I_{\underline{a}}(0),\cdots,S_l^{n_1-1}I_{\underline{a}}(0),I_{\underline{a}}(1),\cdots,S_l^{n_2-1}I_{\underline{a}}(1)].$$

Theorem 6.13. *Let $(I/O)_{\underline{a}\ (p,\underline{N}-p)}$ be the finite-sized Input/output matrix of $\underline{a} \in F(U_{\underline{N}}^*,Y)$. Then there exists a natural partial realization of $\underline{a}$ if and only if the following conditions hold:*
rank $(I/O)_{\underline{a}\ (p,\underline{N}-p)}$ = rank $(I/O)_{\underline{a}\ (p,\underline{N}-p-1)}$ = rank $(I/O)_{\underline{a}\ (p+1,\underline{N}-p)}$ *for some $p \in N$.*

Theorem 6.14. *Let a partial time-invariant input response $\underline{a} \in F(U_{\underline{N}}^*,Y)$ satisfy the condition of Theorem (6.13), then the real time standard system $\sigma_s = ((\mathbf{R}^n, F_s), g_s, h_s, h^0)$ that realizes $\underline{a}$ can be obtained by the following algorithm. Set $n := $ rank $(I/O)_{\underline{a}\ (p,\underline{N}-p)}$, where $(I/O)_{\underline{a}\ (p,\underline{N}-p)}$ is the finite Input/output matrix of $\underline{a} \in F(U_{\underline{N}}^*,Y)$.*
1) Select the linearly independent vectors $\{S_l^j(\chi(u_i)); 1 \le i \le m,\ 0 \le j \le m_i-1\}$ from $(I/O)_{\underline{a}\ (p,\underline{N}-p)}$ in order of the numerical value.
2) Let the state space be $\mathbf{R}^n$. Let the map $g_s : U \to \mathbf{R}^n$ be $g_s(u_i) := \mathbf{e}_{m_1+\cdots+m_{i-1}+1}$ for $u_i \in U$ such that $1 \le i \le m$ and $F_s^j g_s(u_i) := \mathbf{e}_{m_1+\cdots+m_{i-1}+1+j}$ for $1 \le j \le m_i - 1$. And let $S_l^{m_i} g_s(u_i) := \sum_{j=1}^{m_1+\cdots+m_i} \alpha_{i,j}\mathbf{e}_j$ for $u_i \in U$.
3) Let the output map $h_s = [a(u_1)-a(1), a(u_1^2)-a(u_1),\ \cdots, a(u_1^{m_1})-a(u_1^{m_1-1}),\cdots, a(u_m) - a(1),\ a(u_m^2) - a(u_m),\ \cdots, a(u_m^{m_m}) - a(u_m^{m_m-1})].$
4) Let F_s be the F_s in Fig-2, where $S_l^{m_i}\chi(u_i) := \sum_{j=1}^{m_1+\cdots+m_i} \alpha_{i,j}\chi(u_j),\ \alpha_{i,j} \in \mathbf{R}$ holds in the sense of $F(N_{\underline{N}-p},Y)$ and $\underline{S_l} : F(N_p,Y) \to F(N_{p-1},Y); a \mapsto \underline{S_l}a[; t \mapsto \underline{a}(t+1)$ for some $p \in N$.

6.4 Real-Time Partial Realization of Pseudo Linear Systems

In general, it is well known that non-linear systems can only be determined by multi-experiments. However, for pseudo linear systems, special single-experiments to mimic multi-experiments were given in the reference [Matsuo and Hasegawa, 2003].

In this section, the results are introduced as previously described.

Problem 6.15. Real time partial realization problem
Let a physical object (equivalently, $a \in F(U^*, Y)$) be a finite dimensional pseudo linear system. Then for any given finite data $\{\underline{a}(\underline{\omega})$; an input $\underline{\omega}$ is finite length $\}$, find a pseudo linear system $\sigma = ((\boldsymbol{R}^n, F), g, h, h^0)$ and an input $\underline{\omega}$ such that $a_\sigma(\omega) = a(\omega)$ for any $\omega \in U^*$.

Definition 6.16. *For a finite dimensional pseudo linear system, if there exists a solution of a real time partial realization problem, then an input $\underline{\omega} \in U^*$ of the solution is said to be a (real time partial) realization signal.*

Lemma 6.17. *Let a given time invariant input response map $a \in F(U^*, Y)$ have the behavior of a pseudo linear system whose state space is less than L dimensional. Then there exists an input of finite length $\underline{\omega} \in U^*$ such that the following algorithm provides a finite Input/output matrix, where $p := max\{L_1, L_2, \cdots, L_m\}$.*
1) Find an integer L_1 such that row vectors $\{\underline{S_l}^i(\chi(u_1)) \in \boldsymbol{R}^{L-1}; 0 \leq i \leq L_1 - 1\}$ are linearly independent and $\{\underline{S_l}^i(\chi(u_1)) \in \boldsymbol{R}^{L-1}; 0 \leq i \leq L_1\}$ are linearly dependent. Namely, feed an input $\omega_1 := u_1^{L_1+L}$ into the plant.
2) Find an integer L_2 such that row vectors $\{\underline{S_l}^i(\chi(u_j)) \in \boldsymbol{R}^{L-1}; 0 \leq i \leq L_j - 1, 1 \leq j \leq 2\}$ are linearly independent and $\{\underline{S_l}^i(\chi(uj)) \in \boldsymbol{R}^{L-1},$
$\underline{S_l}^{L_2}(\chi(u_2)) \in \boldsymbol{R}^{L-1}; 0 \leq i \leq L_j - 1, 1 \leq j \leq 2\}$ are linearly dependent. Namely, feed a further input $\omega_2 := u_1^{L_1+L-1}|u_2$ into the plant.
3) Find an integer L_3 such that row vectors $\{\underline{S_l}^i(\chi(u_j)) \in \boldsymbol{R}^{L-1}; 0 \leq i \leq L_j - 1, 1 \leq j \leq 3\}$ are linearly independent and $\{\underline{S_l}^i(\chi(u_j)) \in \boldsymbol{R} \, L - 1,$
$\underline{S_l}^{L_3}(\chi(u_3)) \in \boldsymbol{R}^{L-1}; 0 \leq i \leq L_j - 1, 1 \leq j \leq 3\}$ are linearly dependent. Namely, feed a further input $\omega_3 := u_1^{L_3+L-1}|u_3$ into the plant.

$\vdots$

$\vdots$

m) Find an integer L_m such that row vectors $\{\underline{S_l}^i(\chi(u_j)) \in \boldsymbol{R}^{L-1}; 0 \leq i \leq L_j - 1, 1 \leq j \leq m\}$ are linearly independent and $\{\underline{S_l}^i(\chi(u_j)) \in \boldsymbol{R}^{L-1},$
$\underline{S_l}^{L_m}(\chi(u_m)) \in \boldsymbol{R}^{L-1}; 0 \leq i \leq L_j - 1, 1 \leq j \leq m\}$ are linearly dependent. Namely, feed a further input $\omega_m := u_1^{L_m+L-1}|u_m$ into the plant.
Let $\omega = \omega_m|\omega_{m-1}|\cdots|\omega_2|\omega_1$.
Making row vectors of a matrix from the row vectors $\{\underline{S_l}^i(\chi(uj)) \in \boldsymbol{R}^{L-1}; 0 \leq i \leq L_j - 1, 1 \leq j \leq m\}$ obtained by the above iterations, we will obtain a finite Input/output matrix $H_{\underline{a}\,(L-1,p)}$.

Theorem 6.18. *Let a given time-invariant input response map $a \in F(U^*, Y)$ have the behavior of a pseudo linear system whose state space is less than L-dimensional. Then there exists a realization signal such that the quasi-reachable standard system $\sigma_s = ((\boldsymbol{R}^n, F_s), g_s, h_s, h^0)$ that realizes a can be obtained by the following algorithm:*
1) Find a finite Input/output matrix $(I/O)_{\underline{a}\,(L-1,p)}$ based upon the algorithm given in Lemma (6.17).
2) Apply the algorithm given in Theorem (6.14) to the above finite Input/output matrix $(I/O)_{\underline{a}\,(L-1,p)}$.

Theorem 6.19. *Let the modified impulse response $I_a(u) \in F(M, Y)$ satisfy the conditions of Theorem (6.13). Then the pseudo linear system $\sigma = ((X, F_s), g_s, h_s, h^0)$ which realizes a can be obtained by the following algorithm:*

1) Select Cn_1 independent vectors on the vectors $\{S_l^s I_a(0) : 0 \leq s \leq p\}$. And select Cn_2 independent vectors in $\{S_l^s I_a(1) : 0 \leq s \leq p\}$.

2) Let the state space be $\mathbf{R}^n$. And let g_s^0 and g_s be as follows: $g_s^0 = \mathbf{e}_1$, $g_s = \mathbf{e}_{n_1+1} - \mathbf{e}_1$, where Ag_s is given by $g_s = 0$ if $S_l I_a(0) = 0$ holds. Moreover,

$n = n_1 + n_2$ and $\mathbf{e}_i = [0, \cdots, 0, \overset{i}{1}, 0, \cdots, 0]^T$ hold.

3) $F_s \in \mathbf{R}^{n \times n}$ is the same as F_s in Definition (6.8).
$S_l^{n_1} I_a(0) = \sum_{i=1}^{n_1} \alpha_{1i} S_l^{i-1} I_a(0)$.
$S_l^{n_2} I_a(1)$
$= \sum_{i=1}^{n_1} \alpha_{2i} S_l^{i-1} I_a(0) + \sum_{i=1}^{n_2} \alpha_{2n_1+i} S_l^{i-1} I_a(1)$.

4) Let h_s be $h_s = [a(0) - a(1), a(0^2) - a(0), \cdots, a(0^{n_1}) - a(0^{n_1-1}), a(1) - a(1), a(0|1) - a(0), \cdots, a(0^{n_1-1}|1) - a(0^{n_1-1})]$.

5) Let h^0 be $h^0 = a(1)D$

6.5 Approximate Realization of Pseudo Linear Systems

In this section, we discuss the approximate realization problems of pseudo linear systems.

We will discuss the approximate realization problem under the assumption that the set U of input's values is a finite set $U = \{u_j : 1 \leq j \leq m\}$ for an finite integer $m \in N$. In the reference [Matsuo and Hasegawa, 2003], we showed that this assumption is not so special. However, for simplicity of our discussion, we assume that the set U of input's values is $U = \{u_1, u_2\}$ or $U = \{u_1, u_2, u_3\}$.

Roughly speaking, the approximate realization of pseudo linear systems can be stated as follows:

< For any given partial data of a pseudo linear system, find a pseudo linear system which approximates the given data. >

In order to make our discussion simple, we assume that the set Y of outout's value is the set $\mathbf{R}$ of real numbers, namely 1-output.

Theorem 6.20. *Algorithm for approximate realization*
Let an input response map $\underline{a}$ be a considered object which is a pseudo linear system. Then an approximate realization $\sigma = ((\mathbf{R}^n, F_s), g_s, h_s, h^0)$ of $\underline{a}$ is given by the following algorithm:

1) Based on the ratio of the square root of eigenvalues for a matrix
$H_{\underline{a}\ (p,\bar{p})}(p, 0, 0) H_{\underline{a}\ (p,\bar{p})}(p, 0, 0)^T$, determine the value n_1 of rank for the matrix $H_{\underline{a}\ (p,\bar{p})}(p, 0, 0)$, where $n_1 \leq p$.
Namely, determine the value n_1 of rank for the matrix $H_{\underline{a}\ (p,\bar{p})}(p, 0, 0)$ such that the ratio of the square root of eigenvalues for the covariance matrix becomes very small. The small ratio indicates the nearness of approximation degree.

2) The CLS method is used as follows:
① Let a matrix $A_1 \in \mathbf{R}^{1 \times (n_1+1)}$ be $A_1 = [\alpha_{11}, \alpha_{12}, \cdots, \alpha_{1n_1}, -1]$.

② *Choose the coefficients $\{\alpha_{1i} : 1 \leq i \leq n_1\}$ such that*
$\sum_{j=1}^{n_1+1} \underline{S}_l^{j-1} \bar{I}_{\underline{a}} \cdot \underline{S}_l^{j-1} \bar{I}_{\underline{a}}$ *takes a minimum value, where* $\{\underline{S}_l^i \bar{I}_{\underline{a}} \in \boldsymbol{R}^{L \times 1} :$
$0 \leq i \leq n_1\}$ *are given by the equation* $[\bar{I}_{\underline{a}}(u_1), \underline{S}_l \bar{I}_{\underline{a}}(u_1), \cdots, \underline{S}_l^{n_1} \bar{I}_{\underline{a}}(u_2)]^T :=$
$A_1^T [A_1 A_1^T]^{-1} A_1 H_{\underline{a}\ (n_1+1,L)}^T (n_1 + 1, 0)$ *and* $H_{\underline{a}\ (n_1,L)}^T (n_1, 0, 0) :=$
$[I_{\underline{a}}(u_1), \cdots, S_l^{n_1-1} I_{\underline{a}}(u_1), S_l^{n_1} I_{\underline{a}}(u_1)]$. *And* $\cdot$ *denotes the inner product of
two vectors.*
③ *Let* $h_{1s} \in \boldsymbol{R}^{1 \times n_1}$ *be* h_{1s}
$= [(I_{\underline{a}}(u_1))(0) - (\bar{I}_{\underline{a}}(u_1))(0), (S_l I_{\underline{a}}(u_1))(0) - (S_l \bar{I}_{\underline{a}}(u_1))(0), \cdots ,$
$(S_l^{n_1-1} I_{\underline{a}}(u_1))(0) - (S_l^{n_1-1} \bar{I}_{\underline{a}}(u_1))(0)]$.

3) *Based on the ratio of the square root of eigenvalues for a matrix*
$H_{\underline{a}\ (n_1+p,\bar{p})}(n_1, p, 0) H_{\underline{a}\ (n_1+p,\bar{p})}(n_1, p, 0)^T$, *determine the value n_2 of rank for
the matrix* $H_{\underline{a}\ (n_1+p,\bar{p})}(n_1, p, 0)$, *where* $n_2 \leq p$.
Namely, determine the value n_2 of rank for the matrix $H_{\underline{a}\ (p,\bar{p})}(n_1, p, 0)$ *such
that the ratio of the square root of eigenvalues for the covariance matrix
becomes very small. The small ratio indicates the nearness of
approximation degree.*

4) *The CLS method is used as follows:*
① *Let a matrix* $A_2 \in \boldsymbol{R}^{1 \times (n_1+n_2+1)}$ *be* $A_2 = [\alpha_{21}, \alpha_{22}, \cdots, \alpha_{2n_1+n_2}, -1]$.
② *Choose the coefficients* $\{\alpha_{2i} : 1 \leq i \leq n_1 + n_2\}$ *such that*
$\sum_{j=1}^{n_1+n_2+1} \underline{S}_l^{j-1} \bar{I}_{\underline{a}} \cdot \underline{S}_l^{j-1} \bar{I}_{\underline{a}}$ *takes a minimum value, where* $\{\underline{S}_l^i \bar{I}_{\underline{a}} \in \boldsymbol{R}^{L \times 1} :$
$0 \leq i \leq n_1 + n_2\}$ *are given by the equation* $[\bar{I}_{\underline{a}}, \underline{S}_l \bar{I}_{\underline{a}}, \cdots, \underline{S}_l^{n_1} \bar{I}_{\underline{a}}]^T :=$
$A_2^T [A_2 A_2^T]^{-1} A_2 H_{\underline{a}\ (n_1+n_2+1,L)}^T (n_1, n_2 + 1, 0)$ *and* $H_{\underline{a}\ (n_1+n_2;1,L)}^T (n_1, n_2 + 1, 0) :=$
$[I_{\underline{a}}(u_1), \cdots, S_l^{n_1-1} I_{\underline{a}}(u_1), I_{\underline{a}}(u_2), \cdots, S_l^{n_2-1} I_{\underline{a}}(u_2), S_l^{n_2} I_{\underline{a}}(u_2)]$.
And $\cdot$ *denotes the inner product of two vectors.*
③ *Let* $h_{2s} \in \boldsymbol{R}^{1 \times n_2}$ *be* h_{2s}
$= [(I_{\underline{a}}(u_2))(0) - (\bar{I}_{\underline{a}}(u_2))(0), (S_l I_{\underline{a}}(u_2))(0) - (S_l \bar{I}_{\underline{a}}(u_2))(0),$
$\cdots, (S_l^{n_2-1} I_{\underline{a}}(u_2))(0) - (S_l^{n_2-1} \bar{I}_{\underline{a}}(u_2))(0)]$.

5) *Based on the ratio of the square root of eigenvalues for a matrix*
$H_{\underline{a}\ (n_1+p,\bar{p})}(n_1, n_2, q) H_{\underline{a}\ (n_1+n_2+q,\bar{q})}(n_1, n_2, q)^T$, *determine the value n_3 of rank
for the matrix* $H_{\underline{a}\ (n_1+n_2+q,\bar{q})}(n_1, n_2, q)$, *where* $n_3 \leq q$.
Namely, determine the value n_3 of rank for the matrix $H_{\underline{a}\ (n_1+n_2+q,\bar{q})}(n_1, n_2, q)$
*such that the ratio of the square root of eigenvalues for the covariance
matrix becomes very small. The small ratio indicates the nearness of
approximation degree.*

6) *The CLS method is used as follows:*
① *Let a matrix* $A_3 \in \boldsymbol{R}^{1 \times (n_1+n_2+n_3+1)}$ *be* $A_3 = [\alpha_{31}, \alpha_{32}, \cdots, \alpha_{3n_1+n_2+n_3}, -1]$.
② *Choose the coefficients* $\{\alpha_{3i} : 1 \leq i \leq n_1 + n_2 + n_3\}$ *such that*
$\sum_{j=1}^{n_1+n_2+n_3} \underline{S}_l^{j-1} \bar{I}_{\underline{a}} \cdot \underline{S}_l^{j-1} \bar{I}_{\underline{a}}$ *takes a minimum value, where* $\{\underline{S}_l^i \bar{I}_{\underline{a}} \in \boldsymbol{R}^{L \times 1} :$
$0 \leq i \leq n_1 + n_2 + n_3\}$ *are given by the equation* $[\bar{I}_{\underline{a}}, \underline{S}_l \bar{I}_{\underline{a}}, \cdots, \underline{S}_l^{n_1} \bar{I}_{\underline{a}}]^T :=$
$A_3^T [A_3 A_3^T]^{-1} A_3 H_{\underline{a}\ (n_1+n_2+n_3+1,L)}^T (n_1, n_2, n_3 + 1)$ *and*
$H_{\underline{a}\ (n_1+n_2+n_3+1,L)}^T (n_1, n_2, n_3 + 1) := [I_{\underline{a}}(u_1), \cdots, S_l^{n_1-1} I_{\underline{a}}(u_1), I_{\underline{a}}(u_2), \cdots,$
$S_l^{n_2-1} I_{\underline{a}}(u_2), S_l^{n_2} I_{\underline{a}}(u_2), I_{\underline{a}}(u_3), \cdots, S_l^{n_3-1} I_{\underline{a}}(u_3), S_l^{n_3} I_{\underline{a}}(u_3),]$.
And $\cdot$ *denotes the inner product of two vectors.*

③ *Let $h_{3s} \in \mathbf{R}^{1 \times n_3}$ be $h_{3s} =$*
$$[(I_{\underline{a}}(u_3))(0) - (\bar{I}_{\underline{a}}(u_3))(0), (S_l I_{\underline{a}}(u_3))(0) - (S_l \bar{I}_{\underline{a}}(u_3))(0),$$
$$\cdots, (S_l^{n_3-1} I_{\underline{a}}(u_3))(0) - (S_l^{n_3-1} \bar{I}_{\underline{a}}(u_3))(0)].$$

$\vdots$

*2*m-1) Based on the ratio of the square root of eigenvalues for a matrix*
$H_{\underline{a}\ (n_1+\cdots+n_m+p,\bar{p})}(n_1, \cdots, n_m, q) H_{\underline{a}\ (n_1+\cdots+n_m+q,\bar{q})}(n_1, \cdots, n_m, q)^T$, *determine*
the value n_m of rank for the matrix $H_{\underline{a}\ (n_1+\cdots+n_m+q,\bar{q})}(n_1, \cdots, n_m, q)$,
where $n_m \leq q$.
Namely, determine the value n_3 of rank for the matrix $H_{\underline{a}\ (n_1+n_2+q,\bar{q})}(n_1, \cdots,$
$n_m, q)$
such that the ratio of the square root of eigenvalues for the covariance
matrix becomes very small. The small ratio indicates the nearness of
approximation degree.

*2*m) The CLS method is used as follows:*
① *Let a matrix $A_m \in \mathbf{R}^{1 \times (n_1+\cdots+n_m+1)}$ be $A_m = [\alpha_{m1}, \alpha_{m2}, \cdots, \alpha_{mn_1+\cdots+n_m},$*
$-1]$.
② *Choose the coefficients $\{\alpha_{mi} : 1 \leq i \leq n_1 + \cdots + n_m\}$ such that*
$\sum_{j=1}^{n_1+\cdots+n_m} \underline{S}_l^{j-1} \bar{I}_{\underline{a}} \cdot \underline{S}_l^{j-1} \bar{I}_{\underline{a}}$ *takes a minimum value, where $\{\underline{S}_l^i \bar{I}_{\underline{a}} \in \mathbf{R}^{L \times 1} :$*
$0 \leq i \leq n_1 + \cdots + n_m\}$ *are given by the equation $[\bar{I}_{\underline{a}}, \underline{S}_l \bar{I}_{\underline{a}}, \cdots, \underline{S}_l^{n_1} \bar{I}_{\underline{a}}]^T :=$*
$A_m^T [A_m A_m^T]^{-1} A_m H_{\underline{a}\ (n_1+\cdots+n_m+1,L)}^T(n_1, \cdots, n_m + 1)$ *and*
$H_{\underline{a}\ (n_1+\cdots+n_m+1,L)}^T(n_1, \cdots, n_m + 1) := [I_{\underline{a}}(u_1), \cdots, S_l^{n_1-1} I_{\underline{a}}(u_1), I_{\underline{a}}(u_2), \cdots,$
$S_l^{n_2-1} I_{\underline{a}}(u_2), S_l^{n_2} I_{\underline{a}}(u_2), I_{\underline{a}}(u_m), \cdots, S_l^{n_m-1} I_{\underline{a}}(u_m), S_l^{n_m} I_{\underline{a}}(u_m),].$
And $\cdot$ denotes the inner product of two vectors.
③ *Let $h_{ms} \in \mathbf{R}^{1 \times n_m}$ be $h_{ms} =$*
$$[(I_{\underline{a}}(u_m))(0) - (\bar{I}_{\underline{a}}(u_m))(0), (S_l I_{\underline{a}}(u_m))(0) - (S_l \bar{I}_{\underline{a}}(u_m))(0),$$
$$\cdots, (S_l^{n_m-1} I_{\underline{a}}(u_m))(0) - (S_l^{n_m-1} \bar{I}_{\underline{a}}(u_m))(0)].$$

*2*m+1) Let $g_s \in F(U, \mathbf{R}$ be $g_s(u_1) := \mathbf{e}_1$, $g_s(u_2) := \mathbf{e}_{n_1+1}, \cdots,$*
$g_s(u_m) := \mathbf{e}_{n_1+\cdots+n_{m-1}+1}.$
Let $F_s \in \mathbf{R}^{n \times n}$ be the same as in Theorem (6.19).
Let $h_s \in \mathbf{R}^{1 \times n}$ be $h_s := [h_{1s}, h_{2s}, \cdots, h_{ms}]$,
where $n := n_1 + n_2 + \cdots + n_m$.

[proof] By 1) and 3), the reduction part in the data can be excluded in the sense of the number of dimensions by using the ratio of matrix norm, which produces a degree of information loss. The matrices A_1 in 2), A_2 in 4), A_3 in 6), $\cdots$ and A_m in 2*m) correspond to the matrix A in Proposition (2.14). Hence, the reduced part of the given finite-sized Input/output matrix were obtained. Therefore, applying Theorem (6.19), we can obtain g_s, F_s and h_s by 2*m+1).

For the real time standard system $\sigma_s = ((\mathbf{R}^n, F_s), g_s^0, g_s, h_s, h_s^0)$, its modified impulse responses $I(u_1)(i) := h_s F_s^i g_s(u_1)$, $I(u_2)(i) := h_s F_s^i g_s(u_2)$ and $I(u_3)(i) := h_s F_s^i g_s(u_3)$ are written by $I(1)_{-}(n_1, n_2, n_3)$ and $I(2)_{-}(n_1, n_2, n_3)$ and $I(3)_{-}(n_1, n_2, n_3)$ respectively.

Example 6.21. Let the signals be the modified impulse responses of the following 3-dimensional pseudo linear system: $\sigma = ((\mathbf{R}^3, F), g, h, h^0)$, where

$$F = \begin{bmatrix} 0 & 0.3 & 0 \\ 1 & 0.6 & 0 \\ 0 & 0 & 0.7 \end{bmatrix}, \ g(u_1) = \mathbf{e}_1, \ g(u_2) = \mathbf{e}_3, \ h = [12, \ -1, \ -15], \ h^0 = 1.$$

Then the approximate realization problem is solved as follows:

covariance matrix	eigenvalues			
	1	2	3	4
$H^T_{\underline{a}\ (3,50)}(3,0)H_{\underline{a}\ (3,50)}(3,0)$	199	41	0	
$H^T_{\underline{a}\ (3,50)}(1,2)H_{\underline{a}\ (3,50)}(1,2)$	744	55	0	
$H^T_{\underline{a}\ (3,50)}(2,1)H_{\underline{a}\ (3,50)}(2,1)$	567	62	19	
$H^T_{\underline{a}\ (4,50)}(2,2)H_{\underline{a}\ (4,50)}(2,2)$	779	65	20	0
covariance matrix	square root of eigenvalues			
$H^T_{\underline{a}\ (3,50)}(3,0)H_{\underline{a}\ (3,50)}(3,0)$	14.1	6.4	0	
$H^T_{\underline{a}\ (3,50)}(1,2)H_{\underline{a}\ (3,50)}(1,2)$	27.3	7.4	0	
$H^T_{\underline{a}\ (3,50)}(2,1)H_{\underline{a}\ (3,50)}(2,1)$	23.8	7.9	4.4	
$H^T_{\underline{a}\ (4,50)}(2,2)H_{\underline{a}\ (4,50)}(2,2)$	28	8	4.5	0

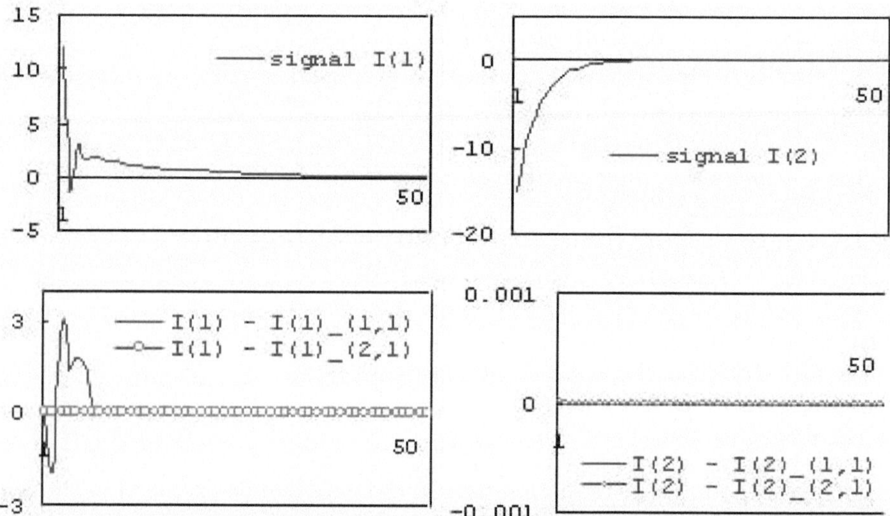

Fig. 6.1. The left are the original modified impulse response $I(1)$ and the difference between it and a modified impulse response $I(1)\_(1,1)$ or $I(1)\_(2,1)$ approximated by a 2 or 3-dimensional pseudo linear system. The right are the original modified impulse response $I(2)$ and the difference between it and a modified impulse response $I(2)\_(1,1)$ or $I(2)\_(2,1)$ approximated by a 2 or 3-dimensional pseudo linear system in Example (6.21).

1) Since the ratio $\frac{4.5}{28} = 0.16$ obtained by the square root of $H^T_{\underline{a}\ (4,50)}(2,2)H_{\underline{a}\ (4,50)}(2,2)$ is large, the approximate 2-dimensional pseudo linear system obtained by the CLS method may not be good.

2) After determining the numbers n_1 and n_2 of dimensions which are 2 and 0, we will continue the approximate realization algorithm by the CLS method.

Therefore, the modified impulse responses $I(1)$ and $I(2)$ of the approximate pseudo linear system obtained by the CLS method is constructed by 2-dimensional space.

The 3-dimensional pseudo linear system $\sigma_2 = ((\boldsymbol{R}^3, \ F_2), \ g_2, \ h_2)$ obtained by the CLS method can be expressed as follows:

$$F_2 = \begin{bmatrix} 0 & 0.3 & 0 \\ 1 & 0.6 & 0 \\ 0 & 0 & 0.7 \end{bmatrix}, \ h_2 = [12, \ -1, \ -15], \ g_2(u_1) = [1, \ 0, \ 0]^T,$$

$g_2(u_2) = [0, \ 0, \ 1]^T, \ h^0 = 1.$

For reference, we list the 2-dimensional pseudo linear system $\sigma_1 = ((\boldsymbol{R}^2, \ F_1), \ g_1, \ h_1, \ h^0)$ which is an approximate pseudo linear system.

$$F_1 = \begin{bmatrix} 0.07 & 0 \\ 0 & 0.7 \end{bmatrix}, \ h_1 = [12, \ -15], \ g_1(u_1) = [1, \ 0]^T, g_1(u_2) = [0, \ 1]^T, \ h^0 = 1.$$

In this example, the original signals are considered as the modified impulse responses of a 3-dimensional pseudo linear system and the desirable modified impulse responses are obtained by the CLS method within our bad expectations. The model obtained by the CLS method is a 2-dimensional pseudo linear system.

For reference, a 3-dimensional pseudo linear system is also given by the CLS method. The system completely reconstructs the original system.

Just as we thought, the following table and Fig. 6.1 truly indicate that the 2-dimensional pseudo linear system obtained by the CLS method is a bad approximation. For reference, the modified impulse responses of the same dimensional pseudo linear system as the original system are shown. Hence, there does not exist a good approximation for the given system.

dimen-ion	ratio of matrices	mean values of square root for sum of			cosine ① and ②	error ratio
		signal ①	signal by CLS ②	error ③	$\cos\theta$	③/①
I(1)_(1,1)	0.18	0.26	0.24	0.11	0.91	0.42
I(2)_(1,1)	0	0.42	0.42	0	1	0
I(1)_(2,1)	0	0.26	0.26	0	1	0
I(2)_(2,1)	0	0.42	0.42	0	1	0

Example 6.22. Let the signals be the modified impulse responses of the following 3-dimensional pseudo linear system: $\sigma = ((\boldsymbol{R}^3, F), g, h, h^0)$, where $F =$

$$\begin{bmatrix} 0.9 & 0.3 & 0.1 \\ 0 & 0.2 & 0 \\ 0 & 0 & -0.2 \end{bmatrix}, \ g(u_1) = \mathbf{e}_1, \ g(u_2) = \mathbf{e}_2, \ g(u_3) = \mathbf{e}_3,$$

$h = [12, \ -1, \ -15], \ h^0 = 1.$

Then the approximate realization problem is solved as follows:

covariance matrix	eigenvalues			
	1	2	3	4
$H^T_{\underline{a}\ (3,50)}(3,0,0)\,H_{\underline{a}\ (3,50)}(3,0,0)$	1869	0	0	
$H^T_{\underline{a}\ (3,50)}(1,2,0)\,H_{\underline{a}\ (3,50)}(1,2,0)$	934	26	0	
$H^T_{\underline{a}\ (4,50)}(1,1,2)\,H_{\underline{a}\ (4,50)}(1,1,2)$	655	268	3.4	0
covariance matrix	square root of eigenvalues			
$H^T_{\underline{a}\ (3,50)}(3,0,0)\,H_{\underline{a}\ (3,50)}(3,0,0)$	43.2	0	0	
$H^T_{\underline{a}\ (3,50)}(1,2,0)\,H_{\underline{a}\ (3,50)}(1,2,0)$	31	5.1	0	
$H^T_{\underline{a}\ (4,50)}(1,1,2)\,H_{\underline{a}\ (4,50)}(1,1,2)$	25.6	16.4	1.8	0

1) Since the ratio $\frac{1.8}{25.6} = 0.07$ obtained by the square root of
$H^T_{\underline{a}\ (4,50)}(1,1,2)\,H_{\underline{a}\ (4,50)}(1,1,2)$ is not so large, the approximate pseudo linear system obtained by the CLS method may not be good.
2) After determining the numbers n_1, n_2 and n_3 of dimensions which are 1, 1 and 0, we will continue the approximate realization algorithm by the CLS method.

Therefore, the modified impulse responses $I(1)$, $I(2)$ and $I(3)$ of an approximate pseudo linear system obtained by the CLS method is constructed for a 2-dimensional space.

The 2-dimensional pseudo linear system $\sigma_2 = ((\boldsymbol{R}^2,\ F_2),\ g_2,\ h_2)$ obtained by the CLS method can be expressed as follows:
$$F_2 = \begin{bmatrix} 0.9 & 0.3 \\ 0 & 0.2 \end{bmatrix},\ h_2 = [12,\ -1],\ g_2(u_1) = [1,\ 0]^T, g_2(u_2) = [0,\ 1]^T,$$
$g_2(u_3) = [-1.07,\ 3.05]^T,\ h^0 = 1.$

The 3-dimensional pseudo linear system $\sigma_3 = ((\boldsymbol{R}^3,\ F_3),\ g_3,\ h_3)$ obtained by the CLS method can be expressed as follows:
$$F_3 = \begin{bmatrix} 0.9 & 0.3 & 0.1 \\ 0 & 0.2 & 0 \\ 0 & 0 & -0.2 \end{bmatrix},\ h_3 = [12,\ -1,\ -15],\ g_3(u_1) = [1,\ 0,\ 0]^T,$$
$g_3(u_2) = [0,\ 1,\ 0]^T,\ g_3(u_3) = [0,\ 0,\ 1]^T,\ h^0 = 1.$

In this example, the original signals are considered as the modified impulse responses of a 3-dimensional pseudo linear system and the desirable modified impulse responses are obtained by the CLS method within our expectations. The model obtained by the CLS method is a 2-dimensional pseudo linear system.

For reference, a 3-dimensional pseudo linear system is also given by the CLS method. The system completely reconstructs the original system.

Just as we thought, the following table and Fig. 6.2 truly indicate that the 2-dimensional pseudo linear system obtained by the CLS method is not a good approximation. For reference, the modified impulse responses of the same dimensional pseudo linear system as the original system are also shown. Contrary to what we expected, there exists a good approximation for the given system.

dimen-ion	ratio of matrices	mean values of square root for sum of		error	cosine ① and ②	error ratio
		signal ①	signal by CLS ②	error ③	$\cos\theta$	③/①
I(1)_(1,1,0)	0	0.55	0.55	0	1	0
I(2)_(1,1,0)	0	0.20	0.20	0	1	0
I(3)_(1,1,0)	0.07	0.31	0.33	0.12	0.93	0.38
I(1)_(1,1,1)	0	0.55	0.55	0	1	0
I(2)_(1,1,1)	0	0.20	0.20	0	1	0
I(3)_(1,1,1)	0	0.31	0.31	0	1	0

Example 6.23. Let the signals be the modified impulse responses of the following 3-dimensional pseudo linear system: $\sigma = ((\boldsymbol{R}^3, F), g, h, h^0)$, where $F =$
$$\begin{bmatrix} 0.9 & 0.7 & 0.1 \\ 0 & 0.1 & 0 \\ 0 & 0 & -0.2 \end{bmatrix}, \ g(u_1) = \mathbf{e}_1, \ g(u_2) = \mathbf{e}_2, \ g(u_3) = \mathbf{e}_3,$$
$h = [12, \ 0.1, \ -15]$, $h^0 = 1$.

Then the approximate realization problem is solved by the following algorithm:

covariance matrix	eigenvalues			
	1	2	3	4
$H^T_{\underline{a}\ (3,50)}(3,0,0)H_{\underline{a}\ (3,50)}(3,0,0)$	1869	0	0	
$H^T_{\underline{a}\ (3,50)}(1,2,0)H_{\underline{a}\ (3,50)}(1,2,0)$	1597	60	0	
$H^T_{\underline{a}\ (3,50)}(1,0,2)H_{\underline{a}\ (3,50)}(1,0,2)$	870	611	0	
$H^T_{\underline{a}\ (4,50)}(1,1,2)H_{\underline{a}\ (4,50)}(1,1,2)$	1165	738	29	0
covariance matrix	square root of eigenvalues			
$H^T_{\underline{a}\ (3,50)}(3,0,0)H_{\underline{a}\ (3,50)}(3,0,0)$	43.2	0	0	
$H^T_{\underline{a}\ (3,50)}(1,2,0)H_{\underline{a}\ (3,50)}(1,2,0)$	40	7.7	0	
$H^T_{\underline{a}\ (3,50)}(1,0,2)H_{\underline{a}\ (3,50)}(1,0,2)$	29.5	25	0	
$H^T_{\underline{a}\ (4,50)}(1,1,2)H_{\underline{a}\ (4,50)}(1,1,2)$	34	27.2	5.4	0

1) Since the ratio $\frac{5.4}{34} = 0.16$ obtained by the square root of $H^T_{\underline{a}\ (4,50)}(1,1,2)H_{\underline{a}\ (4,50)}(1,1,2)$ is not so small, the approximate pseudo linear system obtained by the CLS method may not be good.

2) After determining the numbers n_1, n_2 and n_3 of dimensions which are 1, 0 and 1, we will continue the approximate realization algorithm by the CLS method.

Therefore, the modified impulse responses $I(1)$, $I(2)$ and $I(3)$ of approximate pseudo linear system obtained by the CLS method is constructed for a 2-dimensional space.

The 2-dimensional pseudo linear system $\sigma_2 = ((\boldsymbol{R}^2, F_2), g_2, h_2, h^0)$ obtained by the CLS method can be expressed as follows:
$$F_2 = \begin{bmatrix} 0.9 & 0.1 \\ 0 & -0.7 \end{bmatrix}, \ h_2 = [12, -15], \ g_2(u_1) = [1, \ 0]^T, g_2(u_2) = [0.75, \ 0]^T,$$
$g_2(u_3) = [0, \ 1]^T$, $h^0 = 1$.

For reference, the 3-dimensional pseudo linear system $\sigma_3 = ((\boldsymbol{R}^3, F_3), g_3, h_3, h^0)$ obtained by the CLS method is expressed as follows:

$$F_3 = \begin{bmatrix} 0.9 & 0.7 & 0.1 \\ 0 & 0.1 & 0 \\ 0 & 0 & -0.7 \end{bmatrix}, \quad h_3 = [12,\ 0.1,\ -15],\quad g_3(u_1) = [1,\ 0,\ 0]^T,$$

$g_3(u_2) = [0,\ 1,\ 0]^T$, $g_3(u_3) = [0,\ 0,\ 1]^T$, $h^0 = 1$.

In this example, the original signals are considered as the modified impulse responses of a 3-dimensional pseudo linear system and the desirable modified impulse responses are obtained by the CLS method within our bad expectations. The model obtained by the CLS method is a 2-dimensional pseudo linear system.

For reference, a 3-dimensional pseudo linear system is also given by the CLS method. The system completely reconstructs the original system.

Just as we thought, the following table and Fig. 6.3 truly indicate that the 2-dimensional pseudo linear system obtained by the CLS method is not a good approximation. For reference, the modified impulse responses of the same dimensional pseudo linear system as the original system are also shown. Hence, there does not exist a good approximation for the given system.

dimen-ion	ratio of matrices	mean values of square root for sum of signal ①	mean values of square root for sum of signal by CLS ②	mean values of square root for sum of error ③	cosine ① and ② $\cos\theta$	error ratio ③/①
I(1)_(1,0,1)	0	0.55	0.55	0	1	0
I(2)_(1,0,1)	0.07	0.42	0.41	0.18	0.91	0.42
I(3)_(1,0,1)	0	0.43	0.43	0	1	0
I(1)_(1,1,1)	0	0.55	0.55	0	1	0
I(2)_(1,1,1)	0	0.42	0.42	0	1	0
I(3)_(1,1,1)	0	0.43	0.43	0	1	0

Example 6.24. Let the signals be the modified impulse responses of the following 4-dimensional pseudo linear system: $\sigma = ((\boldsymbol{R}^4, F), g, h, h^0)$, where

$$F = \begin{bmatrix} 0.8 & 0.2 & 0 & 0 \\ 0 & 0.6 & 0 & 0.3 \\ 0 & 0 & -0.7 & 0.2 \\ 0 & 0 & 0 & 0.8 \end{bmatrix}, \quad g(u_1) = \mathbf{e}_1,\ g(u_2) = \mathbf{e}_3,\ g(u_3) = \mathbf{e}_4,$$

$h = [12,\ -1,\ -15,\ 4]$, $h^0 = 1$.

Then the approximate realization problem is solved as follows:

covariance matrix	eigenvalues				
	1	2	3	4	5
$H^T_{\underline{a}\ (2,50)}(2,0,0)\,H_{\underline{a}\ (2,50)}(2,0,0)$	656	0			
$H^T_{\underline{a}\ (3,50)}(1,2,0)\,H_{\underline{a}\ (3,50)}(1,2,0)$	719	338	0		
$H^T_{\underline{a}\ (4,50)}(1,1,2)\,H_{\underline{a}\ (4,50)}(1,1,2)$	616	372	106	0.7	
$H^T_{\underline{a}\ (5,50)}(1,1,3)\,H_{\underline{a}\ (5,50)}(1,1,3)$	694	397	123	1.8	0
covariance matrix	square root of eigenvalues				
$H^T_{\underline{a}\ (4,50)}(1,1,2)\,H_{\underline{a}\ (4,50)}(1,1,2)$	24.8	19.3	10.3	0.8	
$H^T_{\underline{a}\ (5,50)}(1,1,3)\,H_{\underline{a}\ (5,50)}(1,1,3)$	26.3	20	11.1	1.3	0

1) Since the ratio $\frac{0.8}{24.8} = 0.05$ obtained by the square root of $H^T_{\underline{a}\,(4,50)}(1,1,2)H_{\underline{a}\,(4,50)}(1,1,2)$ is not so large, the approximate pseudo linear system obtained by the CLS method may not be so good. The reason is likely to be caused by rapid damping in Fig. 6.4.

2) After determining the numbers n_1, n_2 and n_3 of dimensions which are 1, 1 and 1, we will continue the approximate realization algorithm by the CLS method.

Therefore, the modified impulse responses $I(1)$, $I(2)$ and $I(3)$ of approximate pseudo linear system obtained by the CLS method is constructed for 3-dimensional space.

The 3-dimensional pseudo linear system $\sigma_2 = ((\boldsymbol{R}^3,\ F_2),\ g_2,\ h_2,\ h^0)$ obtained by the CLS method can be expressed as follows:

$$F_2 = \begin{bmatrix} 0.8 & 0 & 0.03 \\ 0 & -0.7 & 0.25 \\ 0 & 0 & 0.95 \end{bmatrix},\ h_2 = [12,\ -15,\ 3.7],\ g_2(u_1) = [1,\ 0,\ 0]^T,$$

$g_2(u_2) = [0,\ 1,\ 0]^T,\ g_2(u_3) = [0,\ 0,\ 1]^T,\ h^0 = 1.$

For reference, a 4-dimensional pseudo linear system $\sigma_3 = ((\boldsymbol{R}^4,\ F_3),\ g_3, h_3, h^0)$ obtained by the CLS method can be expressed as follows:

$$F_3 = \begin{bmatrix} 0.8 & 0 & 0 & 0.06 \\ 0 & -0.7 & 0 & -0.26 \\ 0 & 0 & 0 & -0.48 \\ 0 & 0 & 1 & 1.4 \end{bmatrix},\ h_3 = [12,\ -15,\ 4,\ -0.1],\ g_3(u_1) = [1,\ 0,\ 0,\ 0]^T,$$

$g_3(u_2) = [0,\ 1,\ 0,\ 0]^T,\ g_3(u_3) = [0,\ 0,\ 1,\ 0]^T,\ h^0 = 1.$

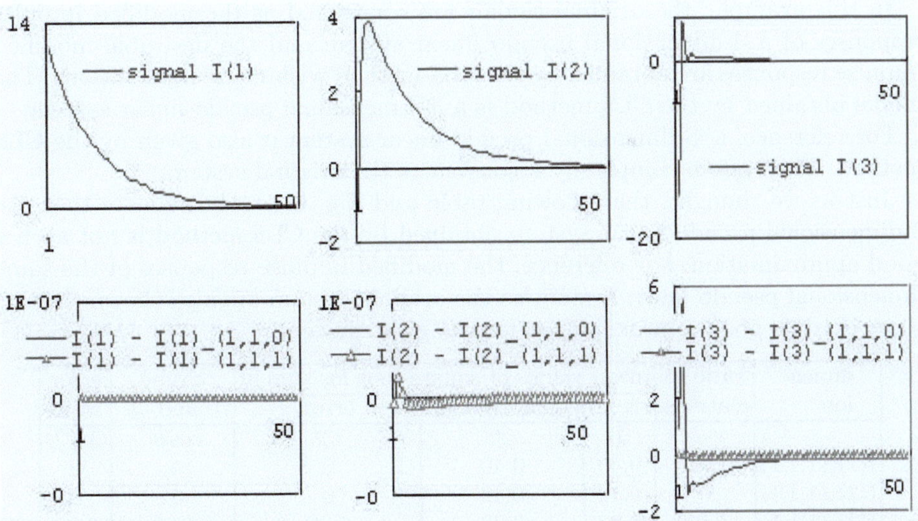

Fig. 6.2. In Example (6.22), the left are the original signal $I(1)$ and the difference between it and the approximate signal $I(1)_-(1,1,0)$ or $I(1)_-(1,1,1)$. The middle are the original signal $I(2)$ and the difference between it and the approximate signal $I(2)_-(1,1,0)$ or $I(2)_-(1,1,1)$. The right are the original signal $I(3)$ and the difference between it and the approximate signal $I(3)_-(1,1,0)$ or $I(3)_-(1,1,1)$.

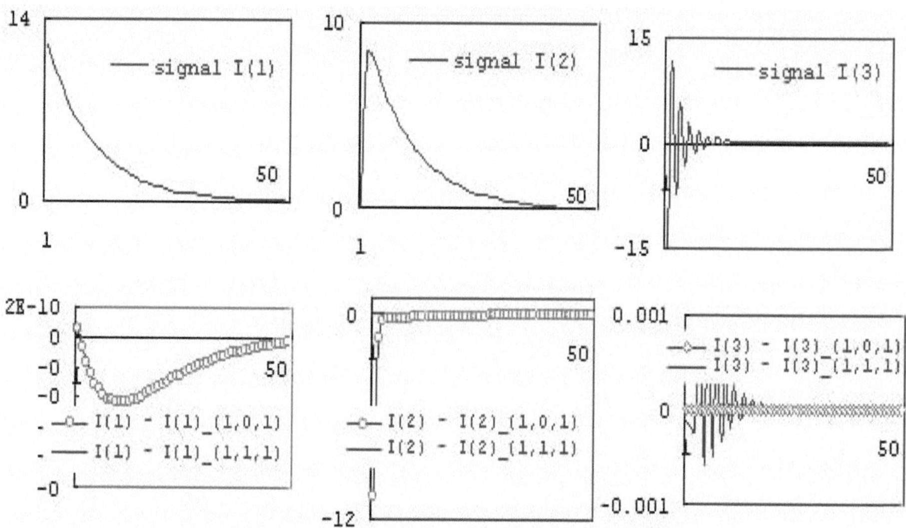

Fig. 6.3. In Example (6.23), the left are the original signal $I(1)$ and the difference between it and the approximate signal $I(1)\_(1,0,1)$ or $I(1)\_(1,1,1)$. The middle are the original signal $I(2)$ and the difference between it and the approximate signal $I(2)\_(1,0,1)$ or $I(2)\_(1,1,1)$. The right are the original signal $I(3)$ and the difference between it and the approximate signal $I(3)\_(1,0,1)$ or $I(3)\_(1,1,1)$.

In this example, the original signals are considered as the modified impulse responses of a 4-dimensional pseudo linear system and the desirable modified impulse responses are obtained by the CLS method within our expectations. The model obtained by the CLS method is a 3-dimensional pseudo linear system.

For reference, a 4-dimensional pseudo linear system is also given by the CLS method. The system completely reconstructs the original system.

Just as we thought, the following table and Fig. 6.4 truly indicate that the 3-dimensional pseudo linear system obtained by the CLS method is not such a good approximation. For reference, the modified impulse responses of the same dimensional pseudo linear system as the original system are also shown. Hence, there exists a good approximation for the given system.

dimen-ion	ratio of matrices	mean values of square root for sum of			cosine ① and ②	error ratio
		signal ①	signal by CLS ②	error ③	$\cos\theta$	③/①
$I(1)\_(1,1,1)$	0	0.40	0.40	0	1	0
$I(2)\_(1,1,1)$	0	0.42	0.42	0	1	0
$I(3)\_(1,1,1)$	0.05	0.23	0.22	0.09	0.92	0.39
$I(1)\_(1,1,2)$	0	0.40	0.40	0	1	0
$I(2)\_(1,1,2)$	0	0.42	0.42	0	1	0
$I(3)\_(1,1,2)$	0	0.23	0.23	0	1	0

Example 6.25. Let the signals be the modified impulse responses of the following 4-dimensional pseudo linear system: $\sigma = ((\mathbf{R}^4, F), g, h, h^0)$, where

$$F = \begin{bmatrix} 0 & 0.2 & 0 & 0 \\ 1 & 0.6 & 0 & 0.3 \\ 0 & 0 & -0.7 & 0.2 \\ 0 & 0 & 0 & 0.8 \end{bmatrix}, \ g(u_1) = \mathbf{e}_1, \ g(u_2) = \mathbf{e}_3, \ g(u_3) = \mathbf{e}_4,$$

$$h = [12, \ -1, \ -15, \ 4], \ h^0 = 1.$$

Then the approximate realization problem is solved as follows:

covariance matrix	eigenvalues				
	1	2	3	4	5
$H^T_{\underline{a}\ (3,50)}(3,0,0)H_{\underline{a}\ (3,50)}(3,0,0)$	156	9.4	0		
$H^T_{\underline{a}\ (4,50)}(2,2,0)H_{\underline{a}\ (4,50)}(2,2,0)$	759	52.4	5	0	
$H^T_{\underline{a}\ (5,50)}(2,1,2)H_{\underline{a}\ (5,50)}(2,1,2)$	566	70	21	0.7	0
covariance matrix	square root of eigenvalues				
$H^T_{\underline{a}\ (5,50)}(2,1,2)H_{\underline{a}\ (5,50)}(2,1,2)$	23.8	8.4	4.6	0.8	0

1) Since the ratio $\frac{0.8}{23.8} = 0.03$ obtained by the square root of
$H^T_{\underline{a}\ (5,50)}(2,1,2)H_{\underline{a}\ (5,50)}(2,1,2)$ is small , the approximate pseudo linear system obtained by the CLS method may be good.
2) After determining the numbers n_1, n_2 and n_3 of dimensions which are 2, 1 and 0, we will continue the approximate realization algorithm by the CLS method.

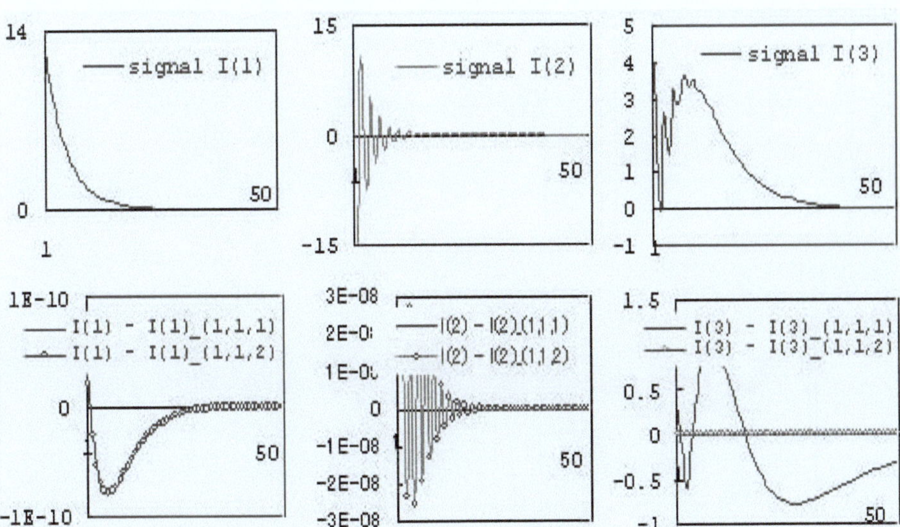

Fig. 6.4. In Example (6.24), the left are the original signal $I(1)$ and the difference between it and the approximate signal $I(1)_-(1,1,1)$ or $I(1)_-(1,1,2)$. The middle are the original signal $I(2)$ and the difference between it and the approximate signal $I(2)_-(1,1,1)$ or $I(2)_-(1,1,2)$. The right are the original signal $I(3)$ and the difference between it and the approximate signal $I(3)_-(1,1,1)$ or $I(3)_-(1,1,2)$.

Therefore, the modified impulse responses $I(1)$, $I(2)$ and $I(3)$ of an approximate pseudo linear system obtained by the CLS method is constructed for a 3-dimensional space.

The 3-dimensional pseudo linear system $\sigma_2 = ((\boldsymbol{R}^3,\ F_2),\ g_2,\ h_2, h^0)$ obtained by the CLS method can be expressed as follows:

$$F_2 = \begin{bmatrix} 0 & 0.2 & 0 \\ 1 & 0.6 & 0 \\ 0 & 0 & -0.7 \end{bmatrix},\ h_2 = [12,\ -1,\ -15],\ g_2(u_1) = [1,\ 0,\ 0]^T,$$

$g_2(u_2) = [0,\ 0,\ 1]^T,\ g_2(u_3) = [0.23,\ 1.7,\ -0.21]^T,\ h^0 = 1.$
For reference, a 4-dimensional pseudo linear system $\sigma_3 = ((\boldsymbol{R}^4,\ F_3),\ g_3,\ h_3, h^0)$ obtained by the CLS method can be expressed as follows:

$$F_3 = \begin{bmatrix} 0 & 0.2 & 0 & 0 \\ 1 & 0.6 & 0 & 0.3 \\ 0 & 0 & -0.7 & 0.2 \\ 0 & 0 & 0 & 0.8 \end{bmatrix},\ h_3 = [12,\ -1,\ -15,\ 4],\ g_3(u_1) = [1,\ 0,\ 0,\ 0]^T,$$

$g_3(u_2) = [0,\ 0,\ 1,\ 0]^T,\ g_3(u_3) = [0,\ 0,\ 0,\ 1]^T,\ h^0 = 1.$

In this example, the original signals are considered as the modified impulse responses of a 4-dimensional pseudo linear system and the desirable modified impulse responses are obtained by the CLS method within our expectations. The model obtained by the CLS method is a 3-dimensional pseudo linear system.

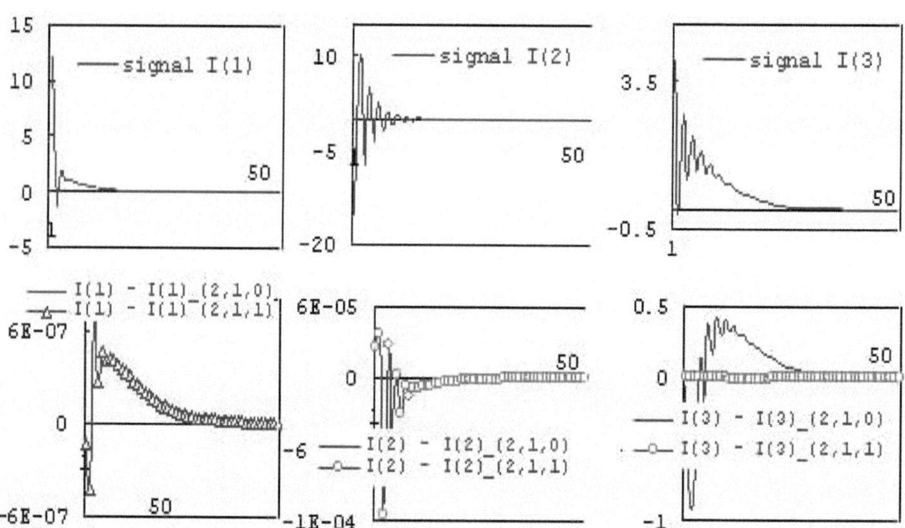

Fig. 6.5. In Example (6.25), the left are the original signal $I(1)$ and the difference between it and the approximate signal $I(1)$_$(2,1,0)$ or $I(1)$_$(2,1,1)$. The middle are the original signal $I(2)$ and the difference between it and the approximate signal $I(2)$_$(2,1,0)$ or $I(2)$_$(2,1,1)$. The right are the original signal $I(3)$ and the difference between it and the approximate signal $I(3)$_$(2,1,0)$ or $I(3)$_$(2,1,1)$.

For reference, a 4-dimensional pseudo linear system is also given by the CLS method. The system completely reconstructs the original system.

Contrary to what we expected, the following table and Fig. 6.5 truly indicate that the 3-dimensional pseudo linear system obtained by the CLS method is not such a good approximation. The result is likely to be caused by rapid damping. For reference, the modified impulse responses of the same dimensional pseudo linear system as the original system are also shown. Hence, there exists a good approximation for the given system.

dimen-ion	ratio of matrices	mean values of square root for sum of		error	cosine $\textcircled{1}$ and $\textcircled{2}$	error ratio
		signal $\textcircled{1}$	signal by CLS $\textcircled{2}$	error $\textcircled{3}$	$\cos\theta$	$\textcircled{3}/\textcircled{1}$
$I(1)\_(2,1,0)$	0	0.24	0.24	0	1	0
$I(2)\_(2,1,0)$	0	0.42	0.42	0	1	0
$I(3)\_(2,1,0)$	0.03	0.122	0.123	0.03	0.97	0.25
$I(1)\_(2,1,1)$	0	0.24	0.24	0	1	0
$I(2)\_(2,1,1)$	0	0.42	0.42	0	1	0
$I(3)\_(2,1,1)$	0	0.23	0.23	0	1	0

Example 6.26. Let the signals be the modified impulse responses of the following 5-dimensional pseudo linear system: $\sigma = ((\boldsymbol{R}^4, F), g, h, h^0)$, where

$$F = \begin{bmatrix} 0 & 0.4 & 0 & 0 & 0 \\ 1 & 0.6 & 0 & 0 & 0.3 \\ 0 & 0 & 0 & -0.7 & 0.2 \\ 0 & 0 & 1 & 0 & 0.8 \\ 0 & 0 & 0 & 0 & 0.8 \end{bmatrix}, \; g(u_1) = \mathbf{e}_1, \; g(u_2) = \mathbf{e}_3, \; g(u_3) = \mathbf{e}_5,$$

$$h = [12, \; -1, \; -15, \; 4, \; -2], \; h^0 = 1.$$

Then the approximate realization problem is solved as follows:

covariance matrix	— eigenvalues					
	1	2	3	4	5	6
$H_{\underline{a}\,(3,50)}^T(3,0,0)H_{\underline{a}\,(3,50)}(3,0,0)$	1158	96	0			
$H_{\underline{a}\,(5,50)}^T(2,3,0)H_{\underline{a}\,(5,50)}(2,3,0)$	884	647	244	54	0	
$H_{\underline{a}\,(6,50)}^T(2,2,2)H_{\underline{a}\,(6,50)}(2,2,2)$	2445	594	354	56	0.0002	0
covariance matrix	square root of eigenvalues					
$H_{\underline{a}\,(6,50)}^T(2,2,2)H_{\underline{a}\,(6,50)}(2,2,2)$	49	24.3	18.8	7.5	0.01	0

1) Since the ratio $\frac{0.8}{23.8} = 0.0002$ obtained by the square root of $H_{\underline{a}\,(6,50)}^T(2,2,2)H_{\underline{a}\,(6,50)}(2,2,2)$ is small , the approximate pseudo linear system obtained by the CLS method may be good.

2) After determining the numbers n_1, n_2 and n_3 of dimensions which are 2, 2 and 0, we will continue the approximate realization algorithm by the CLS method.

Therefore, the modified impulse responses $I(1)$, $I(2)$ and $I(3)$ of an approximate pseudo linear system obtained by the CLS method is constructed for a 4-dimensional space.

The 4-dimensional pseudo linear system $\sigma_2 = ((\mathbf{R}^4, F_2), g_2, h_2, h^0)$ obtained by the CLS method can be expressed as follows:

$$F_2 = \begin{bmatrix} 0 & 0.4 & 0 & 0 \\ 1 & 0.6 & 0 & 0 \\ 0 & 0 & 0 & -0.7 \\ 0 & 0 & 1 & 0 \end{bmatrix}, \ h_2 = [12, -1, -15, 4], \ g_2(u_1) = [1, 0, 0, 0]^T,$$

$g_2(u_2) = [0, 0, 1, 0]^T, \ g_2(u_3) = [0.5, 1, 0.3, -0.6]^T, \ h^0 = 1.$

For reference, a 5-dimensional pseudo linear system $\sigma_3 = ((\mathbf{R}^5, F_3), g_3, h_3, h^0)$ obtained by the CLS method can be expressed as follows:

$$F_3 = \begin{bmatrix} 0 & 0.4 & 0 & 0 & 0 \\ 1 & 0.6 & 0 & 0 & 0.3 \\ 0 & 0 & 0 & -0.7 & 0.2 \\ 0 & 0 & 1 & 0 & 0.8 \\ 0 & 0 & 0 & 0 & 0.8 \end{bmatrix}, \ h_3 = [12, -1, -15, 4, -2], \ g_3(u_1) = [1, 0, 0, 0, 0]^T,$$

$g_3(u_2) = [0, 0, 1, 0, 0]^T, \ g_3(u_3) = [0, 0, 0, 0, 1]^T, \ h^0 = 1.$

In this example, the original signals are considered as the modified impulse responses of a 5-dimensional pseudo linear system and the desirable modified impulse responses are obtained by the CLS method within our expectations. The model obtained by the CLS method is a 4-dimensional pseudo linear system.

For reference, a 5-dimensional pseudo linear system is also given by the CLS method. The system completely reconstructs the original system.

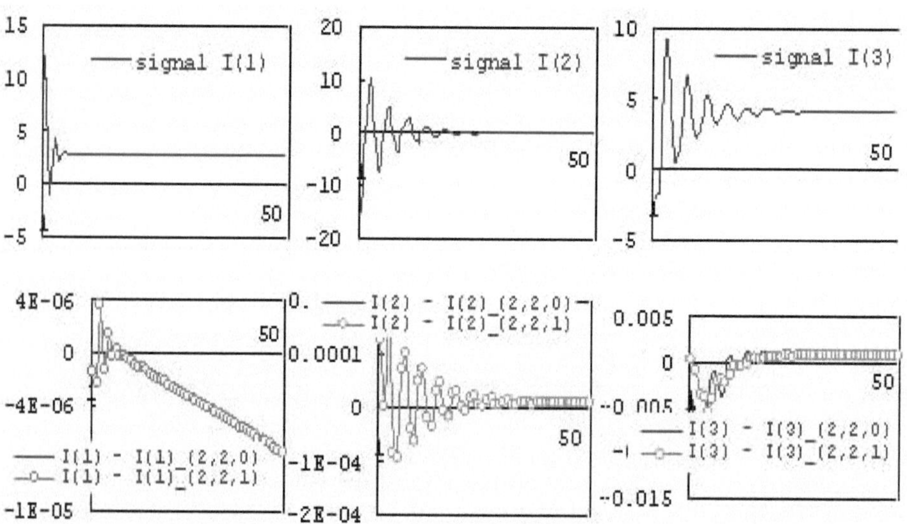

Fig. 6.6. In Example (6.26), the left are the original signal $I(1)$ and the difference between it and the approximate signal $I(1)\_(2,2,0)$ or $I(1)\_(2,2,1)$. The middle are the original signal $I(2)$ and the difference between it and the approximate signal $I(2)\_(2,2,0)$ or $I(2)\_(2,2,1)$. The right are the original signal $I(3)$ and the difference between it and the approximate signal $I(3)\_(2,2,0)$ or $I(3)\_(2,2,1)$.

Just as we expected, the following table and Fig. 6.6 truly indicate that the 4-dimensional pseudo linear system obtained by the CLS method is a very good approximation. For reference, the modified impulse responses of the same dimensional pseudo linear system as the original system are also shown. Hence, there exists a good approximation for the given system.

dimen- ion	ratio of matrices	mean values of square root for sum of		error	cosine ① and ②	error ratio
		signal	signal by CLS	error		
		①	②	③	$\cos\theta$	③/①
$I(1)_-(2,2,0)$	0	0.45	0.45	0	1	0
$I(2)_-(2,2,0)$	0	0.43	0.43	0	1	0
$I(3)_-(2,2,0)$	0.0002	0.61	0.61	0	1	0
$I(1)_-(2,2,1)$	0	0.45	0.45	0	1	0
$I(2)_-(2,2,1)$	0	0.43	0.43	0	1	0
$I(3)_-(2,2,1)$	0	0.61	0.61	0	1	0

Example 6.27. Let the signals be the modified impulse responses of the following 6-dimensional pseudo linear system $\sigma = ((\boldsymbol{R}^6, F), g, h, h^0)$, where

$$F = \begin{bmatrix} 0 & 0.2 & 0 & 0.2 & 0 & 0.1 \\ 1 & 0.7 & 0 & 0.3 & 0 & -0.3 \\ 0 & 0 & 0 & 0.1 & 0 & -0.2 \\ 0 & 0 & 1 & -0.5 & 0 & 0.4 \\ 0 & 0 & 0 & 0 & 0 & 0.3 \\ 0 & 0 & 0 & 0 & 1 & -0.6 \end{bmatrix}, \ g(u_1) = \mathbf{e}_1, \ g(u_2) = \mathbf{e}_3, \ g(u_3) = \mathbf{e}_5,$$

$h = [12, \ -1, \ -15, \ 4, \ -8, \ 1], \ h^0 = 1.$

Then the approximate realization problem is solved as follows:

covariance matrix	eigenvalues						
	1	2	3	4	5	6	7
$H_{\underline{a}\,(3,50)}(3,0,0)H^T_{\underline{a}\,(3,50)}(3,0,0)$	161	16.6	0				
$H_{\underline{a}\,(5,50)}(2,3,0)H^T_{\underline{a}\,(5,50)}(2,3,0)$	413	24	2.4	0.2	0		
$H_{\underline{a}\,(5,50)}(2,2,1)H^T_{\underline{a}\,(5,50)}(2,2,1)$	490	146	22	1.4	0.2		
$H_{\underline{a}\,(6,50)}(2,2,2)H^T_{\underline{a}\,(6,50)}(2,2,2)$	507	288	23	3.5	0.6	0.006	
$H_{\underline{a}\,(7,50)}(2,2,3)H^T_{\underline{a}\,(7,50)}(2,2,3)$	512	442	23	3.6	0.6	0.007	0
covariance matrix	square root of eigenvalues						
$H^T_{\underline{a}\,(6,50)}(2,2,2)H_{\underline{a}\,(6,50)}(2,2,2)$	23	17	4.8	1.9	0.8	0.07	
$H^T_{\underline{a}\,(7,50)}(2,2,3)H_{\underline{a}\,(7,50)}(2,2,3)$	23	21	4.8	1.9	0.8	0.08	0

1) Since the ratios $\frac{0.8}{23} = 0.03$ and $\frac{0.08}{23} = 0.003$ obtained by the square root of $H^T_{\underline{a}\,(7,50)}(2,2,3)H_{\underline{a}\,(7,50)}(2,2,3)$ are small , the approximate pseudo linear system obtained by the CLS method may be good.
However, the ratio $\frac{0.8}{23} = 0.03$ of $H^T_{\underline{a}\,(7,50)}(2,2,3)H_{\underline{a}\,(7,50)}(2,2,3)$ comes from $H^T_{\underline{a}\,(5,50)}(2,3,0)H_{\underline{a}\,(5,50)}(2,3,0)$, hence the reduction from it is not good.
2) After determining the numbers n_1, n_2 and n_3 of dimensions which are 2, 2 and 1, we will continue the approximate realization algorithm by the CLS method.

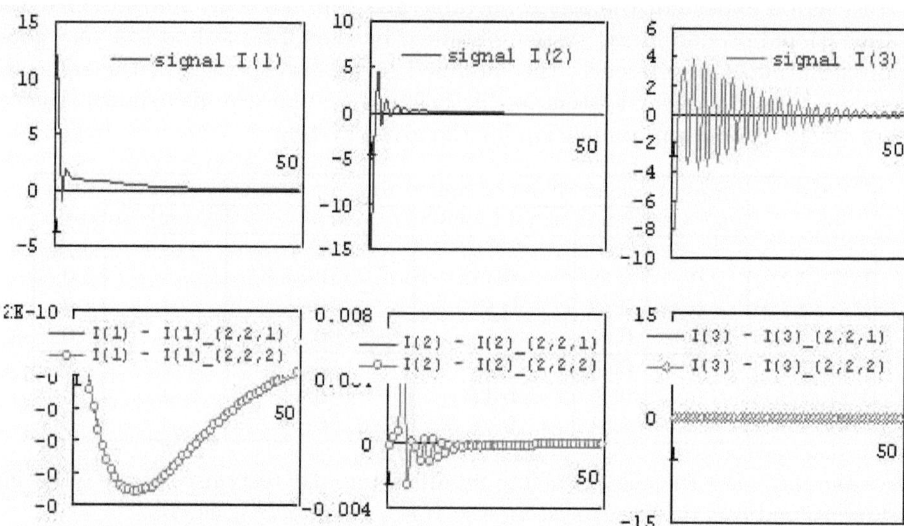

Fig. 6.7. In Example (6.27), the left are the original signal $I(1)$ and the difference between it and the approximate signal $I(1)\_(2,2,1)$ or $I(1)\_(2,2,2)$. The middle are the original signal $I(2)$ and the difference between it and the approximate signal $I(2)\_(2,2,1)$ or $I(2)\_(2,2,2)$. The right are the original signal $I(3)$ and the difference between it and the approximate signal $I(3)\_(2,2,1)$ or $I(3)\_(2,2,2)$.

Therefore, the modified impulse responses $I(1)$, $I(2)$ and $I(3)$ of an approximate pseudo linear system obtained by the CLS method is constructed for a 5-dimensional space.

The 5-dimensional pseudo linear system $\sigma_3 = ((\boldsymbol{R}^5, F_3), g_3, h_3, h^0)$ obtained by the CLS method can be expressed as follows:

$$F_3 = \begin{bmatrix} 0 & 0.2 & 0 & 0.2 & -1.3 \\ 1 & 0.7 & 0 & 0.3 & 3.1 \\ 0 & 0 & 0 & 0.1 & -1.6 \\ 0 & 0 & 1 & -0.5 & -2.7 \\ 0 & 0 & 0 & 0 & -0.94 \end{bmatrix}, \; h_3 = [12, -1, -15, 4, -8], \; g_3(u_1) = [1, 0, 0,$$

$0, 0]^T, g_3(u_2) = [0, 0, 1, 0, 0]^T, \; g_3(u_3) = [0, 0, 0, 0, 1]^T, \; h^0 = 1$.

For reference, a 6-dimensional pseudo linear system $\sigma_4 = ((\boldsymbol{R}^6, F_4), g_4, h_4, h^0)$ obtained by the CLS method can be expressed as follows:

$$F_4 = \begin{bmatrix} 0 & 0.2 & 0 & 0.2 & 0 & 0.1 \\ 1 & 0.7 & 0 & 0.3 & 0 & -0.3 \\ 0 & 0 & 0 & 0.1 & 0 & -0.2 \\ 0 & 0 & 1 & -0.5 & 0 & 0.4 \\ 0 & 0 & 0 & 0 & 0 & 0.3 \\ 0 & 0 & 0 & 0 & 1 & -0.6 \end{bmatrix}, \; h_4 = [12, -1, -15, 4, -8, 1],$$

$g_4(u_1) = [1, 0, 0, 0, 0, 0]^T, g_4(u_2) = [0, 0, 1, 0, 0, 0]^T,$
$g_4(u_3) = [0, 0, 0, 0, 1, 0]^T, \; h^0 = 1$.

In this example, the original signals are considered as the modified impulse responses of a 6-dimensional pseudo linear system and the desirable modified impulse responses are obtained by the CLS method within our expectations. The model obtained by the CLS method is a 5-dimensional pseudo linear system.

For reference, a 6-dimensional pseudo linear system is also given by the CLS method. The system completely reconstructs the original system.

Just as we expected, the following table and Fig. 6.7 truly indicate that the 5-dimensional pseudo linear system obtained by the CLS method is a good approximation. For reference, the modified impulse responses of the same dimensional pseudo linear system as the original system are also shown. Hence, there exists a good approximation for the given system.

dimen-sion	ratio of matrices	mean values of square root for sum of			cosine ① and ②	error ratio
		signal ①	signal by CLS ②	error ③	$\cos\theta$	③/①
I(1)_(2,2,1)	0	0.24	0.24	0	1	0
I(2)_(2,2,1)	0	0.31	0.31	0	1	0
I(3)_(2,2,1)	0.003	0.3	0.3	0.019	0.998	0.06
I(1)_(2,2,2)	0	0.24	0.24	0	1	0
I(2)_(2,2,2)	0	0.31	0.31	0	1	0
I(3)_(2,2,2)	0	0.3	0.3	0	1	0

6.6　Noisy Realization of Pseudo Linear Systems

In this section, we discuss the noisy realization problem of pseudo linear systems.

For noise $\{\bar\gamma(t) : t \in N\}$ added to the unknown pseudo linear system a, we will obtain the observed data $\{\hat\gamma(|\omega|) + \bar\gamma(|\omega|) : \omega \in U^*\}$.

For any given $\{\hat\gamma(|\omega|) + \bar\gamma(|\omega|) : \omega \in U^*\}$, σ which satisfies $a_\sigma(\omega) \approx \hat\gamma(|\omega|) : \omega \in U^*$ is called a noisy realization of a.

Roughly speaking, we can propose the following noisy realization problem:

For any given $\{\hat\gamma(|\omega|) + \bar\gamma(|\omega|) : \omega \in U^*\}$, find a pseudo linear system σ which satisfies $a_\sigma(\omega) \approx \hat\gamma(|\omega|)$ for any $\omega \in U^*$.

In order to make our discussion simple, we assume that the set Y of outout is the set R of real numbers, namely 1-output.

A situation for noisy realization problem 6.28

Let the observed object be a pseudo linear system and noise be added to output. Then we will obtain the data $\{\gamma(t) = \hat\gamma(t) + \bar\gamma(t) : 0 \leq t \leq \underline{N}\}$ for some integer $\underline{N} \in N$, where $\hat\gamma(t)$ is the exact signal which comes from the observed pseudo linear system and $\bar\gamma(t)$ is the noise added at the time of observation.

Problem 6.29. Problem statement of a noisy realization for Pseudo Linear Systems

Let $H_{\underline{a}\ (p,\bar p)}$ be the measured finite-sized Input/output matrix. Then find the cleaned-up Input/output matrix $\hat H_{\underline{a}\ (p,\bar p)}$ such that $H_{\underline{a}\ (p,\bar p)} = \hat H_{\underline{a}\ (p,\bar p)} + \bar H_{\underline{a}\ (p,\bar p)}$ holds.

Namely, find a minimal dimensional pseudo linear system
$\sigma = ((\boldsymbol{R}^n, F_r), g_r, h_r, h^0))$ which realizes $\hat{H}_{\underline{a}\ (p,\bar{p})}$.

Theorem 6.30. *Algorithm for noisy realization*
Let a partial input response map $\underline{a}$ be a considered object which is a pseudo linear system. Then a noisy realization $\sigma = ((\boldsymbol{R}^n, F_s), g_s, h_s, h^0)$ of $\underline{a}$ is given by the following algorithm:

1) Based on the square root of eigenvalues for a matrix
$H_{\underline{a}\ (p,\bar{p})}(p,0,0)H_{\underline{a}\ (p,\bar{p})}(p,0,0)^T$, *determine the value n_1 of rank for the matrix* $H_{\underline{a}\ (p,\bar{p})}(p,0,0)$, *where $n_1 \leq p$.*

Namely, determine the value n_1 of rank for the matrix $H_{\underline{a}\ (p,\bar{p})}(p,0,0)$ such that a set of the square root of eigenvalues for the covariance matrix composed of relatively small and equally-sized numbers is excluded, where the signal part effected by the set may be the noisy part.

2) The CLS method is used as follows:

① Let a matrix $A_1 \in \boldsymbol{R}^{1 \times (n_1+1)}$ be $A_1 = [\alpha_{11}, \alpha_{12}, \cdots, \alpha_{1n_1}, -1]$.

② Choose the coefficients $\{\alpha_{1i} : 1 \leq i \leq n_1\}$ such that
$\sum_{j=1}^{n_1+1} \underline{S}_l^{j-1} \bar{I}_{\underline{a}} \cdot \underline{S}_l^{j-1} \bar{I}_{\underline{a}}$ *takes a minimum value, where $\{\underline{S}_l^i \bar{I}_{\underline{a}} \in \boldsymbol{R}^{L \times 1} :$*
$0 \leq i \leq n_1\}$ are given by the equation $[\bar{I}_{\underline{a}}, \underline{S}_l \bar{I}_{\underline{a}}, \cdots, \underline{S}_l^{n_1} \bar{I}_{\underline{a}}]^T :=$
$A_1^T[A_1 A_1^T]^{-1}A_1 H_{\underline{a}\ T(n_1+1,L)}^T(n_1+1,0,0)$ *and* $H_{\underline{a}\ (n_1+1,L)}(n_1+1,0,0) :=$
$[I_{\underline{a}}(0), \cdots, S_l^{n_1-1}I_{\underline{a}}(0), S_l^{n_1}I_{\underline{a}}(0)]$. *And $\cdot$ denotes the inner product of two vectors.*

③ Let h_{1s} be $h_{1s} = [I_{\underline{a}}(1) - \bar{I}_{\underline{a}}(1), I_{\underline{a}}(2) - \bar{I}_{\underline{a}}(2), \cdots, I_{\underline{a}}(n_1) - \bar{I}_{\underline{a}}(n_1)]$.

3) Based on the square root of eigenvalues for a matrix
$H_{\underline{a}\ (n_1+p,\bar{p})}(n_1,p,0)H_{\underline{a}\ (n_1+p,\bar{p})}(n_1,p,0)^T$, *determine the value n_2 of rank for the matrix $H_{\underline{a}\ (n_1+p,\bar{p})}(n_1,p,0)$, where $n_2 \leq p$.*

Namely, determine the value n_2 of rank for the matrix $H_{\underline{a}\ (p,\bar{p})}(n_1,p,0)$ such that a set of the square root of eigenvalues for the covariance matrix composed of relatively small and equally-sized numbers is excluded, where the signal part effected by the set may be the noisy part.

4) The CLS method is used as follows:

① Let a matrix $A_2 \in \boldsymbol{R}^{1 \times (n_1+n_2+1)}$ be $A_2 = [\alpha_{21}, \alpha_{22}, \cdots, \alpha_{2n_1+n_2}, -1]$.

② Choose the coefficients $\{\alpha_{1i} : 1 \leq i \leq n_1 + n_2\}$ such that
$\sum_{j=1}^{n_1+n_2+1} \underline{S}_l^{j-1} \bar{I}_{\underline{a}} \cdot \underline{S}_l^{j-1} \bar{I}_{\underline{a}}$ *takes a minimum value, where $\{\underline{S}_l^i \bar{I}_{\underline{a}} \in \boldsymbol{R}^{L \times 1} :$*
$0 \leq i \leq n_1 + n_2\}$ are given by the equation $[\bar{I}_{\underline{a}}, \underline{S}_l \bar{I}_{\underline{a}}, \cdots, S_l^{n_1} \bar{I}_{\underline{a}}]^T :=$
$A_2^T[A_2 A_2^T]^{-1}A_2 H_{\underline{a}\ (n_1+n_2+1,L)}^T(n_1,n_2+1,0)$ *and* $H_{\underline{a}\ (n_1+n_2;1,L)}^T(n_1,n_2+1,0)$
$:= [I_{\underline{a}}(u_1), \cdots, S_l^{n_1-1}I_{\underline{a}}(u_1), I_{\underline{a}}(u_2), \cdots, S_l^{n_2-1}I_{\underline{a}}(u_2), S_l^{n_2}I_{\underline{a}}(u_2)]$.
And $\cdot$ denotes the inner product of two vectors.

③ Let h_{2s} be $h_{2s} = [(I_{\underline{a}}(u_2))(0) - (\bar{I}_{\underline{a}}(u_2))(0), (S_l I_{\underline{a}}(u_2))(0) - (S_l \bar{I}_{\underline{a}}(u_2))(0),
$\cdots, (S_l^{n_2-1}I_{\underline{a}}(u_2))(0) - (S_l^{n_2-1}\bar{I}_{\underline{a}}(u_2))(0)]$.

5) Based on the square root of eigenvalues for a matrix
$H_{\underline{a}\ (n_1+n_2+p,\bar{p})}(n_1,n_2,q)H_{\underline{a}\ (n_1+n_2+q,\bar{q})}(n_1,n_2,q)^T$, *determine the value n_3 of rank for the matrix $H_{\underline{a}\ (n_1+n_2+q,\bar{q})}(n_1,n_2,q)$, where $n_3 \leq q$.*

Namely, determine the value n_3 of rank for the matrix $H_{\underline{a}\ (n_1+n_2+q,\bar{q})}(n_1,n_2,q)$ such that a set of the square root of eigenvalues for the covariance matrix *composed of relatively small and equally-sized numbers is excluded, where the signal part effected by the set may be the noisy part.*

6) The CLS method is used as follows:

① Let a matrix $A_3 \in \mathbf{R}^{1 \times (n_1+n_2+n_3+1)}$ be

$A_3 = [\alpha_{21}, \alpha_{22}, \cdots, \alpha_{2n_1+n_2+n_3}, -1].$

*② Choose the coefficients $\{\alpha_{3i} : 1 \leq i \leq n_1 + n_2 + n_3\}$ such that $\sum_{j=1}^{n_1+n_2+n_3} \underline{S}_l^{j-1}\bar{I}_{\underline{a}} \cdot \underline{S}_l^{j-1}\bar{I}_{\underline{a}}$ takes a minimum value, where $\{\underline{S}_l^i\bar{I}_{\underline{a}} \in \mathbf{R}^{L \times 1} : 0 \leq i \leq n_1 + n_2 + n_3\}$ are given by the equation $[\bar{I}_{\underline{a}}, \underline{S}_l\bar{I}_{\underline{a}}, \cdots, \underline{S}_l^{n_1}\bar{I}_{\underline{a}}]^T :=$
$A_3^T [A_3 A_3^T]^{-1} A_3 H_{\underline{a}\,(n_1+n_2+n_3+1,L)}^T (n_1, n_2, n_3 + 1)$ and*

$H_{\underline{a}\,(n_1+n_2+n_3+1,L)}^T (n_1, n_2, n_3 + 1) := [I_{\underline{a}}(u_1), \cdots, S_l^{n_1-1} I_{\underline{a}}(u_1), I_{\underline{a}}(u_2),$
$\cdots, S_l^{n_2-1} I_{\underline{a}}(u_2), S_l^{n_2} I_{\underline{a}}(u_2), I_{\underline{a}}(u_3), \cdots, S_l^{n_3-1} I_{\underline{a}}(u_3), S_l^{n_3} I_{\underline{a}}(u_3)].$

And . denotes the inner product of two vectors.

*③ Let h_{3s} be $h_{3s} = [(I_{\underline{a}}(u_3))(0) - (\bar{I}_{\underline{a}}(u_3))(0), (S_l I_{\underline{a}}(u_3))(0) - (S_l \bar{I}_{\underline{a}}(u_3))(0),$
$\cdots, (S_l^{n_3-1} I_{\underline{a}}(u_3))(0) - (S_l^{n_3-1} \bar{I}_{\underline{a}}(u_3))(0)].$*

$\vdots$

*2*m-1) Based on the square root of eigenvalues for a matrix*
$H_{\underline{a}\,(n_1+\cdots+n_{m-1}+p,\bar{p})}(n_1, \cdots, n_{m-1}, q) H_{\underline{a}\,(n_1+\cdots+n_{m-1}+q,\bar{q})}(n_1, \cdots, n_{m-1}, q)^T,$
determine the value n_m of rank for the matrix
$H_{\underline{a}\,(n_1+\cdots+n_{m-1}+p,q)}(n_1, \cdots, n_{m-1}, q),$ *where $n_m \leq q$.*
Namely, determine the value n_3 of rank for the matrix
$H_{\underline{a}\,(n_1+\cdots,n_{m-1}+q,\bar{q})}(n_1, \cdots, n_{m-1}, q)$ *such that a set of the square root of eigenvalues for the covariance matrix*
composed of relatively small and equally-sized numbers is excluded, where the signal part effected by the set may be the noisy part.

*2*m) The CLS method is used as follows:*

① Let a matrix $A_m \in \mathbf{R}^{1 \times (n_1+\cdots+n_m+1)}$ be

$A_m = [\alpha_{m1}, \alpha_{m2}, \cdots, \alpha_{mn_1+\cdots+n_m}, -1].$

*② Choose the coefficients $\{\alpha_{mi} : 1 \leq i \leq n_1 + \cdots + n_m\}$ such that $\sum_{j=1}^{n_1+\cdots+n_m} \underline{S}_l^{j-1}\bar{I}_{\underline{a}} \cdot \underline{S}_l^{j-1}\bar{I}_{\underline{a}}$ takes a minimum value, where $\{\underline{S}_l^i\bar{I}_{\underline{a}} \in \mathbf{R}^{L \times 1} : 0 \leq i \leq n_1 + \cdots + n_m\}$ are given by the equation $[\bar{I}_{\underline{a}}, \underline{S}_l\bar{I}_{\underline{a}}, \cdots, \underline{S}_l^{n_1}\bar{I}_{\underline{a}}]^T$
$:= A_m^T [A_m A_m^T]^{-1} A_m H_{\underline{a}\,(n_1+\cdots+n_m+1,L)}^T (n_1, \cdots, n_m + 1)$ and*

$H_{\underline{a}\,(n_1+\cdots+n_m+1,L)}^T (n_1, \cdots, n_m + 1) := [I_{\underline{a}}(u_1), \cdots, S_l^{n_1-1} I_{\underline{a}}(u_1), I_{\underline{a}}(u_2),$
$\cdots, S_l^{n_2-1} I_{\underline{a}}(u_2), S_l^{n_2} I_{\underline{a}}(u_2), \cdots, I_{\underline{a}}(u_m), \cdots, S_l^{n_m-1} I_{\underline{a}}(u_m), S_l^{n_m} I_{\underline{a}}(u_m)].$

And . denotes the inner product of two vectors.

*③ Let h_{ms} be $h_{ms} = [(I_{\underline{a}}(u_m))(0) - (\bar{I}_{\underline{a}}(u_m))(0), (S_l I_{\underline{a}}(u_m))(0) - (S_l \bar{I}_{\underline{a}}(u_m))$
$(0), \cdots, (S_l^{n_m-1} I_{\underline{a}}(u_m))(0) - (S_l^{n_m-1} \bar{I}_{\underline{a}}(u_m))(0)].$*

*2*m+1) Let $g_s \in F(U, \mathbf{R}$ be $g_s(u_1) := \mathbf{e}_1, g_s(u_2) := \mathbf{e}_{n_1+1}, \cdots,$*
$g_s(u_m) := \mathbf{e}_{n_1+\cdots+n_{m-1}+1}.$
Let $F_s \in \mathbf{R}^{n \times n}$ be the same as in Theorem (6.19).
Let $h_s \in \mathbf{R}^{1 \times n}$ be $h_s := [h_{1s}, h_{2s}, \cdots, h_{ms}],$
where $n := n_1 + n_2 + \cdots + n_m.$

For the real time standard system $\sigma_r = ((\mathbf{R}^n, F_r), g_r^0, g_r, h_r, h_r^0)$, its modified impulse responses $I(u_1)(i) := h_r F_r^i g_r(u_1)$, $I(u_2)(i) := h_r F_r^i g_r(u_2)$ and $I(u_3)(i) := h_r F_r^i g_r(u_3)$ are written by $I(1)\_(n_1, n_2, n_3)$ and $I(2)\_(n_1, n_2, n_3)$ and $I(3)\_(n_1, n_2, n_3)$ respectively.

[proof] By 1) and 3), the noisy part in the data can be excluded in the sense of the number of dimensions by checking what part is the noisy part and by using the ratio of Input/output matrix norm, which implies the noise to signal ratio. The matrices A_1 in 2), A_2 in 4), A_3 in 6), $\cdots$ and A_m in 2*m) correspond to the matrix A in Proposition (2.14). Hence, the noisy part of the given finite-sized Input/output matrix is excluded. Therefore, applying Theorem (6.19), we can obtain g_s, F_s and h_s by 2*m+1).

Remark 1: A determination method of the degree of n in the linear system $\sigma = ((\mathbf{R}^n, F_s), g, h_s)$ can be found in the Principal Component Method. The method is popular.

Remark 2: Let S and N be the norm of a signal and noise. Then the selected ratio of matrices in the algorithm may be considered as $\frac{N}{S+N}$.

Remark 3: This noisy realization method is very new.

Remark 4: For the noisy case, the AIC method is famous for determining linear systems including dimensions of the state spaces.

Example 6.31. Let signals be the modified impulse responses of the following 3-dimensional pseudo linear system $\sigma = ((R^3, F), g, h, h^0)$,

where $F = \begin{bmatrix} 0 & 0.3 & 0 \\ 1 & 0.6 & 0 \\ 0 & 0 & 0.7 \end{bmatrix}$, $h = [12, \ -1, \ -15]$, $g(u_1) = [1, \ 0, \ 0]^T$,

$g(u_2) = [0, \ 0, \ 1]^T$, $h^0 = 1$.

Let added noises be given in Fig. 6.8.
Then the noisy realization problem is solved as follows:

covariance matrix	eigenvalues				
	1	2	3	4	5
$H_{\underline{a}\ (3,50)}(3,0)H_{\underline{a}\ (3,50)}^T(3,0)$	208	44	3		
$H_{\underline{a}\ (4,50)}(4,0)H_{\underline{a}\ (4,50)}^T(4,0)$	216	56	3.2	1.6	
$H_{\underline{a}\ (4,50)}(2,2)H_{\underline{a}\ (4,50)}^T(2,2)$	792	76	21	3.7	
$H_{\underline{a}\ (5,50)}(2,3)H_{\underline{a}\ (5,50)}^T(2,3)$	894	77	21	4.4	1.9
covariance matrix	square root of eigenvalues				
	1	2	3	4	5
$H_{\underline{a}\ (4,50)}(4,0)H_{\underline{a}\ (4,50)}^T(4,0)$	14.7	7.5	1.8	1.3	
$H_{\underline{a}\ (5,50)}(2,3)H_{\underline{a}\ (5,50)}^T(2,3)$	30	8.8	4.6	2.1	1.4

1) A set {1.8, 1.3} is composed of relatively small and equally-sized numbers in the square root of eigenvalues for $H_{\underline{a}\ (4,50)}(4,0)H_{\underline{a}\ (4,50)}^T(4,0)$.
2) After determining the number n_1 of dimensions which is 2, we will continue the noisy realization algorithm by the CLS method.

Therefore, the modified impulse response $I(1)$ of a pseudo linear system obtained by the CLS method is constructed for a 2-dimensional space.

3) Since a set $\{2.1, 1.4\}$ is composed of relatively small and equally-sized numbers in the square root of $H_{\underline{a}\,(5,50)}(2,3)H_{\underline{a}\,(5,50)}^{T}(2,3)$, the approximate pseudo linear system obtained by the CLS method may be realized by adding another 1-dimensional space.

4) After determining the number n_2 of dimensions which is 1, we execute the noisy realization algorithm by the CLS method.

Therefore, the modified impulse responses $I(1)$ and $I(2)$ of a approximate pseudo linear system obtained by the CLS method is realized by a (2,1)-dimensional pseudo linear system.

The 3-dimensional pseudo linear system obtained by the CLS method is expressed as follows:

$$\sigma_n = ((\mathbf{R}^3,\ F_n),\ g_n,\ h_n,\ h^0),\ \text{where } F_n = \begin{bmatrix} 0 & 0.36 & -0.07 \\ 1 & 0.49 & 0.05 \\ 0 & 0 & 0.67 \end{bmatrix},\ g_n(u_1) = \mathbf{e}_1,$$

$g_n(u_2) = \mathbf{e}_3$, $h_n = [-12,\ -1.2,\ -15]$ and $h^0 = 1$.

The obtained modified impulse responses $I(1)$ and $I(2)$ are illustrated in Fig. 6.8.

In this example, the original signal $I(1)$ is characterized as the modified impulse responses of a 2-dimensional linear space and the original signal $I(2)$ is characterized as the modified impulse response of an added 1-dimensional linear

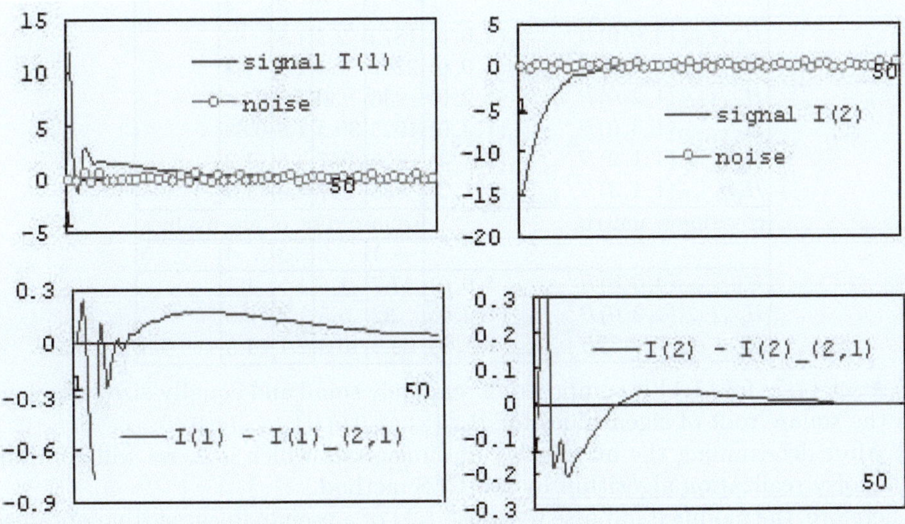

Fig. 6.8. In Example (6.31), the left are the original signal $I(1)$ with noise and the difference between $I(1)$ and the obtained signal $I(1)_-(2,1)$. The right are the original signal $I(2)$ with noise and the difference between $I(2)$ and the obtained signal $I(2)_-(2,1)$.

space. The desirable modified impulse responses are attempted to be obtained by the CLS method. The model obtained by the CLS method is a (2,1)-dimensional pseudo linear system which has the same number of dimensions as the number of the original system.

Just as we expected, the following table and Fig. 6.8 indicate that the model obtained by the CLS method is a good (2,1)-dimensional system for the original (2,1)-dimensional system.

dimen-ion	ratio of matrices	mean values of square root for sum of			cosine	error
		signal	signal by CLS	error	① and ②	ratio
		①	②	③	$\cos\theta$	③/①
I(1)_(2,1)	0.13	0.266	0.267	0.02	0.997	0.07
I(2)_(2,1)	0.08	0.42	0.422	0.01	0.999	0.02

Example 6.32. Let signals be the modified impulse responses of the following 3-dimensional pseudo linear system $\sigma = ((R^3, F), g, h, h^0)$,

where $F = \begin{bmatrix} 0.9 & 0.3 & 0.1 \\ 0 & 0.2 & 0 \\ 0 & 0 & -0.2 \end{bmatrix}$, $h = [12, -1, -15]$, $g(u_1) = [1, 0, 0]^T$,

$g(u_2) = [0, 1, 0]^T$, $g(u_3) = [0, 0, 1]^T$, $h^0 = 1$.

Let added noises be given in Fig. 6.9.

Then the noisy realization problem is solved as follows:

covariance matrix	eigenvalues				
	1	2	3	4	5
$H_{\underline{a}\,(3,50)}(3,0,0)H^T_{\underline{a}\,(3,50)}(3,0,0)$	1872	2.3	1.8		
$H_{\underline{a}\,(4,50)}(4,0,0)H^T_{\underline{a}\,(4,50)}(4,0,0)$	2272	2.4	2.2	1.3	
$H_{\underline{a}\,(3,50)}(1,2,0)H^T_{\underline{a}\,(3,50)}(1,2,0)$	936	30	1.3		
$H_{\underline{a}\,(4,50)}(1,3,0)H^T_{\underline{a}\,(4,50)}(1,3,0)$	1025	30.4	1.8	0.83	
$H_{\underline{a}\,(4,50)}(1,1,2)H^T_{\underline{a}\,(4,50)}(1,1,2)$	855	267	4.6	1.2	
$H_{\underline{a}\,(5,50)}(1,1,3)H^T_{\underline{a}\,(5,50)}(1,1,3)$	858	267	4.6	1.9	0.6
covariance matrix	square root of eigenvalues				
	1	2	3	4	5
$H_{\underline{a}\,(4,50)}(4,0,0)H^T_{\underline{a}\,(4,50)}(4,0,0)$	47.7	1.5	1.5	1.1	
$H_{\underline{a}\,(4,50)}(1,3,0)H^T_{\underline{a}\,(4,50)}(1,3,0)$	32	5.5	1.3	0.9	
$H_{\underline{a}\,(5,50)}(1,1,3)H^T_{\underline{a}\,(5,50)}(1,1,3)$	29.3	16.3	2.1	1.3	0.8

1) A set $\{1.5,\ 1.5,\ 1.1\}$ is composed of relatively small and equally-sized numbers in the square root of eigenvalues for $H_{\underline{a}\,(4,50)}(4,0.0)H^T_{\underline{a}\,(4,50)}(4,0,0)$.
2) After determining the number n_1 of dimensions which is 1, we will continue the noisy realization algorithm by the CLS method.
Therefore, the modified impulse response $I(1)$ of a pseudo linear system obtained by the CLS method is constructed for a 1-dimensional space.
3) A set $\{1.3,\ 0.9\}$ is composed of relatively small and equally-sized numbers in the square root of eigenvalues for $H_{\underline{a}\,(4,50)}(1,3,0)H^T_{\underline{a}\,(4,50)}(1,3,0)$.
4) After determining the number n_2 of dimensions which is 1, we execute the noisy realization algorithm by the CLS method.

5) A set $\{1.3, 0.8\}$ is composed of relatively small and equally-sized numbers in the square root of eigenvalues for $H_{\underline{a}\,(5,50)}(1,1,3)H_{\underline{a}\,(5,50)}^T(1,1,3)$.

6) After determining the number n_3 of dimensions which is 1, we execute the noisy realization algorithm by the CLS method.

Therefore, the modified impulse responses $I(1)$, $I(2)$ and $I(3)$ of an approximate pseudo linear system obtained by the CLS method is realized by a (1,1,1)-dimensional pseudo linear system.

The 3-dimensional pseudo linear system obtained by the CLS method is expressed as follows:

$$\sigma_n = ((\boldsymbol{R}^3,\ F_n),\ g_n,\ h_n,\ h^0),\ \text{where } F_n = \begin{bmatrix} 0.9 & 0.32 & 0.08 \\ 0 & 0.14 & 0.06 \\ 0 & 0 & -0.21 \end{bmatrix},\ g_n(u_1) = \mathbf{e}_1,$$

$g_n(u_2) = \mathbf{e}_2$, $g_n(u_3) = \mathbf{e}_3$, $h_n = [11.9, -1.3, -14.9]$ and $h^0 = 1$.

The obtained modified impulse responses $I(1)$, $I(2)$ and $I(3)$ are illustrated in Fig. 6.9.

In this example, the original signals $I(1)$, $I(2)$ and $I(3)$ are characterized as the modified impulse responses of a 1-dimensional linear space respectively. The desirable modified impulse responses are attempted to be obtained by the CLS method. The model obtained by the CLS method is a (1,1,1)-dimensional pseudo linear system which has the same number of dimensions as the number of the original system.

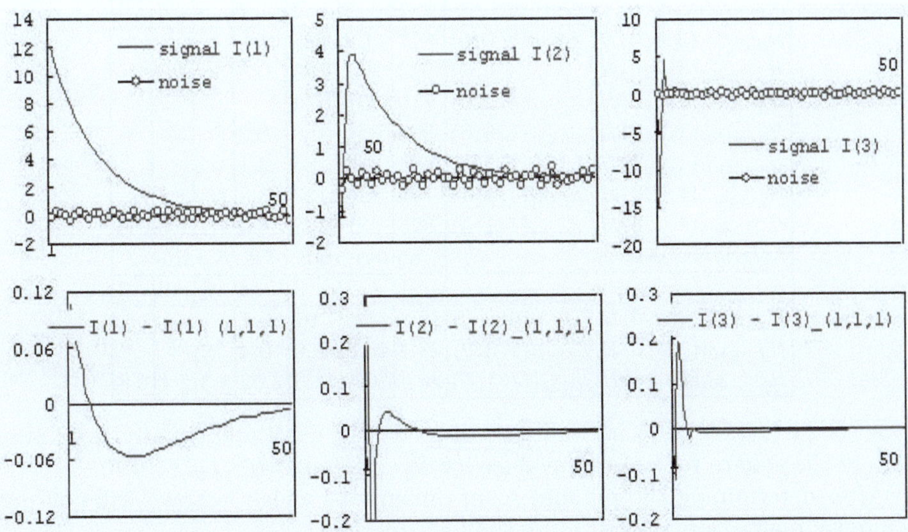

Fig. 6.9. In Example (6.32), the left are the original signal $I(1)$ with noise and the difference between $I(1)$ and the obtained signal $I(1)$_$(1,1,1)$. The middle are the original signal $I(2)$ with noise and the difference between $I(2)$ and the obtained signal $I(2)$_$(1,1,1)$. The right are the original signal $I(3)$ with noise and the difference between $I(3)$ and the obtained signal $I(3)$_$(1,1,1)$.

Just as we expected, the following table and Fig. 6.9 indicate that the model obtained by the CLS method is a good (1,1,1)-dimensional noisy realization system for the original (1,1,1)-dimensional system.

dimen-sion	ratio of matrices	mean values of square root for sum of			cosine	error ratio
		signal ①	signal by CLS ②	error ③	① and ② $\cos\theta$	③/①
I(1)₋(1,1,1)	0.03	0.55	0.55	0.005	0.9999	0.01
I(2)₋(1,1,1)	0.03	0.201	0.203	0.008	0.999	0.03
I(3)₋(1,1,1)	0.04	0.313	0.311	0.004	0.9999	0.01

Example 6.33. Let signals be the modified impulse responses of the following 4-dimensional pseudo linear system $\sigma = ((R^4, F), g, h, h^0)$,

where $F = \begin{bmatrix} 0.8 & 0.2 & 0 & 0 \\ 0 & 0.6 & 0 & 0.3 \\ 0 & 0 & -0.7 & 0.2 \\ 0 & 0 & 0 & 0.8 \end{bmatrix}$, $h = [12, -1, -15, 4]$, $g(u_1) = [1, 0, 0, 0]^T$,

$g(u_2) = [0, 0, 1, 0]^T$, $g(u_3) = [0, 0, 0, 1]^T$, $h^0 = 1$.

Let added noises be given in Fig. 6.10.

Then the noisy realization problem is solved by the following algorithm:

covariance matrix	eigenvalues						
	1	2	3	4	5	6	7
$H_{\underline{a}\,(3,40)}(3,0,0)H^T_{\underline{a}\,(3,40)}(3,0,0)$	817	1.2	0.8				
$H_{\underline{a}\,(4,40)}(4,0,0)H^T_{\underline{a}\,(4,40)}(4,0,0)$	922	1.2	1.03	0.5			
$H_{\underline{a}\,(3,40)}(1,2,0)H^T_{\underline{a}\,(3,40)}(1,2,0)$	710	337	0.4				
$H_{\underline{a}\,(4,40)}(1,3,0)H^T_{\underline{a}\,(4,40)}(1,3,0)$	803	344	0.8	0.2			
$H_{\underline{a}\,(5,40)}(1,1,3)H^T_{\underline{a}\,(5,40)}(1,1,3)$	694	393	124	3.1	1.4		
$H_{\underline{a}\,(6,40)}(1,1,4)H^T_{\underline{a}\,(6,40)}(1,1,4)$	771	424	127	3.8	1.5	1.1	
$H_{\underline{a}\,(7,40)}(1,1,5)H^T_{\underline{a}\,(7,40)}(1,1,5)$	875	431	128	4.1	1.7	1.3	0.6
covariance matrix	square root of eigenvalues						
	1	2	3	4	5	6	7
$H_{\underline{a}\,(4,40)}(4,0,0)H^T_{\underline{a}\,(4,40)}(4,0,0)$	30.3	1.1	1.01	0.7			
$H_{\underline{a}\,(4,40)}(1,3,0)H^T_{\underline{a}\,(4,40)}(1,3,0)$	28.3	18.5	0.9	0.4			
$H_{\underline{a}\,(7,40)}(1,1,5)H^T_{\underline{a}\,(7,40)}(1,1,5)$	29.6	20.7	11.3	2.02	1.3	1.1	0.8

1) A set $\{1.1,\ 1.01,\ 0.7\}$ is composed of relatively small and equally-sized numbers in the square root of eigenvalues for $H_{\underline{a}\,(4,40)}(4,0.0)H^T_{\underline{a}\,(4,40)}(4,0,0)$.
2) After determining the number n_1 of dimensions which is 1, we will continue the noisy realization algorithm by the CLS method.
Therefore, the modified impulse response $I(1)$ of a pseudo linear system obtained by the CLS method is constructed for a 1-dimensional space.
3) A set $\{0.9,\ 0.4\}$ is composed of relatively small and equally-sized numbers in the square root of eigenvalues for $H_{\underline{a}\,(4,40)}(1,3,0)H^T_{\underline{a}\,(4,40)}(1,3,0)$.
4) After determining the number n_2 of dimensions which is 1, we execute the noisy realization algorithm by the CLS method.

5) A set $\{1.3,\ 1.1,\ 0.8\}$ is composed of relatively small and equally-sized numbers in the square root of eigenvalues for $H_{\underline{a}\,(7,40)}(1,1,5)H_{\underline{a}\,(7,40)}^{T}(1,1,5)$.
6) After determining the number n_3 of dimensions which is 2, we execute the noisy realization algorithm by the CLS method.

Therefore, the modified impulse responses $I(1)$, $I(2)$ and $I(3)$ of an approximate pseudo linear system obtained by the CLS method is realized by a (1,1,2)-dimensional pseudo linear system.

The 4-dimensional pseudo linear system obtained by the CLS method is expressed as follows:

$$\sigma_n = ((\boldsymbol{R}^4,\ F_n),\ g_n,\ h_n,\ h^0),\ \text{where } F_n = \begin{bmatrix} 0.8 & 0 & 0 & 0.06 \\ 0 & -0.7 & 0 & -0.3 \\ 0 & 0 & 0 & -0.59 \\ 0 & 0 & 1 & 1.5 \end{bmatrix},\ g_n(u_1) = \boldsymbol{e}_1,$$

$g_n(u_2) = \boldsymbol{e}_2$, $g_n(u_3) = \boldsymbol{e}_3$, $h_n = [12.1,\ -15.1,\ 4.1,\ -0.04]$ and $h^0 = 1$.
The obtained modified impulse responses $I(1)$, $I(2)$ and $I(3)$ are illustrated in Fig. 6.10.

In this example, the original signals $I(1)$ and $I(2)$ are characterized as the modified impulse responses of a 1-dimensional linear space respectively. The original signal $I(3)$ is characterized as the modified impulse responses of a 2-dimensional linear space. The desirable modified impulse responses are

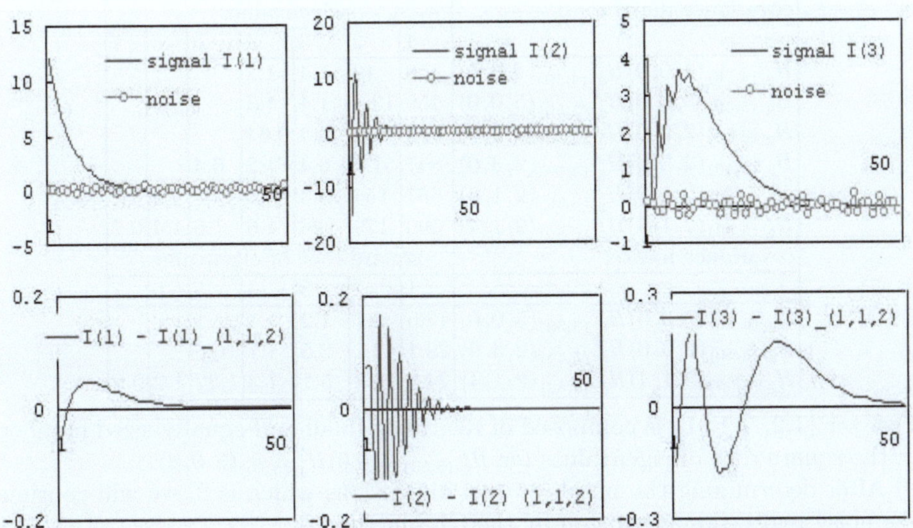

Fig. 6.10. In Example (6.33), the left are the original signal $I(1)$ with noise and the difference between $I(1)$ and the obtained signal $I(1)$_$(1,1,2)$. The middle are the original signal $I(2)$ with noise and the difference between $I(2)$ and the obtained signal $I(2)$_$(1,1,2)$. The right are the original signal $I(3)$ with noise and the difference between $I(3)$ and the obtained signal $I(3)$_$(1,1,2)$.

attempted to be obtained by the CLS method. The model obtained by the CLS method is a (1,1,2)-dimensional pseudo linear system which has the same number of dimensions as the number of the original system.

Just as we expected, the following table and Fig. 6.10 indicate that the model obtained by the CLS method is a good (1,1,2)-dimensional noisy realization system for the original (1,1,2)-dimensional system.

dimen-ion	ratio of matrices	mean values of square root for sum of			cosine ① and ②	error ratio
		signal ①	signal by CLS ②	error ③	$\cos\theta$	③/①
I(1)_(1,1,2)	0.04	0.4	0.399	0.003	0.9999	0.008
I(2)_(1,1,2)	0.03	0.42	0.417	0.008	0.999	0.02
I(3)_(1,1,2)	0.04	0.233	0.233	0.01	0.999	0.04

Example 6.34. Let signals be the modified impulse responses of the following 4-dimensional pseudo linear system $\sigma = ((R^4, F), g, h, h^0)$,

where $F = \begin{bmatrix} 0 & 0.2 & 0 & 0 \\ 1 & 0.6 & 0 & 0.3 \\ 0 & 0 & -0.7 & 0.2 \\ 0 & 0 & 0 & 0.8 \end{bmatrix}$, $h = [12,\ -1,\ -15,\ 6]$, $g(u_1) = [1,\ 0,\ 0,\ 0]^T$,

$g(u_2) = [0,\ 0,\ 1,\ 0]^T$, $g(u_3) = [0,\ 0,\ 0,\ 1]^T$, $h^0 = 1$.
Let added noises be given in Fig. 6.11.

Then the noisy realization problem is solved by the following algorithm:

covariance matrix	eigenvalues						
	1	2	3	4	5	6	7
$H_{\underline{a}\ (4,50)}(4,0,0)H_{\underline{a}\ (4,50)}^T(4,0,0)$	158	12	1.4	1.2			
$H_{\underline{a}\ (5,50)}(5,0,0)H_{\underline{a}\ (5,50)}^T(5,0,0)$	159	13.4	1.4	1.3	1		
$H_{\underline{a}\ (4,50)}(2,2,0)H_{\underline{a}\ (4,50)}^T(2,2,0)$	754	51	6.4	0.64			
$H_{\underline{a}\ (5,50)}(2,3,0)H_{\underline{a}\ (5,50)}^T(2,3,0)$	847	51.4	6.4	1.2	0.4		
$H_{\underline{a}\ (5,50)}(2,1,3)H_{\underline{a}\ (5,50)}^T(2,1,3)$	587	154	25.3	1,5	1.48	0.9	
$H_{\underline{a}\ (6,50)}(2,1,4)H_{\underline{a}\ (6,50)}^T(2,1,4)$	590	179	26	1.6	1.5	1.4	0.7
covariance matrix	square root of eigenvalues						
	1	2	3	4	5	6	7
$H_{\underline{a}\ (5,50)}(5,0,0)H_{\underline{a}\ (5,50)}^T(5,0,0)$	12.6	3.7	1.2	1.1	1		
$H_{\underline{a}\ (5,50)}(2,3,0)H_{\underline{a}\ (5,50)}^T(2,3,0)$	29.1	7.2	2.5	1.1	0.6		
$H_{\underline{a}\ (6,50)}(2,1,4)H_{\underline{a}\ (6,50)}^T(2,1,4)$	24.3	13.3	5.1	1.3	1.22	1.2	0.9

1) A set {1.2, 1.1, 1} is composed of relatively small and equally-sized numbers in the square root of eigenvalues for $H_{\underline{a}\ (5,50)}(5,0.0)H_{\underline{a}\ (5,50)}^T(5,0,0)$.
2) After determining the number n_1 of dimensions which is 2, we will continue the noisy realization algorithm by the CLS method.
Therefore, the modified impulse response $I(1)$ of a pseudo linear system obtained by the CLS method is constructed for a 2-dimensional space.
3) A set {1.1, 0.6} is composed of relatively small and equally-sized numbers in the square root of eigenvalues for $H_{\underline{a}\ (5,50)}(2,3,0)H_{\underline{a}\ (5,50)}^T(2,3,0)$.
4) After determining the number n_2 of dimensions which is 1, we execute the noisy realization algorithm by the CLS method.

5) A set $\{1.3,\ 1.22,\ 1.2,\ 0.9\}$ is composed of relatively small and equally-sized numbers in the square root of eigenvalues for $H_{\underline{a}\ (6,50)}(2,1,4)H_{\underline{a}\ (6,50)}^{T}(2,1,4)$.

6) After determining the number n_3 of dimensions which is 0, we execute the noisy realization algorithm by the CLS method. The noisy realization by the CLS method may not be so good. The reason is likely to be caused by rapid damping in Fig. 6.11.

Therefore, the modified impulse responses $I(1)$, $I(2)$ and $I(3)$ of an approximate pseudo linear system obtained by the CLS method is realized by a (2,1,0)-dimensional pseudo linear system.

The 4-dimensional pseudo linear system obtained by the CLS method is expressed as follows:

$$\sigma_n = ((\boldsymbol{R}^3,\ F_n),\ g_n,\ h_n,\ h^0),\ \text{where } F_n = \begin{bmatrix} 0 & 0.2 & 0.04 & \\ 1 & 0.62 & 0 & 2.72 \\ 0 & 0 & -0.67 & \end{bmatrix},$$

$g_n(u_1) = \mathbf{e}_1,\ g_n(u_2) = \mathbf{e}_3,\ g_n(u_3) = [0.57,\ 2.4,\ -0.1]^T,$
$h_n = [11.95,\ -1.12,\ -15.1]$ and $h^0 = 1.$

The obtained modified impulse responses $I(1)$, $I(2)$ and $I(3)$ are illustrated in Fig. 6.11.

In this example, the original signal $I(1)$ is characterized as the modified impulse response of a 2-dimensional linear space. The original signals $I(2)$ and

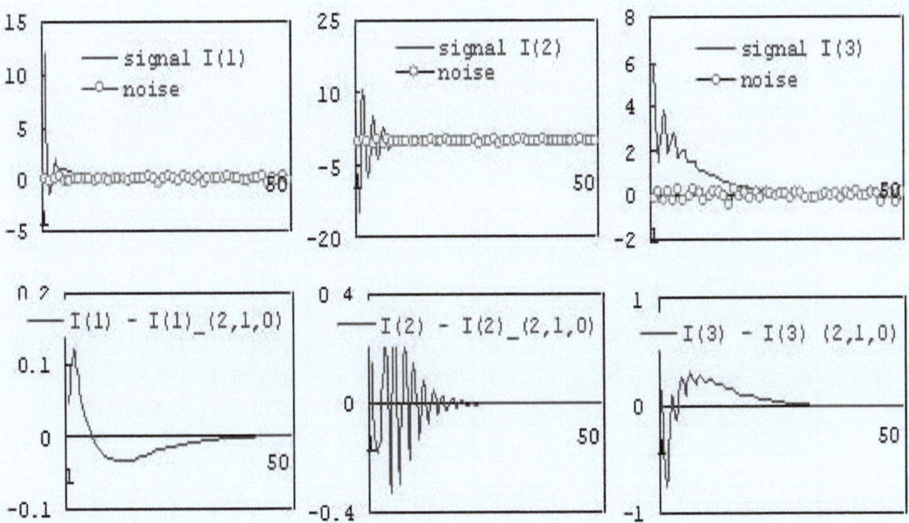

Fig. 6.11. In Example (6.34), the left are the original signal $I(1)$ with noise and the difference between $I(1)$ and the obtained signal $I(1)$_$(2,1,0)$. The middle are the original signal $I(2)$ with noise and the difference between $I(2)$ and the obtained signal $I(2)$_$(2,1,0)$. The right are the original signal $I(3)$ with noise and the difference between $I(3)$ and the obtained signal $I(3)$_$(2,1,0)$.

$I(3)$ are characterized as the modified impulse responses of a 1-dimensional linear space respectively. The desirable modified impulse responses are attempted to be obtained by the CLS method. The model obtained by the CLS method is a (2,1,0)-dimensional pseudo linear system which has a different number of dimensions from the number of the original system.

Just as we thought, the following table and Fig. 6.11 indicate that the model obtained by the CLS method is not such a good (2,1,0)-dimensional noisy realization system for the original (2,1,1)-dimensional system.

dimen-ion	ratio of matrices	mean values of square root for sum of		error	cosine ① and ②	error ratio
		signal ①	signal by CLS ②	error ③	$\cos\theta$	③/①
I(1)-(2,1,0)	0.12	0.246	0.245	0.004	0.9999	0.02
I(2)-(2,1,0)	0.03	0.42	0.419	0.01	0.9999	0.04
I(3)-(2,1,0)	0.04	0.181	0.174	0.02	0.994	0.11

Example 6.35. Let signals be the modified impulse responses of the following 5-dimensional socalled linear system $\sigma = ((R^5, F),\ g,\ h,\ h^0)$,

$$\text{where } F = \begin{bmatrix} 0 & 0.4 & 0 & 0 & 0.7 \\ 1 & 0.6 & 0 & 0 & 0.3 \\ 0 & 0 & 0 & -0.7 & 0.2 \\ 0 & 0 & 1 & 0 & 0.8 \\ 0 & 0 & 0 & 0 & -0.8 \end{bmatrix},\ h = [12,\ -1,\ -15,\ 4,\ -2],$$

$g(u_1) = [1,\ 0,\ 0,\ 0,\ 0]^T,\ g(u_2) = [0,\ 0,\ 1,\ 0,\ 0]^T,\ g(u_1) = [0,\ 0,\ 0,\ 0,\ 1]^T,$
$h^0 = 1.$

Let added noises be given in Fig. 6.12.

Then the noisy realization problem is solved by the following algorithm:

covariance matrix	eigenvalues							
	1	2	3	4	5	6	7	8
$H_{\underline{a}\ (4,50)}(4,0,0)H_{\underline{a}\ (4,50)}^T(4,0,0)$	1514	103	1.2	0.9				
$H_{\underline{a}\ (5,50)}(5,0,0)H_{\underline{a}\ (5,50)}^T(5,0,0)$	1882	105	1.3	1.04	0.8			
$H_{\underline{a}\ (6,50)}(2,4,0)H_{\underline{a}\ (6,50)}^T(2,4,0)$	892	651	364	53	2.4	0.3		
$H_{\underline{a}\ (7,50)}(2,5,0)H_{\underline{a}\ (7,50)}^T(2,5,0)$	964	693	365	53	2.5	1.4	0.1	
$H_{\underline{a}\ (7,50)}(2,2,3)H_{\underline{a}\ (7,50)}^T(2,2,3)$	1213	673	329	214	51	1	0.6	
$H_{\underline{a}\ (8,50)}(2,2,4)H_{\underline{a}\ (8,50)}^T(2,2,4)$	1296	676	398	276	51	1.8	0.8	0.4

covariance matrix	eigenvalues							
	1	2	3	4	5	6	7	8
$H_{\underline{a}\ (5,50)}(5,0,0)H_{\underline{a}\ (5,50)}^T(5,0,0)$	43.5	10.2	1.1	1	0.9			
$H_{\underline{a}\ (7,50)}(2,5,0)H_{\underline{a}\ (7,50)}^T(2,5,0)$	31	26.3	19.1	7.3	1.6	1.2	0.3	
$H_{\underline{a}\ (8,50)}(2,2,4)H_{\underline{a}\ (8,50)}^T(2,2,4)$	36	26	19.9	16.6	7.1	1.3	0.9	0.6

1) A set {1.1, 1, 0.9} is composed of relatively small and equally-sized numbers in the square root of eigenvalues for $H_{\underline{a}\ (5,50)}(5,0.0)H_{\underline{a}\ (5,50)}^T(5,0,0)$.
2) After determining the number n_1 of dimensions which is 2, we will continue the noisy realization algorithm by the CLS method.

Therefore, the modified impulse response $I(1)$ of a pseudo linear system obtained by the CLS method is constructed for a 2-dimensional space.

3) A set $\{1.6,\ 1.2,\ 0.3\}$ is composed of relatively small and equally-sized numbers in the square root of eigenvalues for $H_{\underline{a}\,(7,50)}(2,5,0)H_{\underline{a}\,(7,50)}^{T}(2,5,0)$.

4) After determining the number n_2 of dimensions which is 2, we execute the noisy realization algorithm by the CLS method.

5) A set $\{1.3,\ 0.9,\ 0.6\}$ is composed of relatively small and equally-sized numbers in the square root of eigenvalues for $H_{\underline{a}\,(8,50)}(2,2,4)H_{\underline{a}\,(8,50)}^{T}(2,2,4)$.

6) After determining the number n_3 of dimensions which is 1, we execute the noisy realization algorithm by the CLS method.

Therefore, the modified impulse responses $I(1)$, $I(2)$ and $I(3)$ of an approximate pseudo linear system obtained by the CLS method is realized by a (2,2,1)-dimensional pseudo linear system.

The 5-dimensional pseudo linear system obtained by the CLS method may be expressed as follows:

$$\sigma_n = ((\boldsymbol{R}^5,\ F_n),\ g_n,\ h_n,\ h^0),\ \text{where } F_n = \begin{bmatrix} 0 & 0.4 & 0 & -0.02 & 0.7 \\ 1 & 0.6 & 0 & 0.02 & 0.3 \\ 0 & 0 & 0 & -0.7 & 0.2 \\ 0 & 0 & 1 & 0 & 0.8 \\ 0 & 0 & 0 & 0 & -0.8 \end{bmatrix},$$

$g_n(u_1) = \mathbf{e}_1$, $g_n(u_2) = \mathbf{e}_3, g_n(u_3) = \mathbf{e}_5,$, $h_n = [11.9,\ -0.86,\ -15.3,\ 4.9,\ -2.2]$ and $h^0 = 1$.

The obtained modified impulse responses $I(1)$, $I(2)$ and $I(3)$ are illustrated in Fig. 6.12.

In this example, the original signals $I(1)$ and $I(2)$ are characterized as the modified impulse responses of a 2-dimensional linear space respectively. The original signal $I(3)$ is characterized as the modified impulse responses of a 1-dimensional linear space. The desirable modified impulse responses are attempted to be obtained by the CLS method. The model obtained by the CLS method is a (2,2,1)-dimensional pseudo linear system which has the same number of dimensions as the number of the original system.

Just as we expected, the following table and Fig. 6.12 indicate that the model obtained by the CLS method is a good (2,2,1)-dimensional system for the original (2,2,1)-dimensional system.

dimen- ion	ratio of matrices	mean values of square root for sum of			cosine	error ratio
		signal	signal by CLS	error	① and ②	
		①	②	③	$\cos\theta$	③/①
$I(1)$_(2,2,1)	0.04	0.45	0.448	0.004	0.9999	0.01
$I(2)$_(2,2,1)	0.06	0.435	0.439	0.02	0.999	0.04
$I(3)$_(2,2,1)	0.02	0.355	0.367	0.02	0.999	0.05

Example 6.36. Let signals be the modified impulse responses of the following 6-dimensional pseudo linear system $\sigma = ((R^6, F),\ g,\ h,\ h^0)$,

where $F = \begin{bmatrix} 0 & -0.3 & 0 & 0.2 & 0 & 0.1 \\ 1 & 0.7 & 0 & 0.3 & 0 & -0.3 \\ 0 & 0 & 0 & 0.1 & 0 & -0.2 \\ 0 & 0 & 1 & -0.5 & 0 & 0.5 \\ 0 & 0 & 0 & 0 & 0 & -0.8 \\ 0 & 0 & 0 & 0 & 1 & -0.7 \end{bmatrix}$, $h = [12, -1, -15, 4, -8, 4]$,

$g(u_1) = [1, 0, 0, 0, 0, 0]^T$, $g(u_2) = [0, 0, 1, 0, 0, 0]^T$,
$g(u_3) = [0, 0, 0, 0, 1, 0]^T$, $h^0 = 1$.

Let added noises be given in Fig. 6.13.

Then the noisy realization problem is solved as follows:

covariance matrix	eigenvalues								
	1	2	3	4	5	6	7	8	9
$H_{\underline{a}\ (4,35)}(4,0,0)H^T_{\underline{a}\ (4,35)}(4,0,0)$	193	41	1.3	0.7					
$H_{\underline{a}\ (5,35)}(5,0,0)H^T_{\underline{a}\ (5,35)}(5,0,0)$	194	41	1.3	0.96	0.7				
$H_{\underline{a}\ (6,35)}(2,4,0)H^T_{\underline{a}\ (6,35)}(2,4,0)$	405	37.7	30	4	1.2	0.4			
$H_{\underline{a}\ (7,35)}(2,5,0)H^T_{\underline{a}\ (7,35)}(2,5,0)$	409	38	30	4	1.2	0.8	0.2		
$H_{\underline{a}\ (8,35)}(2,2,4)H^T_{\underline{a}\ (8,35)}(2,2,4)$	1199	600	266	35	20	2.3	1	0.2	
$H_{\underline{a}\ (9,35)}(2,2,5)H^T_{\underline{a}\ (9,35)}(2,2,5)$	1200	799	269	34	20	3	1	0.4	0.1
covariance matrix	square root of eigenvalues								
	1	2	3	4	5	6	7	8	9
$H_{\underline{a}\ (5,35)}(5,0,0)H^T_{\underline{a}\ (5,35)}(5,0,0)$	13.9	6.4	1.1	0.97	0.8				
$H_{\underline{a}\ (7,35)}(2,5,0)H^T_{\underline{a}\ (6,35)}(2,5,0)$	20	6.2	5.5	2	1.1	0.9	0.4		
$H_{\underline{a}\ (9,35)}(2,2,5)H^T_{\underline{a}\ (9,35)}(2,2,5)$	34.6	28.3	16.4	5.8	4.5	1.7	1	0.6	0.3

1) A set $\{1.1, 0.97, 0.8\}$ is composed of relatively small and equally-sized numbers in the square root of eigenvalues for $H_{\underline{a}\ (5,35)}(5,0.0)H^T_{\underline{a}\ (5,35)}(5,0,0)$.
2) After determining the number n_1 of dimensions which is 2, we will continue the noisy realization algorithm by the CLS method.
Therefore, the modified impulse response $I(1)$ of a pseudo linear system obtained by the CLS method is constructed for a 2-dimensional space.
3) A set $\{1.1, 0.9, 0.4\}$ may be composed of relatively small and equally-sized numbers in the square root of eigenvalues for $H_{\underline{a}\ (7,35)}(2,5,0)H^T_{\underline{a}\ (7,35)}(2,5,0)$.
4) After determining the number n_2 of dimensions which is 2, we execute the noisy realization algorithm by the CLS method.
5) A set $\{1, 0.6, 0.5\}$ may be composed of relatively small and equally-sized numbers in the square root of eigenvalues for $H_{\underline{a}\ (9,35)}(2,2,5)H^T_{\underline{a}\ (9,35)}(2,2,5)$.
6) After determining the number n_3 of dimensions which is 2, we execute the noisy realization algorithm by the CLS method.

Therefore, the modified impulse responses $I(1)$, $I(2)$ and $I(3)$ of an approximate pseudo linear system obtained by the CLS method is realized by a (2,2,2)-dimensional pseudo linear system.

The 6-dimensional pseudo linear system obtained by the CLS method is expressed as follows:

We list the following 6-dimensional pseudo linear system σ_n

$$= ((\boldsymbol{R}^6,\ F_n),\ g_n,\ h_n,\ h^0),\ \text{where}\ F_n = \begin{bmatrix} 0 & -0.32 & 0 & 0.23 & 0 & 0.09 \\ 1 & 0.68 & 0 & 0.35 & 0 & -0.35 \\ 0 & 0 & 0 & 0.06 & 0 & -0.23 \\ 0 & 0 & 1 & -0.71 & 0 & 0.36 \\ 0 & 0 & 0 & 0 & 0 & -0.8 \\ 0 & 0 & 0 & 0 & 1 & -0.7 \end{bmatrix},$$

$g_n(u_1) = \mathbf{e}_1, g_n(u_2) = \mathbf{e}_3, g_n(u_3) = \mathbf{e}_5, h_n = [11.9, -0.91, -14.8, 4.2, -8.1, 3.9]$ and $h^0 = 1$.

In this example, the original signal $I(1)$, $I(2)$ and $I(3)$ are characterized as the modified impulse responses of a 2-dimensional linear space respectively. The desirable modified impulse responses are attempted to be obtained by the CLS method. The model obtained by the CLS method is a (2,2,2)-dimensional pseudo linear system which has the same number of dimensions as the number of the original system.

Just as we expected, the following table and Fig. 6.13 indicate that the model obtained by the CLS method is a good (2,2,2)-dimensional noisy realization system for the original (2,2,2)-dimensional system.

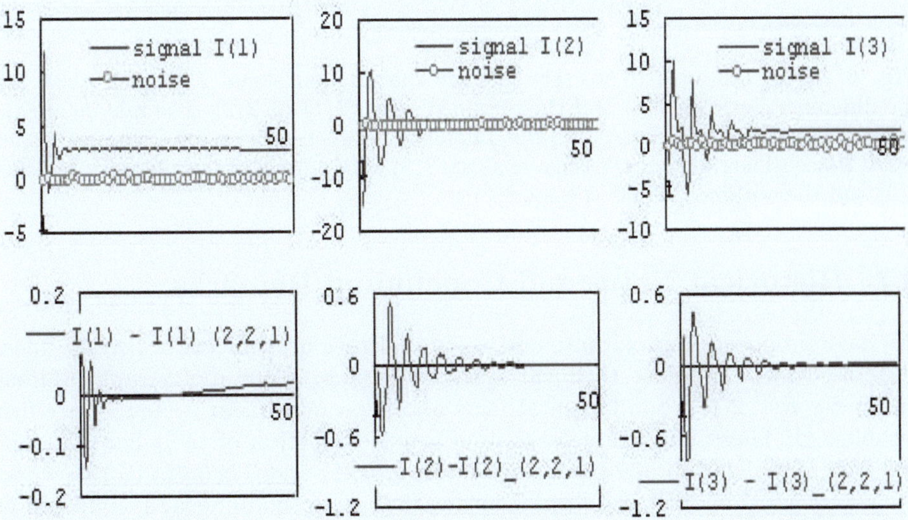

Fig. 6.12. In Example (6.35), the left are the original signal $I(1)$ with noise and the difference between $I(1)$ and the obtained signal $I(1)\_(2,2,1)$. The middle are the original signal $I(2)$ with noise and the difference between $I(2)$ and the obtained signal $I(2)\_(2,2,1)$. The right are the original signal $I(3)$ with noise and the difference between $I(3)$ and the obtained signal $I(3)\_(2,2,1)$.

dimen-sion	ratio of matrices	mean values of square root for sum of			cosine ① and ②	error ratio
		signal ①	signal by CLS ②	error ③	$\cos\theta$	③/①
I(1)₋(2,2,2)	0.14	0.262	0.262	0.006	0.999	0.02
I(2)₋(2,2,2)	0.04	0.3136	0.313	0.02	0.999	0.05
I(3)₋(2,2,2)	0.02	0.445	0.448	0.02	0.999	0.04

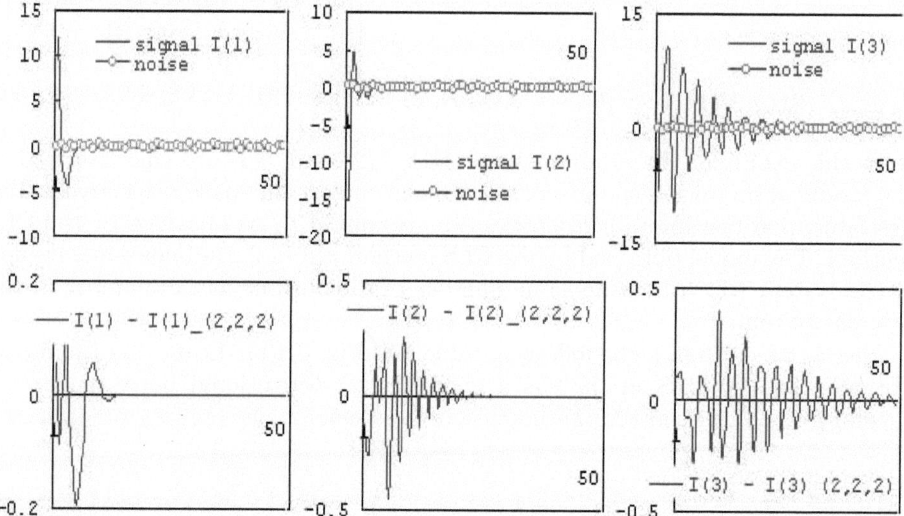

Fig. 6.13. In Example (6.36), the left are the original signal $I(1)$ with noise and the difference between $I(1)$ and the obtained signal $I(1)_-(2,2,2)$. The middle are the original signal $I(2)$ with noise and the difference between $I(2)$ and the obtained signal $I(2)_-(2,2,2)$. The right are the original signal $I(3)$ with noise and the difference between $I(3)$ and the obtained signal $I(3)_-(2,2,2)$.

6.7   Historical Notes and Concluding Remarks

We have proposed approximate and noisy realization problems of pseudo linear systems, which are close to linear systems. In a previous monograph [Matsuo and hasegawa, 2003], fundamental facts about pseudo linear systems were first established. These important facts were a representation of their behavior and the partial realization algorithm in Definition (6.1) and Theorem (6.19), where the representation of their behavior means that any pseudo linear system can be completely characterized by the modified impulse responses. The approximate realization problem was attempted to solve by presenting an approximate realization algorithm. The algorithm is made up of a ratio of input/output matrix norm and the CLS method, i.e., the constrained least square method. Through the introduction of the ratio of matrix norm which is the square norm, we can decrease the dimensional number of state spaces while considering information loss in mind. By using the CLS method, we can make full efforts to characterize the relation of a linear combination.

By applying this algorithm to several examples of pseudo linear systems, we have shown that this algorithm is practical and useful. In the case that the ratio of input/output matrix norm is within some percent, we have shown that this approximate realization algorithm produces good results with the exception of pseudo linear systems whose modified impulse responses have small values and rapid damping as in Examples (6.24) and (6.25). Our several examples show that the changing relations among the ratio of matrix norm and the error to signal ratio are proportional relations and the ratio is 0.01 for the Input/output matrix norm while the error to signal ranges from 0.02 to 0.06. This approximate realization algorithm appears to be very promising.

We treated noisy realization problems and attempted to solve them by presenting a noisy realization algorithm. The algorithm is composed by making a set of singular values of a matrix and applying the CLS method, i.e., the constrained least square method. By producing a set of singular values of a matrix, we can determine the dimensional number of state spaces by drawing a distinction between a noiseless part and a noisy part in the given signal. By using the CLS method, we can make full efforts to characterize a relation of a linear combination in the noiseless part.

By applying this algorithm to several examples of pseudo linear systems, we have shown that this algorithm is practical and useful. In the case that we can make a set composed of relatively small and equally-sized numbers in the square root of eigenvalues for an Input/output matrix, we have shown that this noisy realization algorithm produces good results with the exception of pseudo linear systems whose modified impulse responses have small values and rapid damping as in Examples (6.34). Our several examples in noisy realizations show that the changing relations among the ratio of matrix norm and the error to signal ratio are proportional relations and the ratio is 0.01 for the Input/output matrix norm while the error to signal ratio ranges from 0.002 to 0.03. This noisy realization algorithm also appears to be very promising.

As we mentioned before, concrete discussions of approximate and noisy realization for non linear systems are very new.

7 Approximate and Noisy Realization of Affine Dynamical Systems

In this chapter, we will discuss approximate and noisy realization problems of affine dynamical systems, which realize any input response map, equivalently, as an input/output map with causality. Affine dynamical systems were proposed and the realization problem of the systems were solved in the reference [Matsuo & Hasegawa, 2003]. We characterized the finite-dimensionality of affine dynamical systems. We obtained the same results as ones established in linear system theory.

A criterion for canonical finite-dimensional affine dynamical systems was given. There uniquely exists a quasi-reachable standard system in the isomorphic class of finite-dimensional canonical affine dynamical systems. We obtained a criterion for the behavior of finite-dimensional affine dynamical systems. We also gave a procedure on how to obtain the quasi-reachable standard system from an input response map.

For our discussion of approximate and noisy realization problems of affine dynamical systems, we will need a partial realization algorithm of the systems which was not stated in the reference. We will obtain a partial realization algorithm in section 7.3. The proofs about the results of the algorithm will be stated in the appendix of this chapter.

7.1 Basic Facts about Affine Dynamical Systems

Definition 7.1. *Affine Dynamical Systems*
1) A system given by the following system equation is written as a collection
$\sigma = ((X, F), g, h, h^0)$ and it is said to be an affine dynamical system.

$$\begin{cases} x(t+1) = F(\omega(t+1))x(t) + g(\omega(t+1)) \\ x(0) \quad = 0 \\ \gamma(t) \quad = h^0 + hx(t) \end{cases}$$

for any $t \in N$, $x(t) \in X$, $\gamma(t) \in Y$, where X is a linear space over the field
R that may be called a state space, F is a map $F : U \rightarrow L(X); u \mapsto F(u)$,
a map $g : U \rightarrow X$, a linear map $h : X \rightarrow Y$ and $h^0 \in Y$.

Y. Hasegawa: Approxi. & Noisy Reali. of Discrete-Time Dyn. Sys., LNCIS 376, pp. 165–204, 2008.
springerlink.com

2) *The input response map* $a_\sigma : U^* \to Y ; \omega \mapsto a_\sigma(\omega)$
$= h^0 + h(\sum_{j=1}^{|\omega|}(F(\omega(|\omega|))F(\omega(|\omega|-1))\cdots F(\omega(|\omega|-j))g(\omega(j))))$ *is said to be a behavior of* σ.
An affine dynamical system σ *which satisfies* $a_\sigma = a$ *is said to be a realization of an input response map* a.

3) *An affine dynamical system* σ *is said to be quasi-reachable if the linear hull of the reachable set* $\{\sum_{j=1}^{|\omega|}(F(\omega(|\omega|))F(\omega(|\omega|-1))\cdots F(\omega(|\omega|-j))g(\omega(j));$
$\omega \in U^*\}$ *is equal to* X *and an affine dynamical system* σ *is said to be distinguishable if* $h(F(\omega(|\omega|))F(\omega(|\omega|-1))\cdots F(\omega(|\omega|-j))x_1$
$= h(F(\omega(|\omega|))F(\omega(|\omega|-1))\cdots F(\omega(|\omega|-j))x_2$ *implies* $x_1 = x_2$ *for any* $\omega \in U^*$.

4) *An affine dynamical system* σ *is said to be canonical if* σ *is quasi-reachable and distinguishable.*

Remark 1: It is meant for σ to be a faithful model for the input response map a such that σ realizes a.

Remark 2: Notice that a canonical affine dynamical system:
$\sigma = ((X, F), g, h, h^0)$ is a system which has the most reduced state set X among systems that have the behavior a_σ.

In order to show intuitively that affine dynamical systems are general dynamical systems, we will state a relation between affine dynamical systems and inhomogeneous bilinear systems.

We will consider the following dynamical system:

$$\begin{cases} x(t+1) = (A + \sum_{i=1}^{m} \mathbf{N}_i \cdot \omega_i(t+1)x(t) + \sum_{i=1}^{m} \bar{g} \cdot \omega_i(t+1) \\ x(0) \quad = 0 \\ \gamma(t) \quad = h^0 + hx(t) \end{cases}$$

$\omega_i(t) \in \mathbf{R}$, $x(t)$, $\bar{g} \in \mathbf{R}^n$, $\mathbf{A}$, $\mathbf{N}_i \in \mathbf{R}^{n \times n}$ and $\gamma(t) \in \mathbf{R}$.
Let $F(\omega(t+1)) = \mathbf{A} + \sum_{i=1}^{m} \omega_i(t+1)\mathbf{N}_i$, $g(\omega(t+1)) = \sum_{i=1}^{m} \bar{g}\omega_i(t+1)$.
Then the above dynamical system is an affine dynamical system. Therefore, the inhomogeneous bilinear system is an example of our affine dynamical systems.

Definition 7.2. *Let* $\sigma_1 = ((X_1, F_1), g_1, h_1, h^0)$ *and* $\sigma_2 = ((X_2, F_2), g_2, h_2, h^0)$ *be affine dynamical systems. Then a linear operator* $T : X_1 \to X_2$ *is called an affine dynamical system morphism* $T : \sigma_1 \to \sigma_2$ *if* T *is a linear map* $: X_1 \to X_2$ *that satisfies* $fF_1(u) = F_2(u)f$, $fg_1 = g_2$ *and* $h_1 = h_2f$.
A bijective affine dynamical system morphism $T : \sigma_1 \to \sigma_2$ *is called an isomorphism.*

Corollary 7.3. *Let* σ_1 *and* σ_2 *be affine dynamical systems and* $T : \sigma_1 \to \sigma_2$ *be an affine dynamical system morphism. Then* $a_{\sigma_1} = a_{\sigma_2}$ *holds.*

Example 7.4. Let $U^+ := U \setminus 1$ and $V(U^+) := \{\lambda = \sum_{\omega \in U^+} \lambda(\omega)\mathbf{e}_\omega (\text{finite sum});$
$\lambda(\omega) \in \mathbf{R}\}$, where $\mathbf{e}_\omega(\bar{\omega}) = 1$ for $\omega = \bar{\omega}$ and $\mathbf{e}_\omega(\bar{\omega}) = 0$ for $\omega \neq \omega$.

Let ψ be a map : $U \to L(V(U^+)); u \mapsto \psi(u)[; \mathbf{e}_\omega \mapsto \mathbf{e}_{u|\omega} - \mathbf{e}_u]$.

And let a map $e : U \to V(U^+); u \mapsto \mathbf{e_u}$, where $e(1) = 0$. And we consider a linear map $a_l : V(U^+) \to Y; \mathbf{e}_\omega \mapsto a(\omega) - a(1)$ for any input response map $a \in F(U^*, Y)$. Then $((V(U^+), \psi), e, a_l, a(1))$ is a quasi-reachable affine dynamical system that realizes $a \in F(U^*, Y)$.

Example 7.5. Let $a \in F(U^*, Y)$ be any input response map and S_l be defined by $S_l(u)a : U^* \to Y; \omega \mapsto a(\omega|u)$. Then $S_l(u) \in L(F(U^*, Y)$ for any $u \in U$. Let a map $\xi : U \to F(U^*, Y)$ be $u \mapsto \xi(u)[; \omega \mapsto a(\omega|u) - a(\omega)]$. And let 1 be a linear map : $F(U^*, Y) \to Y; a \mapsto a(1)$. Then $((F(U^*, Y), S_l), \xi, 1, a(1))$ is a distinguishable affine dynamical system that realizes $a \in F(U^*, Y)$.

Remark: Examples (7.4) and (7.5) imply that there exist many affine dynamical systems that realize a given input response map $a \in F(U^*, Y)$. However, there is no relation between them. Therefore, we introduce canonical affine dynamical systems, and we will make a clear relation between them.

Theorem 7.6. *For any input response $a \in F(U^*, Y)$, there exist the following two canonical affine dynamical systems that realize it.*
1) $((V(U^+)/_{=a}, \tilde{\psi}), \tilde{e}, \tilde{a}_l, a(1))$,
  where $V(U^+)/_{=a}$ is a quotient space derived by equivalence relation:
  $\sum_\omega \lambda(\omega)\mathbf{e}_\omega = \sum_{\bar\omega} \lambda(\bar\omega)\mathbf{e}_{\bar\omega} \Longleftrightarrow$
  $\quad \sum_\omega \lambda(\omega)(a(\omega) - a(1)) = \sum_{\bar\omega} \lambda(\bar\omega)(a(\bar\omega) - a(1))$,
  $\tilde{\psi}$ is given by a map :$U \to L(V(U^+)/_{=a}); u \mapsto \tilde{\psi}(u)[; \lambda \mapsto$
  $\sum_\omega \lambda(\omega)(\mathbf{e}_{u|\omega} - \mathbf{e_u})$, *$\tilde{e}$ is given by $\tilde{e} : U \to V(U^+)/_{=a}; u \mapsto [e_u]$ and*
  $\tilde{a}_l$ is given by $\tilde{a}_l : V(U^+)/_{=a} \to Y; [\lambda] \mapsto \tilde{a}_l([\lambda]) = \sum_\omega \lambda(\omega)(a(\omega) - a(1))$.
2) $((\ll S_l(U^*)a - a \gg, S_l), \xi, 1, a(1))$,
  where $S_l(U^*)a - a = \{S_l(\omega)a - a; \omega \in U^*\}$ and $\ll S_l(U^*)a - a \gg$ denotes the smallest linear space which contains $S_l(U^*)a - a$.

We conclude that there exists a canonical affine dynamical system that realize any input response map in Theorem (7.6). Next, we will insist on the uniqueness of the systems that have the same behavior.

Theorem 7.7. *Realization Theorem*
For any input response map $a \in F(U^*, Y)$, there exist at least two canonical affine dynamical systems that realize it.
  Let $\sigma_1 = ((X_1, F_1), g_1, h_1, h^0)$ and $\sigma_2 = ((X_2, F_2), g_2, h_2, h^0)$ be canonical affine dynamical systems that realize any $a \in F(U^*, Y)$, then there exists a unique isomorphism $T : \sigma_1 \to \sigma_2$.

7.2 Finite dimensional Affine Dynamical Systems

Based on Realization Theorem (7.7), we clarified the finite-dimensionality of the systems. Therefore, we obtained the same results as obtained in the linear systems by R. E. Kalman.

As previously described, we introduce finite dimensional affine dynamical systems needed for our approximate and noisy realization problems.

Firstly, we assume that the set U of input's values is finite, and we show that the assumption of finiteness is not so special. Namely, affine dynamical systems with an assumption include biaffine systems as a subclass. Biaffine systems were discussed by Tarn and Nonoyama [1979].

The following results were obtained for the systems. It is given as a criterion for canonical finite dimensional affine dynamical systems. We give a criterion for the behavior of finite dimensional affine dynamical systems. The companion form for canonical finite-dimensional affine dynamical systems is also given. Moreover, a procedure to obtain the companion form from a given input/output map is obtained.

Therefore, it is obvious that the theory of these affine dynamical systems is the extension of the linear system theory established by Kalman et al for the non-linear case.

An affine dynamical system is different from a state-affine system in [Sontag, 1979a]. Our system is introduced on the basis of Theorem (2.6) and Definition (2.7) in [Matsuo and Hasegawa 2003], which is the representation theorem for any input/output map with causality. Hence, our systems are more general than state-affine systems.

If the state space X of an affine dynamical system $\sigma = ((X, F), g, h, h^0)$ is finite dimensional (n-dimensional), then σ is said to be a finite dimensional (n-dimensional) affine dynamical system.

There is the following fact about n dimensional linear space in [Halmos, 1958]. Fact: [Every n dimensional linear space over the field $\boldsymbol{R}$ is isomorphic to $\boldsymbol{R}^n$. Moreover, every linear operator from $\boldsymbol{R}^n$ to $\boldsymbol{R}^m$ is isomorphic to a matrix $F \in \boldsymbol{R}^{m \times n}$.]

Therefore, without loss of generality, a n dimensional affine dynamicai system can be represented by $\sigma = ((X, F), g, h, h^0)$,

where, F is a map : $U \rightarrow \boldsymbol{R}^{n \times n}$, g is a map : $U \rightarrow \boldsymbol{R}^n$ and $h \in \boldsymbol{R}^{p \times n}$ and $h^0 \in \boldsymbol{R}^p$.

According to the above discussion, we can treat an n-dimensional affine dynamical system $\sigma = ((X, F), g, h, h^0)$ which is easily embodied by computer programs or electrical circuits.

From now on, we assume that the set U of input's values is finite. Let $U = \{u_1, u_2, \cdots, u_m\}$. Now, we show that the assumption is not so special.

Biaffine Systems 7.8

We will consider the following system:

$$\begin{cases} x(t+1) = (A + \sum_{i=1}^{m} N_i \cdot \omega_i(t+1))x(t) + \sum_{i=1}^{m} \mathbf{b}_i \cdot \omega_i(t+1) + \mathbf{a} \\ x(0) \quad = 0 \\ \gamma(t) \quad = h^0 + hx(t) \end{cases}$$

$\omega_i(t) \in \mathbf{R}$, $x(t)$, $\mathbf{b}_i$ and $\mathbf{a} \in \mathbf{R}^n$, $N_i \in \mathbf{R}^{n \times n}$ and $\gamma(t) \in Y$.

Transferring time in input, we will conclude that the above system is a biaffine system as treated in [Tarn and Nonoyama, 1979],
where maps $\tilde{F} : \mathbf{R}^m \to \mathbf{R}^{n \times n}$ and $g : \mathbf{R}^m \to \mathbf{R}^{n \times n}$ are affine, namely,
$\tilde{F}(\sum_{i=1}^{m} \omega_i(t+1)\mathbf{e_i}) = A + \sum_{i=1}^{m} N_i \omega_i(t+1)$,
$\tilde{g}((\sum_{i=1}^{m} \omega_i(t+1)\mathbf{e_i}) = \mathbf{a} + \sum_{i=1}^{m} \mathbf{b}_i \omega_i(t+1)$.
Then we can obtain an affine dynamical system $\sigma = ((R^n, F), g, h, h^0)$,
where F and g are given by the following relations:
$F(0) = A, \ F(\mathbf{e_i}) = A + N_i (1 \leq i \leq m)$,
$g(0) = \mathbf{a}, \ g(\mathbf{e_i}) = \mathbf{a} + b_i (1 \leq i \leq m)$.

And U is given by $U = \{\mathbf{0}, \mathbf{e_1}, \mathbf{e_2}, \cdots, \mathbf{e_m}\}$ and $e_i = [0, 0, \cdots, 0, \overset{i}{1}, 0, \cdots, 0]^T$, where T denotes the transpose.

Therefore, we can conclude that the assumption for the set U to be finite is not so special.

Proposition 7.9. *Let $\sigma = ((\mathbf{R}^n, F), g, h, h^0)$ be an affine dynamical system. σ is canonical if and only if*
1) rank $[(g(u_1), g(u_2), \cdots, g(u_m), F(u_1)g(u_1), \cdots, F(u_1)g(u_m), \cdots,$
$F^{n-1}(u_m)g(u_1), \cdots, F^{n-1}(u_m)g(u_m)] = n$.
2) rank $[h^T, (hF(u_1)^T, (hF(u_2)^T, \cdots, (hF(u_m))^T, \cdots,$
$(hF^2(u_1))^T, \cdots, (hF^2(u_m))^T, (hF^{n-1}(u_1)g(u_m))^T, \cdots, (hF^{n-1}(u_m)$
$g(u_m))^T] = n$.

Definition 7.10. *Let the input value's set U be $U := \{u_i; 1 \leq i \leq m\}$ and let a map $\| \ \| : U \to N$ be $u_i \mapsto \|u_i\| = i$. And let a numerical value $\|\omega\|$ of an input $\omega \in U^*$ be $\|\omega\| = \|\omega(|\omega|)\| + \|\omega(|\omega| - 1)\| \times m + \cdots + \|\omega(1)\| \times m^{|\omega|-1}$ and $\|1\| = 0$.*
Then we can define a totally ordered relation by this numerical value in U^*.
Namely, $\omega_1 \leq \omega_2 \iff \|\omega_1\| \leq \|\omega_2\|$.

Definition 7.11. *Let $\sigma_s = ((\mathbf{R}^n, F_s), g_s, h_s, h^0)$ be a canonical affine dynamical system. If input sequences $\{\omega_i \in U^*; 1 \leq i \leq n\}$ satisfy the following conditions, then σ_s is said to be a quasi-reachable standard system.*
1) $\mathbf{e_i} = \sum_{j=1}^{i} F_s(\omega_j(|\omega_j|))F_s(\omega_j(|\omega_j| - 1)F_s(\omega_j(|\omega_j| - j)g_s(\omega_j(j))$
2) $1 = \omega_1 < \omega_2 < \cdots < \omega_n$ and $|\omega_i| \leq i - 1$ for $i(1 \leq i \leq n)$ hold.
3) $\sum_{j=1}^{|\omega|} F_s(\omega(|\omega|))F_s(\omega(|\omega| - 1)F_s(\omega(|\omega| - j)g_s(\omega(j)) = \sum_{i=1}^{j} \alpha_i \mathbf{e_i}, \alpha_i \in \mathbf{R}$ holds
for any input sequence $\omega \in U^*$ such that $\omega_j < \omega < \omega_{j+1}(1 \leq i \leq n - 1)$.

Theorem 7.12. *For any canonical affine dynamical system $\sigma = ((\mathbf{R}^n, F), g, h, h^0)$, there exists a unique quasi-reachable standard system $\sigma_s = ((\mathbf{R}^n, F_s), g_s, h_s, h^0)$ which is isomorphic to it.*

Definition 7.13. *For any input response map $a \in F(U^*, Y)$, there uniquely exists a linear operator $A : V(U^+) \to F(U^*, Y)$ such that A satisfies $S_l(u)A = A\psi(u)$ for any $u \in U$. Hence, $A(e_\omega)(\bar{\omega}) = a(\bar{\omega}|\omega) - a(\bar{\omega})$ holds for any $\omega, \bar{\omega} \in U^*$.*
Therefore, for any $\omega, \bar{\omega} \in U^*$, we can consider the following infinite matrix H_a^A.
The H_a^A is called a Hankel matrix of a. The column vector of H_a^A may be written by $S_l(\omega)a - a$.

$$H_a^A = \begin{array}{c} \\ \\ \overline{\omega} \end{array} \left(\begin{array}{ccc} & & \overset{\omega}{\vdots} \\ & & \vdots \\ & & \vdots \\ \cdots & \cdots & a(\overline{\omega}\,|\omega) - a(\overline{\omega}) \end{array} \right)$$

Theorem 7.14. *Theorem for existence criterion*
For an input response map $a \in F(U^*, Y)$, the following conditions are equivalent:
1) a is a behavior of an n-dimensional canonical affine dynamical system.
2) $\{S_l(\omega)a - a : \omega \in U^*\}$ have n linearly independent vectors.
3) rank of H_a^A is n,
where $S_l(\omega)a - a \in F(U^*, Y)$ is defined by $S_l(\omega)a - a : U^* \to Y; \overline{\omega} \mapsto a(\overline{\omega}|\omega) - a(\overline{\omega})$.

Theorem 7.15. *Theorem for a realization procedure*
Let an input response map $a \in F(U^*, Y)$ satisfy the condition of Theorem (7.14).
Then the quasi-reachable standard system $\sigma_s = ((\mathbf{R}^n, F_s), g_s, h_s, h^0)$ which real-
izes it can be obtained by the following procedure:
1) Select n linearly independent vectors $\{S_l(\omega_i)a - a : (1 \leq i \leq n)\}$ from
$\{S_l(\omega)a - a : \omega \in U^*, |\omega| \leq n - 1\}$ in order of the numerical value of U^*.
2) Let the state space be $\mathbf{R}^n$. For the set $\{\omega_j : |\omega_j| = 1\}$ of input sequence,
set $g_s(\omega_j) = \mathbf{e_j}$. Moreover, let $g_s(\omega_j) = \sum_{i=1}^{j} \alpha_i \mathbf{e_i}$ for any $\omega \in U^*$ such that
$\omega_j < \omega < \omega_{j+1}$ and $|\omega_j| = |\omega_{j+1}| = 1$.
3) Let $h_s = [a(\omega_1) - a(1), a(\omega_2) - a(1), \cdots, a(\omega_n) - a(1)]$.
4) For any $i(1 \leq i \leq n)$, let $_if_j$ in $F_s(u_i) = [_if_1, _if_2, \cdots, _if_n] \in \mathbf{R}^{n \times n}$ be $_if_j =$
$[_if_{j1}, _if_{j2}, \cdots, _if_{jn}]^T$,
where $S_l(u_i)(S_l(\omega_j)a - a) = \sum_{k=1}^{n} {_if_{jk}} (S_l(\omega_k)a - a)$ holds for any $j(1 \leq j \leq n)$.
5) Set $h^0 = a(1)$.

7.3 Partial Realization Theory of Affine Dynamical Systems

Here we consider a partial realization problem by multi-experiment. Let $\underline{a}$ be
an $\underline{N}$ sized input response map($\in F(U_{\underline{N}}^*, Y)$), where $\underline{N} \in N$ and $U_{\underline{N}}^* := \{\omega \in U^*; |\omega| \leq \underline{N}\}$. The $\underline{a}$ is said to be a partial input response map. A finite dimen-
sional affine dynamical system $\sigma = ((X, F), g, , h, h^0)$ is called a partial realiza-
tion of $\underline{a}$ if $h^0 + h(\sum_{j=1}^{|\omega|}(F(\omega(|\omega|))F(\omega(|\omega|-1)) \cdots F(\omega(|\omega|-j))g(\omega(j))) = \underline{a}(\omega)$
holds for any $\omega \in U_{\underline{N}}^*$.

A partial realization problem of affine dynamical systems can be stated as
follows:

$<$ For any given $\underline{a} \in F(U_{\underline{N}}^*, Y)$, find a partial realization σ of $\underline{a}$ such that the
dimensions of state space $\overline{X}$ of σ is minimum, where the σ is said to be a minimal partial realization of $\underline{a}$. Moreover, show when the minimal realizations are
isomorphic.$>$

Since the partial realization problem of affine dynamical systems have not yet been discussed untill now, we will state facts about it. The proof is given in an appendix of this chapter.

Proposition 7.16. *For any given $\underline{a} \in F(U_N^*, Y)$, there always exists a minimal partial realization of it.*

[proof]. For any $\omega \notin U_N^*$, set $\underline{a}(\omega) = 0$. Then $\underline{a} \in F(U^*, Y)$, and Theorem (7.14) implies that there exists a finite dimensional partial realization of $\underline{a}$. Therefore, there exists a minimal partial realization.

Minimal partial realizations are, in general, not unique modulo isomorphisms. Therefore, we introduce a natural partial realization, and we show that natural partial realizations exist if and only if they are isomorphic.

Definition 7.17. *For an affine dynamical system $\sigma = ((X, F), g, h, h^0)$ and some $p \in N$, if $X = \ll \{\sum_{j=1}^{|\omega|}(F(\omega(|\omega|))F(\omega(|\omega|-1))\cdots F(\omega(|\omega|-j))g(\omega(j)); \omega \in U_p^*\} \gg$, then σ is said to be p-quasi-reachable,*
where $\ll S \gg$ denotes the smallest linear space which contains a set S.
Let q be some integer. If $hF(\omega(|\omega|))F(\omega(|\omega|-1))\cdots F(\omega(1))x = 0$ implies x=0 for any $\omega \in U_q^*$, then σ is said to be q-distinguishable.
For a given $\underline{a} \in F(U_L^*, Y)$, if there exist p and $q \in N$ such that $p + q < L$ and σ is p-quasi-reachable and q-distinguishable then σ is said to be a natural partial realization of $\underline{a}$.
For a partial input response map $\underline{a} \in F(U_L^*, Y)$, the following matrix $H_{\underline{a}(p,L-p)}^A$ is said to be a finite-sized Hankel matrix of $\underline{a}$.
The column vector of $H_{\underline{a}(p,L-p)}^A$ may be written by $S_l(\omega)\underline{a} - \underline{a}$.

$$
H_{\underline{a}(p,L-p)}^A = \begin{array}{c} \\ \bar{\omega} \end{array}
\begin{pmatrix}
 & & \vdots & \\
 & & \vdots & \\
 & & \vdots & \\
\cdots & \cdots & \underline{a}(\bar{\omega}|\omega) - \underline{a}(\bar{\omega}) &
\end{pmatrix}
\begin{array}{c} \omega \\ \\ \\ \end{array}
$$

where $\omega \in U_p^*$ and $\bar{\omega} \in U_{L-p}^*$.
In discussion of approximate and noisy realization of affine dynamical systems, the notation of $H_{\underline{a}(p,L-p)}^A(|||\omega_1|||, |||\omega_2|||, |||\omega_3|||, |||\omega_4|||)$ is used as follows:
$H_{\underline{a}(p,L-p)}^A(|||\omega_1|||, |||\omega_2|||, |||\omega_3|||, |||\omega_4|||) := [\underline{S}_l(\omega_1)\underline{a} - \underline{a}, \underline{S}_l(\omega_2)\underline{a} - \underline{a}, \underline{S}_l(\omega_3)\underline{a} - \underline{a}, \underline{S}_l(\omega_4)\underline{a} - \underline{a}]$.

Theorem 7.18. *Let $H_{\underline{a}(p,L-p)}^L$ be the finite Hankel matrix of $\underline{a} \in F(U_L^*, Y)$. Then there exists a natural partial realization of $\underline{a}$ if and only if the following conditions hold:*
rank $H_{\underline{a}(p,L-p)}^L$ = rank $H_{\underline{a}(p,L-p-1)}^L$ =rank $H_{\underline{a}(p+1,L-p-1)}^L$ *for some $p \in N$.*

[proof]. See (7-A.9) in Appendix 7.5.A.

Theorem 7.19. *There exists a natural partial realization of a given partial input response map $\underline{a} \in F(U_L^*, Y)$ if and only if the minimal partial realization of $\underline{a}$ are unique modulo isomorphisms.*

[proof]. See (7-A.11) in Appendix 7.5.A.

Theorem 7.20. *Let a partial input response $\underline{a} \in F(U_L^*, Y)$ satisfy the condition of Theorem (4.26), then the quasi-reachable standard system $\sigma_s = ((X, F_s), g_s, h_s, h^0$ which realizes $\underline{a}$ can be obtained by the following algorithm.*

Set $n := \operatorname{rank} H^L_{\underline{a}\,(p, L-p)}$, where $H^L_{\underline{a}\,(p, L-p)}$ is the finite Hankel-matrix of $\underline{a} \in F(U_L^*, Y)$.

1) Select the linearly independent vectors $\{S_l(\omega_i)\underline{a} - \underline{a} \in F(U_{L-p}^*, Y); 1 \leq i \leq n\}$ from $H^L_{\underline{a}\,(p, L-p)}$ in order of their numerical value.

2) Let the state space be $\boldsymbol{R}^n$, the map $g_s : U \to X$ be $g_s(u_i) = \mathbf{e}_i$, where $\mathbf{e}_i := [0, \cdots, 0, \overset{i}{1}, 0, \cdots, 0]^T$.

3) Let the output map $h_s = [\underline{a}(\omega_1) - \underline{a}(1), \underline{a}(\omega_2) - \underline{a}(1), \underline{a}(\omega_3) - \underline{a}(1), \cdots, \underline{a}(\omega_n) - \underline{a}(1)]$.

4) Let $_i\mathbf{f_j} \in \boldsymbol{R}^n$ in $F_s(u_i) := [_i\mathbf{f_1} \ _i\mathbf{f_2} \ \cdots \ _i\mathbf{f_n}]$ be $_i\mathbf{f_j} := [_if_{j,1}, _if_{j,2} \ \cdots \ _if_{j,n}]^T$ for $1 \leq i \leq n$, where $_i\mathbf{f_j}$ is given by the following:
$S_l(u_i)(S_l(\omega_j)\underline{a} - \underline{a}) = \sum_{k=1}^n \ _if_{j,k}(S_l(\omega_k)\underline{a} - \underline{a}), \ _if_{j,k} \in \boldsymbol{R}$ in the sense of $F(U_{L-p}^*, Y)$ and $S_l(\omega) : F(U_s^*, Y) \to F(U_{s-|\omega|}^*, Y) \ ; \ \underline{a} \mapsto S_l(\omega)\underline{a}[; \bar{\omega} \mapsto \underline{a}(\bar{\omega}|\omega)]$.

[proof]. See (7-A.12) in Appendix 7.5.A.

7.4 Approximate Realization of Affine Dynamical Systems

In this section, we discuss approximate realization problems of affine dynamical systems.

We will discuss an approximate realization problem under the assumption that the set U of input values is a finite set $U = \{u_j : 1 \leq j \leq m\}$ for a finite integer $m \in N$. In the reference [Matsuo and Hasegawa, 2003], we showed that this assumption are not so special. However, for simplicity of our discussion, we assume that the set U of input values is $U = \{u_1, u_2\}$ or $U = \{u_1, \ u_2, \ u_3\}$.

Roughly speaking, the approximate realization of affine dynamical systems can be stated as follows:

< For any given partial data of an affine dynamical system, find an affine dynamical system which approximates the given data. >

In order to make our discussion simple, we assume that the set Y of outout is the set $\boldsymbol{R}$ of real numbers, namely 1-output.

Theorem 7.21. *Algorithm for approximate realization*
Let an input response map $\underline{a}$ be a considered object which is an affine dynamical system. Then an approximate realization $\sigma = ((\boldsymbol{R}^n, F_s), g_s, h_s, h^0)$ of $\underline{a}$ is given by the following algorithm:

1) *Based on the ratio of the square root of eigenvalues for a matrix*
$H_{\underline{a}\ (n,\bar{p})}(|||\omega_1|||, \cdots, |||\omega_n|||) H_{\underline{a}\ (n,\bar{p})}(|||\omega_1|||, \cdots, |||\omega_n|||)^T$, *determine the value n of rank for the matrix, where* $|||\omega_1|||, |||\omega_2|||, \cdots$ *and* $|||\omega_n|||$ *are selected in the order of numerical value of input and* $\{\underline{S_l}(\omega_i)\underline{a} - \underline{a}; 1 \le i \le n, \omega_i \in U^*\}$ *is a set of independent vectors.*
Namely, determine the value n of rank for the matrix
$H_{\underline{a}\ (n,\bar{p})}(|||\omega_1|||, \cdots, |||\omega_n|||)$ *such that the ratio of the square root of eigenvalues for the covariance matrix becomes very small. The small ratio means the nearness of approximation degree.*
2) *In order to determine* g_s, *the CLS method is used as follows:*
   ① *In particular, set* $g_s(\omega_i) := \mathbf{e}_i$ *for* $\omega_i \in U$. *Namely,* $g_s(\omega_1) := \mathbf{e}_1$,
$g_s(\omega_2) := \mathbf{e}_2, \cdots, g_s(\omega_k) := \mathbf{e}_k$ *for some* $k \in N$.
For $u \in U$ *such that* $u \notin \{\omega_i; 1 \le i \le n\}$ *and* $\omega_r < u$,
$g_s(u) = \sum_{j=1}^r b_{u,j}(\underline{S_l}(\omega_j)\underline{a} - \underline{a})$ *is obtained as follows:*
Let a matrix $A_u \in \mathbf{R}^{1 \times (r+1)}$ *be* $A_u := [b_{u,1}, b_{u,2}, \cdots, b_{u,r}, -1]$.
Choose the coefficients $\{b_{u,j} : 1 \le j \le r\}$ *such that*
$\sum_{j=1}^r (\underline{S_l}(\omega_j)\underline{\bar{a}} - \underline{\bar{a}}) \cdot (\underline{S_l}(\omega_j)\underline{\bar{a}} - \underline{\bar{a}})) + (\underline{S_l}(u)\underline{\bar{a}} - \underline{\bar{a}}) \cdot (\underline{S_l}(u)\underline{\bar{a}} - \underline{\bar{a}})$ *take*
a minimum value, where $\{\underline{S_l}(\omega_j)\underline{\bar{a}} - \underline{\bar{a}} \in \mathbf{R}^{L \times 1} : 0 \le j \le r\}$ *are given by*
the equation $[\underline{S_l}(\omega_1)\underline{\bar{a}} - \underline{\bar{a}}, \underline{S_l}(\omega_2)\underline{\bar{a}} - \underline{\bar{a}}, \cdots, \underline{S_l}(\omega_r)\underline{\bar{a}} - \underline{\bar{a}}, \underline{S_l}(u)\underline{\bar{a}} - \underline{\bar{a}}]^T :=$
$A_u^T [A_u A_u^T]^{-1} A_u H_{\underline{a}\ (r+1,L)}^T (|||\omega_1|||, |||\omega_2|||, \cdots, |||\omega_r|||, |||u|||)$
and $H_{\underline{a}\ (|||u|||+1,L)}^T (|||\omega_1|||, |||\omega_2|||, \cdots, |||\omega_r|||, |||u|||) :=$
$[\underline{S_l}(\omega_1)\underline{a} - \underline{a}, \cdots, \underline{S_l}(\omega_r)\underline{a} - \underline{a}, \underline{S_l}(u)\underline{a} - \underline{a}]$. *And* $\cdot$ *denotes the inner product of two vectors.*
3) *In order to obtain* F_s, *the CLS method is used as follows:*
   ① *Let* $_i\mathbf{f}_j \in \mathbf{R}^n$ *in* $F_s(u_i) := [_i\mathbf{f}_1\ _i\mathbf{f}_2\ \cdots\ _i\mathbf{f}_n]$ *be* $_i\mathbf{f}_j := [_if_{j,1}\ _if_{j,2}\ \cdots\ _if_{j,n}]^T$
for $1 \le i \le n$, *where* $_if_{j,k}$ *is given by the following:*
$\underline{S_l}(u_i)(\underline{S_l}(\omega_j)\underline{a} - \underline{a}) = \sum_{k=1}^n {}_if_{j,k}(\underline{S_l}(\omega_k)\underline{a} - \underline{a})$, $_if_{j,k} \in \mathbf{R}$ *in the sense of*
$F(U_{L-p}^*, Y)$.
We cannot directly obtain the coefficients $\{_i\mathbf{f}_j; 1 \le i \le 3, 1 \le j \le n\}$.
Firstly, we will determine coefficients $\{_i\bar{f}_{j,k}; 1 \le k \le n\}$ *from the equation*
$\underline{S_l}(u_i)\underline{S_l}(\omega_j)\underline{a} - \underline{a} = \sum_{k=1}^n {}_i\bar{f}_{j,k}(\underline{S_l}(\omega_k)\underline{a} - \underline{a})$ *by using the CLS method.*
② *For* i $(1 \le i \le 3)$, j $(1 \le j \le n)$ *and for the maximum number*
r $(1 \le r \le n)$ *such that* $\omega_r, \omega_j \in \{\omega_j; 1 \le j \le n\}$ *and* $|||\omega_r||| < |||u_i|\omega_j|||$,
let a matrix $_iA_j \in \mathbf{R}^{1 \times (r+1)}$ *be* $_iA_j := [_i\bar{f}_{j,1}, {}_i\bar{f}_{j,2}, \cdots, {}_i\bar{f}_{j,r}, -1]$.
Choose the coefficients $_i\bar{f}_{j,k} = 0$ *for* k $(r+1 \le k \le n)$ *and* $\{_i\bar{f}_{j,k} : 1 \le k \le r\}$
such that $\sum_{i=1}^r (\underline{S_l}(\omega_i)\underline{\bar{a}} - \underline{\bar{a}}) \cdot (\underline{S_l}(\omega_i)\underline{\bar{a}} - \underline{\bar{a}}) + (\underline{S_l}(u_i|\omega_j)\underline{\bar{a}} - \underline{\bar{a}}) \cdot (\underline{S_l}(u_i|\omega_j)\underline{\bar{a}} - \underline{\bar{a}})$
takes a minimum value, where $\{\underline{S_l}(\omega_i)\underline{\bar{a}} - \underline{\bar{a}} \in \mathbf{R}^{L \times 1} : 0 \le i \le n\}$
and $\underline{S_l}(u_i|\omega_j)\underline{\bar{a}} - \underline{\bar{a}} \in \mathbf{R}^{L \times 1}$ *are given by the equation*
$[\underline{S_l}(\omega_1)\underline{\bar{a}} - \underline{\bar{a}}, \underline{S_l}(\omega_2)\underline{\bar{a}} - \underline{\bar{a}}, \cdots, \underline{S_l}(\omega_r)\underline{\bar{a}} - \underline{\bar{a}}, \underline{S_l}(u_i|\omega_j)\underline{\bar{a}} - \underline{\bar{a}}]^T :=$
$_iA_j^T [_iA_j\ _iA_j^T]^{-1}\ _iA_j\ H_{\underline{a}\ (r+1,L)}^T (|||\omega_1|||, |||\omega_2|||, \cdots, |||\omega_r|||, |||u_i|\omega_j|||)$ *and*
$H_{\underline{a}\ (r+1,L)}^T (|||\omega_1|||, \cdots, |||\omega_r|||, |||u_i|\omega_j|||) :=$
$[\underline{S_l}(\omega_1)\underline{a} - \underline{a}, \cdots, \underline{S_l}(\omega_r)\underline{a} - \underline{a}, \underline{S_l}(u_i|\omega_j)\underline{a} - \underline{a}]$.
And $\cdot$ *denotes the inner product of two vectors.*
③ *Next, using the equations*
$\underline{S_l}(u_i)(\underline{S_l}(\omega_j)\underline{a} - \underline{a})$

$$= \underline{S_l}(u_i)\underline{S_l}(\omega_j)\underline{a} - \underline{a} - \underline{S_l}(u_i)\underline{a} + \underline{a}$$
$$= \sum_{k=1}^{n} {}_i\bar{f}_{j,k}(\underline{S_l}(\omega_k)\underline{a} - \underline{a}) - \sum_{j=1}^{r_{u_i}} b_{u_i,j}(\underline{S_l}(u_j)\underline{a} - \underline{a}),$$
we obtain $\underline{S_l}(u_i)(\underline{S_l}(\omega_j)\underline{a} - \underline{a})$
$$= \sum_{k=1}^{n} {}_i\bar{f}_{j,k}(\underline{S_l}(\omega_k)\underline{a} - \underline{a}) - \sum_{j=1}^{r_{u_i}} b_{u_i,i}(\underline{S_l}(u_j)\underline{a} - \underline{a})$$
$$= \sum_{j=1}^{r_{u_i}} ({}_i\bar{f}_{j,k} - b_{u_i,j})(\underline{S_l}(\omega_j)\underline{a} - \underline{a}) + \sum_{j=r_{u_i}+1}^{n} {}_i\bar{f}_{j,k}(\underline{S_l}(\omega_j)\underline{a} - \underline{a}).$$

On the other hand, the equation

$\underline{S_l}(u_i)(\underline{S_l}(\omega_j)\underline{a} - \underline{a}) = \sum_{k=1}^{n} {}_i f_{j,k}(\underline{S_l}(\omega_k)\underline{a} - \underline{a})$ *holds. Therefore, we obtain the following equation:*

$$\sum_{k=1}^{n} {}_i f_{j,k}(\underline{S_l}(\omega_k)\underline{a} - \underline{a})$$
$$= \sum_{j=1}^{r_{u_i}} ({}_i\bar{f}_{j,k} - b_{u_i,j})\underline{S_l}(\omega_j)\underline{a} - \underline{a} + \sum_{j=r_{u_i}+1}^{n} {}_i\bar{f}_{j,k}(\underline{S_l}(\omega_j)\underline{a} - \underline{a}).$$

Comparing the coefficients, we obtain the following:

$$\sum_{j=1}^{r_{u_i}} ({}_i f_{j,k} - {}_i\bar{f}_{j,k} + b_{u_i,j})(\underline{S_l}(\omega_j)\underline{a} - \underline{a}) + \sum_{j=r_{u_i}+1}^{n} ({}_i f_{j,k} - {}_i\bar{f}_{j,k})(\underline{S_l}(\omega_j)\underline{a} - \underline{a})$$
$$= 0.$$

Finally, coefficients ${}_i f_{j,k}$ *of* ${}_i\mathbf{f}_j$ *in* F_s *are obtained as follows:*

${}_i f_{j,k} = {}_i\bar{f}_{j,k} - b_{u_i,j}$ *for* $1 \le j \le r_{u_i}$,

${}_i f_{j,k} = {}_i\bar{f}_{j,k}$ *for* $r_{u_i} + 1 \le j \le n$.

4) In order to determine $h_s \in \mathbf{R}^{1 \times n}$, *the CLS method is used as follows:*

① *For the first* ω_{r_1+1}, $\omega_{r_1} \in \{\omega_i ; 1 \le i \le n\}$ *such that* $\omega_{r_1+1} > \omega_{r_1}$ *and* $|||\omega_{r_1+1}||| - |||\omega_{r_1}||| > 1$ *when starting out from* ω_1,

set $\underline{S_l}(\lambda_1)\underline{a} - \underline{a} := \sum_{i=1}^{r_1} b_{\lambda_1,i}\mathbf{e}_i$ *for the obtained equation* $\underline{S_l}(\lambda_1)\underline{a} - \underline{a} = \sum_{i=1}^{r_1} b_{\lambda_1,i}(\underline{S_l}(\omega_i)\underline{a} - \underline{a})$ *for* λ_1 *such that* $|||\lambda_1||| = |||\omega_{r_1}||| + 1$.

Let a matrix $A_{\lambda_1} \in \mathbf{R}^{1 \times (r_1+1)}$ *be* $A_{\lambda_1} := [b_{\lambda_1,1}, b_{\lambda_1,2}, \cdots, b_{\lambda_1,r_1}, -1]$.

Choose the coefficients $\{b_{\lambda_1,i} : 1 \le i \le r_1\}$ *such that*

$\sum_{i=1}^{r_1} (\underline{S_l}(\omega_i)\bar{\underline{a}} - \bar{\underline{a}}) \cdot (\underline{S_l}(\omega_i)\bar{\underline{a}} - \bar{\underline{a}}) + (\underline{S_l}(\lambda_1)\bar{\underline{a}} - \bar{\underline{a}}) \cdot (\underline{S_l}(\lambda_1)\bar{\underline{a}} - \bar{\underline{a}})$ *take a minimum value, where* $\{\underline{S_l}(\omega_i)\bar{\underline{a}} - \bar{\underline{a}} \in \mathbf{R}^{L \times 1} : 0 \le i \le r_1\}$ *are given by the equation* $[\underline{S_l}(\omega_1)\bar{\underline{a}} - \bar{\underline{a}}, \underline{S_l}(\omega_2)\bar{\underline{a}} - \bar{\underline{a}}, \cdots, \underline{S_l}(\omega_{r_1})\bar{\underline{a}} - \bar{\underline{a}}, \underline{S_l}(\lambda_1)\bar{\underline{a}} - \bar{\underline{a}}]^T :=$ $A_{\lambda_1}^T [A_{\lambda_1} A_{\lambda_1}^T]^{-1} A_{\lambda_1} H_{\underline{a}}^T{}_{(r_1+1,L)}(|||\omega_1|||, |||\omega_2|||, \cdots, |||\omega_{r_1}|||, |||\lambda_1|||)$ *and* $H_{\underline{a}}^T{}_{(|||\lambda_1|||+1,L)}(|||\omega_1|||, |||\omega_2|||, \cdots, |||\omega_{r_1}|||, |||\lambda_1|||) :=$ $[\underline{S_l}(\omega_1)\underline{a} - \underline{a}, \cdots, \underline{S_l}(\omega_{r_1})\underline{a} - \underline{a}, \underline{S_l}(\lambda_1)\underline{a} - \underline{a}]$. *And* $\cdot$ *denotes the inner product of two vectors.*

Then let $h_{\lambda_1 s} \in \mathbf{R}^{1 \times r_1}$ *be*

$h_{\lambda_1 s} := [\underline{a}(\omega_1) - \underline{a}(1) - (\bar{\underline{a}}(\omega_1) - \bar{\underline{a}}(1)), \underline{a}(\omega_2) - \underline{a}(1) - (\bar{\underline{a}}(\omega_2) - \bar{\underline{a}}(1)), \cdots,$
$\quad \underline{a}(\omega_{r_1}) - \underline{a}(1) - (\bar{\underline{a}}(\omega_{r_1}) - \bar{\underline{a}}(1))]$.

② *For the first* ω_{r_2+1}, $\omega_{r_2} \in \{\omega_i ; 1 \le i \le n\}$ *such that* $\omega_{r_2+1} > \omega_{r_2}$ *and* $|||\omega_{r_2+1}||| - |||\omega_{r_2}||| > 1$ *when starting out from* ω_{r_1+1},

set $\underline{S_l}(\lambda_2)\underline{a} - \underline{a} := \sum_{i=1}^{r_2} b_{\lambda_2,i}\mathbf{e}_i$ *for the obtained equation* $\underline{S_l}(\lambda_2)\underline{a} - \underline{a} = \sum_{i=1}^{r_2} b_{\lambda_2,i}(\underline{S_l}(\omega_i)\underline{a} - \underline{a})$ *for* λ_2 *such that* $|||\lambda_2||| = |||\omega_{r_2}||| + 1$.

Let a matrix $A_{\lambda_2} \in \mathbf{R}^{1 \times (r_2+1)}$ *be* $A_{\lambda_2} := [b_{\lambda_2,1}, b_{\lambda_2,2}, \cdots, b_{\lambda_2,r_2}, -1]$.

Choose the coefficients $\{b_{\lambda_2,i} : 1 \le i \le r_2\}$ *such that*

$\sum_{i=1}^{r_2} (\underline{S_l}(\omega_i)\bar{\underline{a}} - \bar{\underline{a}}) \cdot (\underline{S_l}(\omega_i)\bar{\underline{a}} - \bar{\underline{a}}) + (\underline{S_l}(\lambda_2)\bar{\underline{a}} - \bar{\underline{a}}) \cdot (\underline{S_l}(\lambda_2)\bar{\underline{a}} - \bar{\underline{a}})$ *take a minimum value, where* $\{\underline{S_l}(\omega_i)\bar{\underline{a}} - \bar{\underline{a}} \in \mathbf{R}^{L \times 1} : 0 \le i \le r_2\}$ *are given by the equation* $[\underline{S_l}(\omega_1)\bar{\underline{a}} - \bar{\underline{a}}, \underline{S_l}(\omega_2)\bar{\underline{a}} - \bar{\underline{a}}, \cdots, \underline{S_l}(\omega_{r_2})\bar{\underline{a}} - \bar{\underline{a}}, \underline{S_l}(\lambda_2)\bar{\underline{a}} - \bar{\underline{a}}]^T :=$ $A_{\lambda_2}^T [A_{\lambda_2} A_{\lambda_2}^T]^{-1} A_{\lambda_2} H_{\underline{a}}^T{}_{(r_2+1,L)}(|||\omega_1|||, |||\omega_2|||, \cdots, |||\omega_{r_2}|||, |||\lambda_2|||)$ *and* $H_{\underline{a}}^T{}_{(|||\lambda_2|||+1,L)}(|||\omega_1|||, |||\omega_2|||, \cdots, |||\omega_{r_2}|||, |||\lambda_2|||) :=$

$[\underline{S_l}(\omega_1)\underline{a} - \underline{a}, \cdots , \underline{S_l}(\omega_{r_2})\underline{a} - \underline{a}, \underline{S_l}(\lambda_2)\underline{a} - \underline{a}]$. *And* **.** *denotes the inner product of two vectors.*

Then let $h_{\lambda_2 s} \in \mathbf{R}^{1 \times r_2}$ *be*

$h_{\lambda_2 s} := [\underline{a}(\omega_{r_1+1}) - \underline{a}(1) - (\underline{\bar{a}}(\omega_{r_1+1}) - \underline{\bar{a}}(1)),$
$\underline{a}(\omega_{r_1+2}) - \underline{a}(1) - (\underline{\bar{a}}(\omega_{r_1+2}) - \underline{\bar{a}}(1)), \cdots , \underline{a}(\omega_{r_2}) - \underline{a}(1) - (\underline{\bar{a}}(\omega_{r_2}) - \underline{\bar{a}}(1))].$

$\vdots$

(t) For $\omega \in U^*$ *such that* $\||\omega\|| = \||\omega_n\|| + 1$, *let a matrix* $A_\omega \in \mathbf{R}^{1 \times (n+1)}$ *be* $A_\omega := [b_{\omega,1}, b_{\omega,2}, \cdots , b_{\omega,n}, -1]$.

Choose the coefficients $\{b_{\omega,i} : 1 \leq i \leq n\}$ *such that*
$\sum_{i=1}^{n}(\underline{S_l}(\omega_i)\underline{\bar{a}} - \underline{\bar{a}}) \boldsymbol{\cdot} (\underline{S_l}(\omega_i)\underline{\bar{a}} - \underline{\bar{a}}) + (\underline{S_l}(\omega)\underline{\bar{a}} - \underline{\bar{a}}) \boldsymbol{\cdot} (\underline{S_l}(\omega)\underline{\bar{a}} - \underline{\bar{a}})$ *takes a minimum value, where* $\{\underline{S_l}(\omega_i)\underline{\bar{a}} - \underline{\bar{a}} : 0 \leq i \leq n\}$ *and* $\underline{S_l}(\omega)\underline{\bar{a}} - \underline{\bar{a}}$ *are given by the following equation:*

$[\underline{S_l}(\omega_1)\underline{\bar{a}} - \underline{\bar{a}}, \underline{S_l}(\omega_2)\underline{\bar{a}} - \underline{\bar{a}}, \cdots , \underline{S_l}(\omega_n)\underline{\bar{a}} - \underline{\bar{a}}, \underline{S_l}(\omega)\underline{\bar{a}} - \underline{\bar{a}}]^T :=$
$A_\omega^T [A_\omega A_\omega^T]^{-1} A_\omega H_{\underline{a}}^T{}_{(n+1,L)}(\||\omega_1\||, \||\omega_2\||, \cdots , \||\omega_n\||, \||\omega\||)$
and $H_{\underline{a}}^T{}_{(n+1,L)}(\||\omega_1\||, \||\omega_2\||, \cdots , \||u\||) :=$
$[\underline{S_l}(\omega_1)\underline{a} - \underline{a}, \cdots , \underline{S_l}(\omega_n)\underline{a} - \underline{a}, \underline{S_l}(\omega)\underline{a} - \underline{a}]$, *and* **.** *denotes the inner product of two vectors.*

Then let $h_{\omega s}$ *be*
$h_{\omega s} := [\underline{a}(\omega_{r_t+1}) - \underline{a}(1) - (\underline{\bar{a}}(\omega_{r_t+1}) - \underline{\bar{a}}(1)),$
$\underline{a}(\omega_{r_t+2}) - \underline{a}(1) - (\underline{\bar{a}}(\omega_{r_t+2}) - \underline{\bar{a}}(1)), \cdots , \underline{a}(\omega_n) - \underline{a}(1) - (\underline{\bar{a}}(\omega_n) - \underline{\bar{a}}(1))].$
Finally, let $h_s \in \mathbf{R}^{1 \times n}$ *be*
$h_s := [h_{\lambda_1 s}, h_{\lambda_2 s}, \cdots , h_{\omega s}].$

[proof]. In 1), the number of dimensions is determined by considering the ratio of Hankel matrix norm, which means a degree of information loss. According to Theorem (7.20), an affine dynamical system $\sigma = ((\mathbf{R}^n, F_s), g_s, h_s, h^0)$ is obtained as follows. In 2), g_s is obtained directly or by using the CLS method for A_u corresponding to the matrix A in Proposition (2.14). In 3), F_s is obtained by using the CLS method for $_iA_j$ corresponding to the matrix A in Proposition (2.14). In 4), h_s is obtained by using the CLS method for A_{λ_1}, A_{λ_2}, A_{λ_ω} corresponding to the matrix A in Proposition (2.14).

In the figures of this chapter, we use a notation $Signal\_n\text{-}d$ as an input response map obtained by a n-dimensional affine dynamical system.

In the examples of this chapter, a notation $H_{\underline{a}}^T{}_{(r,40)}(1, \cdots , r)$ is used in place of $H_{\underline{a}}^T{}_{(r,40)}(1, 2, 3, \cdots , r-1, r)$.

Example 7.22. Let the signals be the input response map of the following 3-dimensional affine dynamical system: $\sigma = ((\mathbf{R}^3, F), g, h, h^0)$, where $F(u_1) =$
$$\begin{bmatrix} -1 & -0.3 & -0.5 \\ 0 & 0 & -0.8 \\ 0 & -0.3 & -0.7 \end{bmatrix}, \ F(u_2) = \begin{bmatrix} -0.6 & 0 & 0.7 \\ 0 & 1 & 0 \\ -0.7 & 0 & 0.6 \end{bmatrix}, \ F(u_3) = \begin{bmatrix} 0 & -0.5 & 0 \\ 0 & 0.1 & 0.3 \\ 0 & 0 & 0.1 \end{bmatrix},$$
$g(u_1) = \mathbf{e}_1, \ g(u_2) = \mathbf{e}_2, \ g(u_3) = \mathbf{e}_3, \ h = [6, \ -3, \ -6], \ h^0 = 1.$

Then the approximate realization problem is solved as follows:

covariance matrix	eigenvalues						
	1	2	3	4	5	$\cdots$	12
$H^T_{\underline{a}\,(2,40)}(1,2)H_{\underline{a}\,(2,40)}(1,2)$	177	103					
$H^T_{\underline{a}\,(3,40)}(1,2,3)H_{\underline{a}\,(3,40)}(1,2,3)$	238	123	72				
$H^T_{\underline{a}\,(4,40)}(1,\cdots,4)H_{\underline{a}\,(4,40)}(1,\cdots,4)$	238	123	72	0			
$H^T_{\underline{a}\,(5,40)}(1,\cdots,5)H_{\underline{a}\,(5,40)}(1,\cdots,5)$	272	178	123	0	0		
$H^T_{\underline{a}\,(12,40)}(1,\cdots,12)H_{\underline{a}\,(12,40)}(1,\cdots,12)$	1394	558	326	0	0	$\cdots$	0
covariance matrix	square root of eigenvalues						
$H^T_{\underline{a}\,(2,40)}(1,2)H_{\underline{a}\,(2,40)}(1,2)$	13.3	10.1					
$H^T_{\underline{a}\,(3,40)}(1,2,3)H_{\underline{a}\,(3,40)}(1,2,3)$	15.4	11	8.5				
$H^T_{\underline{a}\,(4,40)}(1,\cdots,4)H_{\underline{a}\,(4,40)}(1,\cdots,4)$	15.4	11	8.5	0			
$H^T_{\underline{a}\,(5,40)}(1,\cdots,5)H_{\underline{a}\,(5,40)}(1,\cdots,5)$	16.5	13.3	11	0	0		
$H^T_{\underline{a}\,(12,40)}(1,\cdots,12)H_{\underline{a}\,(12,40)}(1,\cdots,12)$	37.3	23.6	18	0	0	$\cdots$	0

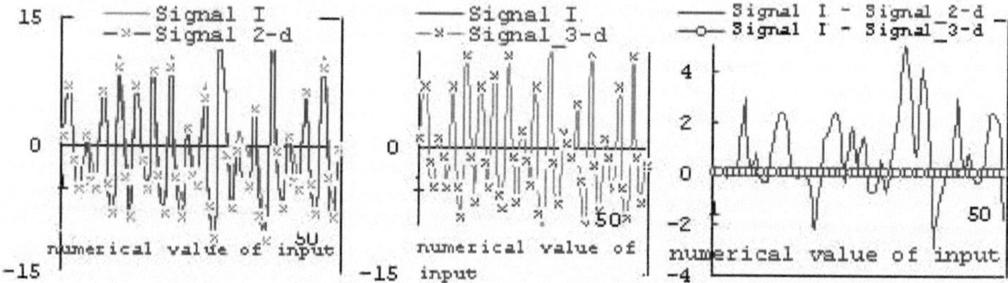

Fig. 7.1. The left is the original input response map and the behavior of a 2-dimensional affine dynamical system obtained by the CLS method. The middle is the original input response map and the behavior of a 3-dimensional affine dynamical system obtained by the CLS method. The right is the difference between the original one and the behavior of the 2-dimensional affine dynamical system obtained by the CLS method or the 3-dimensional affine dynamical system obtained by the CLS method in Example (7.22).

1) Since the ratio $\frac{8.5}{15.4} = 0.55$ obtained by the square root of $H^T_{\underline{a}\,(3,40)}(1,2,3)H_{\underline{a}\,(3,40)}(1,2,3)$ is large, the approximate 2-dimensional affine dynamical system obtained by the CLS method may not be good.

2) After determining the independent vectors $\underline{S}_l(u_1)\underline{a}-\underline{a}$ and $\underline{S}_l(u_2)\underline{a}-\underline{a}$ whose numerical value of input are 1 and 2, we will continue the approximate realization algorithm by the CLS method.

Therefore, an approximate 2-dimensional affine dynamical system $\sigma_1 = ((\boldsymbol{R}^2,\ F_1),\ g_1,\ h_1,\ h^0)$ obtained by the CLS method is constructed as follows:

$$F_1(u_1) = \begin{bmatrix} -1 & -0.17 \\ 0 & -0.12 \end{bmatrix},\ F_1(u_2) = \begin{bmatrix} -0.42 & 0 \\ 0.02 & 1 \end{bmatrix},\ F_1(u_3) = \begin{bmatrix} 0 & -0.64 \\ 0 & -0.31 \end{bmatrix},$$

$g_1(u_1) = \mathbf{e}_1,\ g_1(u_2) = \mathbf{e}_2,\ g_1(u_3) = [-0.5,\ 1.2],\ h_1 = [5.91,\ -2.3],\ h^0 = 1.$

For reference, a 3-dimensional affine dynamical system $\sigma_2 = ((\boldsymbol{R}^3,\ F_2),\ g_2,\ h_2,h^0)$ obtained by the CLS method can be expressed as follows:

$$F_2(u_1) = \begin{bmatrix} -1 & -0.3 & -0.5 \\ 0 & 0 & -0.8 \\ 0 & -0.3 & 0.7 \end{bmatrix}, \quad F_2(u_2) = \begin{bmatrix} -0.6 & 0 & 0.7 \\ 0 & 1 & 0 \\ -0.7 & 0 & 0.6 \end{bmatrix},$$

$$F_2(u_3) = \begin{bmatrix} 0 & -0.5 & 0 \\ 0 & 0.1 & 0.3 \\ 0 & 0 & 0.1 \end{bmatrix}, \quad g_2(u_1) = \mathbf{e}_1, \ g_2(u_2) = \mathbf{e}_2, \ g_2(u_3) = \mathbf{e}_3,$$

$$h_2 = [6, \ -3, \ -6], \ h^0 = 1.$$

In this example, the original signals are considered as the input response map of a 3-dimensional affine dynamical system and the desirable input response map is obtained by the CLS method with our bad feeling. The model obtained by the CLS method is a 2-dimensional affine dynamical system.

For reference, a 3-dimensional affine dynamical system is also given by the CLS method. The system completely reconstructs the original system.

Just as we thought, the following table and Fig. 7.1 truly indicate that the 2-dimensional affine dynamical system obtained by the CLS method is a bad approximation. For reference, the input response map of the same dimensional affine dynamical system as the original system is shown. Hence, there does not exist a good approximation for the given system.

dimen-ion	ratio of matrices	mean values of square root for sum of		error	cosine ① and ②	error ratio
		signal ①	signal by CLS ②	③	$\cos\theta$	③/①
$a_{1,2}$	0.55	0.883	0.926	0.24	0.97	0.27
$a_{1,2,3}$	0	0.883	0.883	0	1	0

Example 7.23. Let the signals be the input response map of the following 3-dimensional affine dynamical system: $\sigma = ((\mathbf{R}^3, F), g, h, h^0)$, where $F(u_1) =$

$$\begin{bmatrix} -0.6 & -0.3 & 0 \\ 1 & 0.1 & 0.1 \\ 0 & 1 & 0 \end{bmatrix}, \quad F(u_2) = \begin{bmatrix} -0.9 & -0.1 & 0 \\ 0.5 & 0.3 & 0.05 \\ 0 & 0.5 & 0 \end{bmatrix}, \quad F(u_3) = \begin{bmatrix} 0.5 & 0 & 0 \\ 1 & 0.3 & 0 \\ 1 & 0 & 0.05 \end{bmatrix},$$

$$g(u_1) = \mathbf{e}_1, \ g(u_2) = \mathbf{e}_2, \ g(u_3) = \mathbf{e}_3, \ h = [6, \ -4, \ 2], \ h^0 = 1.$$

Then the approximate realization problem is solved as follows:

covariance matrix	eigenvalues						
	1	2	3	4	5	$\cdots$	12
$H^T_{\underline{a}\,(2,40)}(1,2)\,H_{\underline{a}\,(2,40)}(1,2)$	542	38					
$H^T_{\underline{a}\,(3,40)}(1,2,3)\,H_{\underline{a}\,(3,40)}(1,2,3)$	542	41	1.3				
$H^T_{\underline{a}\,(4,40)}(1,\cdots,4)\,H_{\underline{a}\,(4,40)}(1,\cdots,4)$	676	68	1.3	0			
$H^T_{\underline{a}\,(5,40)}(1,\cdots,5)\,H_{\underline{a}\,(5,40)}(1,\cdots,5)$	1004	155	1.3	0	0		
$H^T_{\underline{a}\,(12,40)}(1,\cdots,12)\,H_{\underline{a}\,(12,40)}(1,\cdots,12)$	2057	272	9.7	0	0	$\cdots$	0
covariance matrix	square root of eigenvalues						
$H^T_{\underline{a}\,(2,40)}(1,2)\,H_{\underline{a}\,(2,40)}(1,2)$	23.3	6.2					
$H^T_{\underline{a}\,(3,40)}(1,2,3)\,H_{\underline{a}\,(3,40)}(1,2,3)$	23.3	6.4	1.1				
$H^T_{\underline{a}\,(4,40)}(1,\cdots,4)\,H_{\underline{a}\,(4,40)}(1,\cdots,4)$	26	8.2	1.1	0			
$H^T_{\underline{a}\,(5,40)}(1,\cdots,5)\,H_{\underline{a}\,(5,40)}(1,\cdots,5)$	31.7	12.4	1.1	0	0		
$H^T_{\underline{a}\,(12,40)}(1,\cdots,12)\,H_{\underline{a}\,(12,40)}(1,\cdots,12)$	45.4	16.5	3.1	0	0	$\cdots$	0

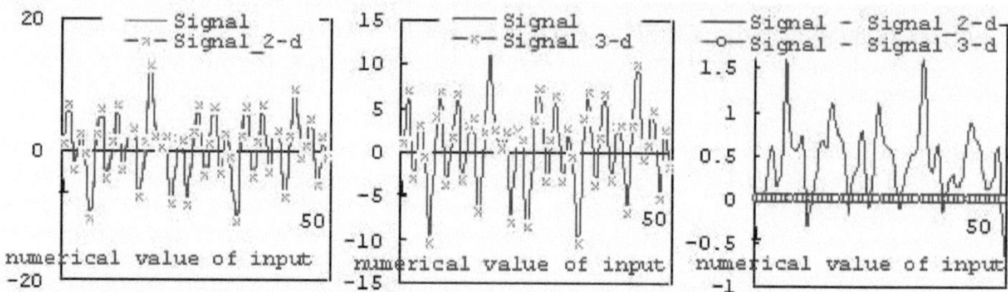

Fig. 7.2. The left is the original input response map and the behavior of a 2-dimensional affine dynamical system obtained by the CLS method. The middle is the original input response map and the behavior of a 3-dimensional affine dynamical system obtained by the CLS method. The right is the difference between the original input response map and the behavior of the 2-dimensional affine dynamical system obtained by the CLS method or the 3-dimensional affine dynamical system obtained by the CLS method in Example (7.23).

1) Since the ratio $\frac{1.1}{23.3} = 0.05$ obtained by the square root of $H_{\underline{a}}^{T}{}_{(3,40)}(1,2,3)$ $\times H_{\underline{a}\,(3,40)}(1,2,3)$ and the ratio $\frac{3.1}{43.4} = 0.06$ obtained by the square root of $H_{\underline{a}}^{T}{}_{(12,40)}(1,\cdots,12)H_{\underline{a}\,(12,40)}(1,\cdots,12)$ are not so small, the approximate 2-dimensional affine dynamical system obtained by the CLS method may not be good.

2) After determining the independent vectors $\underline{S}_{l}(u_1)\underline{a} - \underline{a}$ and $\underline{S}_{l}(u_2)\underline{a} - \underline{a}$ whose numerical value of input are 1 and 2, we will continue an approximate realization algorithm by the CLS method.

Therefore, an approximate 2-dimensional affine dynamical system $\sigma_1 = ((\boldsymbol{R}^2,\ F_1),\ g_1,\ h_1,\ h^0)$ obtained by the CLS method is constructed as follows:

$$F_1(u_1) = \begin{bmatrix} -0.6 & -0.25 \\ 1 & -0.16 \end{bmatrix}, \ F_1(u_2) = \begin{bmatrix} -0.9 & -0.08 \\ 0.5 & 0.18 \end{bmatrix}, \ F_1(u_3) = \begin{bmatrix} 0.55 & 0 \\ 0.8 & 0.31 \end{bmatrix},$$

$g_1(u_1) = \mathbf{e}_1$, $g_1(u_2) = \mathbf{e}_2$, $g_1(u_3) = [0.05,\ -0.27]$, $h_1 = [6.03,\ -4.16]$, $h^0 = 1$.

For reference, a 3-dimensional affine dynamical system $\sigma_2 = ((\boldsymbol{R}^3,\ F_2),\ g_2,\ h_2, h^0)$ obtained by the CLS method can be expressed as follows:

$$F_2(u_1) = \begin{bmatrix} -0.6 & -0.3 & 0 \\ 1 & 0.1 & 0.1 \\ 0 & 1 & 0 \end{bmatrix}, \ F_2(u_2) = \begin{bmatrix} -0.9 & -0.1 & 0 \\ 0.5 & 0.3 & 0.05 \\ 0 & 0.5 & 0 \end{bmatrix},$$

$$F_2(u_3) = \begin{bmatrix} 0.5 & 0 & 0 \\ 1 & 0.3 & 0 \\ 1 & 0 & 0.05 \end{bmatrix}, \ g_2(u_1) = \mathbf{e}_1, \ g_2(u_2) = \mathbf{e}_2, \ g_2(u_3) = \mathbf{e}_3,$$

$h_2 = [6,\ -4,\ 2]$, $h^0 = 1$.

In this example, the original signals are considered as the input response map of a 3-dimensional affine dynamical system and the desirable input response map is obtained by the CLS method based on our thinking. The model obtained by the CLS method is a 2-dimensional affine dynamical system.

For reference, a 3-dimensional affine dynamical system is also given by the CLS method. The system completely reconstructs the original system.

Just as we thought, the following table and Fig. 7.2 truly indicate that the 2-dimensional affine dynamical system obtained by the CLS method is not such a good approximation. For reference, the input response map of the same dimensional affine dynamical system as the original system is shown. Hence, there does not exist a good approximation for the given system.

dimen-ion	ratio of matrices	mean values of square root for sum of			cosine ① and ②	error ratio
		signal ①	signal by CLS ②	error ③	$\cos \theta$	③/①
$a_{1,2}$	0.05	0.733	0.717	0.08	0.99	0.11
$a_{1,2,3}$	0	0.733	0.733	0	1	0

Example 7.24. Let the signals be the input response map of the following 4-dimensional affine dynamical system: $\sigma = ((\boldsymbol{R}^4, F), g, h, h^0)$, where

$$F(u_1) = \begin{bmatrix} 0 & -0.5 & -0.4 & -0.4 \\ 0 & 0.6 & 0 & -0.03 \\ -0.6 & 0 & -0.1 & 0.4 \\ 0 & -0.5 & 0.5 & -0.6 \end{bmatrix}, \quad F(u_2) = \begin{bmatrix} 0.6 & 0 & -0.3 & 0.3 \\ 0 & 0.8 & 0 & 0 \\ -0.3 & 0 & 0.6 & 0.3 \\ 0.3 & 0 & 0.3 & 0.7 \end{bmatrix},$$

$$F(u_3) = \begin{bmatrix} -1 & -0.3 & 0 & -0.7 \\ 0 & 0 & 1 & 0.8 \\ 0 & -0.3 & -0.3 & 0.7 \\ 0.8 & 0.3 & 0.4 & 0.7 \end{bmatrix}, \quad g(u_1) = \mathbf{e}_1, \; g(u_2) = \mathbf{e}_2, \; g(u_3) = \mathbf{e}_3,$$

$h = [12, \; -8, \; 0, \; 1], \; h^0 = 1.$

Then the approximate realization problem is solved as follows:

covariance matrix	eigenvalues						
	1	2	3	4	5	$\cdots$	12
$H_{\underline{a}\,(2,50)}^T(1,2)H_{\underline{a}\,(2,50)}(1,2)$	1312	1063					
$H_{\underline{a}\,(3,50)}^T(1,2,3)H_{\underline{a}\,(3,50)}(1,2,3)$	1733	1230	738				
$H_{\underline{a}\,(4,50)}^T(1,\cdots,4)H_{\underline{a}\,(4,50)}(1,\cdots,4)$	1892	1632	1201	0			
$H_{\underline{a}\,(5,50)}^T(1,\cdots,5)H_{\underline{a}\,(5,50)}(1,\cdots,5)$	4257	1699	1442	40	0		
$H_{\underline{a}\,(12,50)}^T(1,\cdots,12)H_{\underline{a}\,(12,50)}(1,\cdots,12)$	17072	4975	3480	540	0	$\cdots$	0
covariance matrix	square root of eigenvalues						
$H_{\underline{a}\,(2,50)}^T(1,2)H_{\underline{a}\,(2,50)}(1,2)$	36.2	32.6					
$H_{\underline{a}\,(3,50)}^T(1,2,3)H_{\underline{a}\,(3,50)}(1,2,3)$	41.6	35	27.1				
$H_{\underline{a}\,(4,50)}^T(1,\cdots,4)H_{\underline{a}\,(4,50)}(1,\cdots,4)$	43.5	40.4	34.7	0			
$H_{\underline{a}\,(5,50)}^T(1,\cdots,5)H_{\underline{a}\,(5,50)}(1,\cdots,5)$	65.2	41.2	38	6.3	0		
$H_{\underline{a}\,(12,50)}^T(1,\cdots,12)H_{\underline{a}\,(12,50)}(1,\cdots,12)$	130	70.5	59	23.2	0	$\cdots$	0

1) Since the ratio $\frac{6.3}{65.2} = 0.1$ obtained by the square root of $H_{\underline{a}\,(5,50)}^T(1,\cdots,5) \times H_{\underline{a}\,(5,50)}(1,\cdots,5)$ is not so small and the ratio $\frac{29.2}{130} = 0.22$ obtained by the square root of $H_{\underline{a}\,(12,50)}^T(1,\cdots,12)H_{\underline{a}\,(12,50)}(1,\cdots,12)$ is rather large, the approximate 3-dimensional affine dynamical system obtained by the CLS method may not be so good.

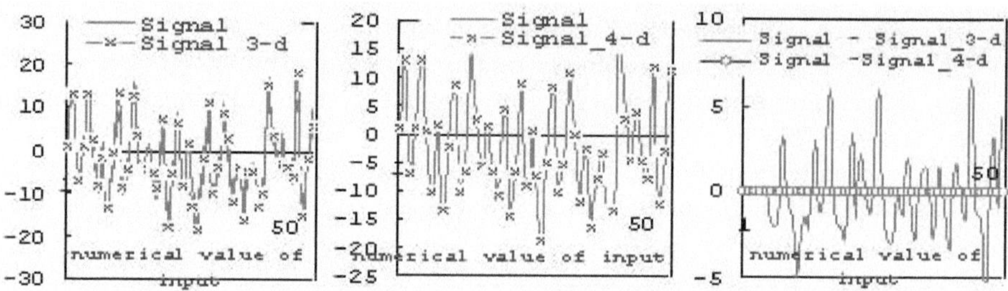

Fig. 7.3. The left is the original input response map and the behavior of a 3-dimensional affine dynamical system obtained by the CLS method. The middle is the original input response map and the behavior of a 4-dimensional affine dynamical system obtained by the CLS method. The right is the difference between the original one and the behavior of the 3-dimensional affine dynamical system obtained by the CLS method or the 4-dimensional affine dynamical system obtained by the CLS method in Example (7.24).

2) After determining the independent vectors $\underline{S}_l(u_1)\underline{a} - \underline{a}$, $\underline{S}_l(u_2)\underline{a} - \underline{a}$ and $\underline{S}_l(u_3)\underline{a} - \underline{a}$ whose numerical value of input are 1, 2 and 3, we will continue the approximate realization algorithm by the CLS method.

Therefore, an approximate 3-dimensional affine dynamical system $\sigma_1 = ((\boldsymbol{R}^3, F_1), g_1, h_1, h^0)$ obtained by the CLS method is constructed as follows:

$$F_1(u_1) = \begin{bmatrix} 0 & -0.93 & 0.25 \\ 0 & 0.43 & 0.29 \\ -0.6 & -0.37 & 0.13 \end{bmatrix}, F_1(u_2) = \begin{bmatrix} 0.93 & 0 & 0.02 \\ 0.19 & 0.8 & 0.19 \\ -0.16 & 0 & 0.79 \end{bmatrix},$$

$$F_1(u_3) = \begin{bmatrix} -0.47 & 0.03 & 0.39 \\ 0.47 & 0.16 & 1.3 \\ 0.8 & -0.09 & -0.05 \end{bmatrix}, g_1(u_1) = \mathbf{e}_1, g_1(u_2) = \mathbf{e}_2,$$

$g_1(u_3) = \mathbf{e}_3$, $h_1 = [12, -8, 0]$, $h^0 = 1$.

For reference, a 4-dimensional affine dynamical system $\sigma_2 = ((\boldsymbol{R}^4, F_2), g_2, h_2, h^0)$ obtained by the CLS method can be expressed as follows:

$$F_2(u_1) = \begin{bmatrix} 0 & 0.5 & -1.4 & 1.16 \\ 0 & 2.27 & -1.67 & 3.36 \\ -0.6 & -0.5 & 0.4 & -1 \\ 0 & -1.67 & 1.67 & -2.77 \end{bmatrix}, F_2(u_2) = \begin{bmatrix} 0 & 0 & -0.9 & -0.06 \\ -1 & 0.8 & -1 & -0.2 \\ 0 & 0 & 0.9 & 0.03 \\ 1 & 0 & 1 & 1 \end{bmatrix},$$

$$F_2(u_3) = \begin{bmatrix} -2.6 & -0.9 & -0.8 & -2.85 \\ -2.7 & -1 & -0.3 & -2.96 \\ 0.8 & 0 & 0.1 & 0.87 \\ 2.7 & 1 & 1.3 & 2.9 \end{bmatrix}, g_2(u_1) = \mathbf{e}_1, g_2(u_2) = \mathbf{e}_2,$$

$g_2(u_3) = \mathbf{e}_3$, $h_2 = [12, -8, 0, 0.5]$, $h^0 = 1$.

In this example, the original signals are considered as the input response map of a 4-dimensional affine dynamical system and the desirable input response map is obtained by the CLS method with our bad feeling. The model obtained by the CLS method is a 3-dimensional affine dynamical system.

For reference, a 4-dimensional affine dynamical system is also given by the CLS method. The system completely reconstructs the original system.

Just as we thought, the following table and Fig. 7.3 truly indicate that the 3-dimensional affine dynamical system obtained by the CLS method is a bad approximation. For reference, the input response map of the same dimensional affine dynamical system as the original system is shown. Hence, there exists only a small good approximation for the given system.

dimen-ion	ratio of matrices	mean values of square root for sum of			cosine $\textcircled{1}$ and $\textcircled{2}$	error ratio
		signal $\textcircled{1}$	signal by CLS $\textcircled{2}$	error $\textcircled{3}$	$\cos\theta$	$\textcircled{3}/\textcircled{1}$
$a_{1,2,3}$	0.1	1.3	1.31	0.38	0.96	0.29
$a_{1,2,3,5}$	0	1.3	1.3	0	1	0

Example 7.25. Let the signals be the input response map of the following 4-dimensional affine dynamical system: $\sigma = ((\boldsymbol{R}^4, F), g, h, h^0)$, where

$$F(u_1) = \begin{bmatrix} -1.5 & -0.3 & -1 & -0.1 \\ 0 & 0 & 1 & 0.6 \\ 0 & 2.3 & -0.3 & 0 \\ 0.5 & 0.3 & 0.42 & 0.4 \end{bmatrix}, \; F(u_2) = \begin{bmatrix} 0.6 & 0 & -0.3 & 0.3 \\ 0 & 1 & 0 & 0 \\ -0.3 & 0 & 0.6 & 0.3 \\ 0.3 & 0 & 0.3 & 0.6 \end{bmatrix},$$

$$F(u_3) = \begin{bmatrix} 0 & -0.5 & 0 & -0.4 \\ 0 & 0.5 & 0 & 0 \\ -1 & 0 & -0.1 & 0 \\ 0 & 0.5 & 0.5 & 0.5 \end{bmatrix}, \; g(u_1) = \mathbf{e}_1, \; g(u_2) = \mathbf{e}_2, \; g(u_3) = \mathbf{e}_3,$$

$h = [1, \; 8, \; -2, \; -1]$, $h^0 = 1$.

Then the approximate realization problem is solved as follows:

covariance matrix	eigenvalues						
	1	2	3	4	5	$\cdots$	12
$H^T_{\underline{a}\,(2,40)}(1,2)H_{\underline{a}\,(2,40)}(1,2)$	2225	332					
$H^T_{\underline{a}\,(3,40)}(1,2,3)H_{\underline{a}\,(3,40)}(1,2,3)$	2441	423	328				
$H^T_{\underline{a}\,(4,40)}(1,\cdots,4)H_{\underline{a}\,(4,40)}(1,\cdots,4)$	2625	473	416	5.6			
$H^T_{\underline{a}\,(5,40)}(1,\cdots,5)H_{\underline{a}\,(5,40)}(1,\cdots,5)$	5741	469	422	7.9	0		
$H^T_{\underline{a}\,(12,40)}(1,\cdots,12)H_{\underline{a}\,(12,40)}(1,\cdots,12)$	17356	4658	989	16	0	$\cdots$	0
covariance matrix	square root of eigenvalues						
$H^T_{\underline{a}\,(2,40)}(1,2)H_{\underline{a}\,(2,40)}(1,2)$	47.2	19.2					
$H^T_{\underline{a}\,(3,40)}(1,2,3)H_{\underline{a}\,(3,40)}(1,2,3)$	49.4	20.6	18.1				
$H^T_{\underline{a}\,(4,40)}(1,\cdots,4)H_{\underline{a}\,(4,40)}(1,\cdots,4)$	51.2	21.7	20.4	2.4			
$H^T_{\underline{a}\,(5,40)}(1,\cdots,5)H_{\underline{a}\,(5,40)}(1,\cdots,5)$	75.8	21.7	20.5	2.8	0		
$H^T_{\underline{a}\,(12,40)}(1,\cdots,12)H_{\underline{a}\,(12,40)}(1,\cdots,12)$	132	68.2	31.4	4	0	$\cdots$	0

1) Since the ratio $\frac{2.4}{51.2} = 0.05$ obtained by the square root of $H^T_{\underline{a}\,(4,40)}(1,\cdots,4) \times H_{\underline{a}\,(4,40)}(1,\cdots,4)$ and the ratio $\frac{4}{132} = 0.03$ obtained by the square root of $H^T_{\underline{a}\,(12,40)}(1,\cdots,12)H_{\underline{a}\,(12,40)}(1,\cdots,12)$ are small, the approximate 3-dimensional affine dynamical system obtained by the CLS method may be somewhat good.
2) After determining the independent vectors $\underline{S}_l(u_1)\underline{a} - \underline{a}$, $\underline{S}_l(u_2)\underline{a} - \underline{a}$ and $\underline{S}_l(u_3)\underline{a} - \underline{a}$ whose numerical value of input are 1, 2 and 3, we will continue the approximate realization algorithm by the CLS method.

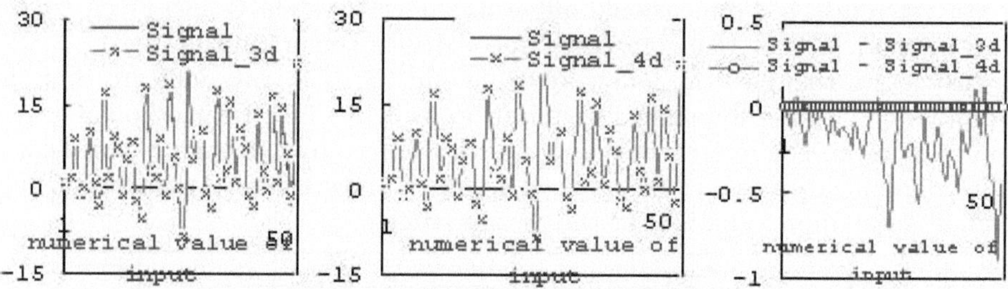

Fig. 7.4. The left is the original input response map and the behavior of a 3-dimensional affine dynamical system obtained by the CLS method. The middle is the original input response map and the behavior of a 4-dimensional affine dynamical system obtained by the CLS method. The right is the difference between the original one and the behavior of the 3-dimensional affine dynamical system obtained by the CLS method or the 4-dimensional affine dynamical system obtained by the CLS method in Example (7.25).

Therefore, an approximate 3-dimensional affine dynamical system $\sigma_1 = ((\mathbf{R}^3, F_1), g_1, h_1, h^0)$ obtained by the CLS method is constructed as follows:

$$F_1(u_1) = \begin{bmatrix} -1.66 & -0.39 & -1.13 \\ 0.07 & 0.04 & 1.06 \\ 0.3 & 2.5 & -0.04 \end{bmatrix}, \ F_1(u_2) = \begin{bmatrix} 0.5 & 0 & -0.4 \\ 0.04 & 1 & 0.04 \\ -0.12 & 0 & 0.8 \end{bmatrix},$$

$$F_1(u_3) = \begin{bmatrix} 0 & -0.7 & -0.15 \\ 0 & 0.64 & 0.07 \\ -1 & 0.3 & 0.2 \end{bmatrix}, \ g_1(u_1) = \mathbf{e}_1, \ g_1(u_2) = \mathbf{e}_2, \ g_1(u_3) = \mathbf{e}_3,$$

$h_1 = [1.1, \ 8, \ -2], \ h^0 = 1$.

For reference, a 4-dimensional affine dynamical system $\sigma_2 = ((\mathbf{R}^4, F_2), g_2, h_2, h^0)$ obtained by the CLS method can be expressed as follows:

$$F_2(u_1) = \begin{bmatrix} -1 & 0 & -0.6 & 0.65 \\ 0 & 0 & 1 & 0.3 \\ 0 & 2.3 & -0.3 & 0 \\ 1 & 0.6 & 0.84 & -0.1 \end{bmatrix}, \ F_2(u_2) = \begin{bmatrix} 0.9 & 0 & 0 & 0 \\ 0 & 1 & 0 & 0 \\ -0.3 & 0 & 0.6 & 0.3 \\ 0.6 & 0 & 0.6 & 0.3 \end{bmatrix},$$

$$F_2(u_3) = \begin{bmatrix} 0 & 0 & 0.5 & 0.05 \\ 0 & 0.5 & 0 & 0 \\ -1 & 0 & -0.1 & 0.5 \\ 0 & 1 & 1 & 0.5 \end{bmatrix}, \ g_2(u_1) = \mathbf{e}_1, \ g_2(u_2) = \mathbf{e}_2, \ g_2(u_3) = \mathbf{e}_3,$$

$h_2 = [1, \ 8, \ -2, \ -1], \ h^0 = 1$.

In this example, the original signals are considered as the input response map of a 4-dimensional affine dynamical system and the desirable input response map is obtained by the CLS method within our expectations. The model obtained by the CLS method is a 3-dimensional affine dynamical system.

For reference, a 4-dimensional affine dynamical system is also given by the CLS method. The system completely reconstructs the original system.

Just as we thought, the following table and Fig. 7.4 truly indicate that the 3-dimensional affine dynamical system obtained by the CLS method is a somewhat good approximation. For reference, the input response map of the same dimensional affine dynamical system as the original system is shown. Hence, there exists a somewhat good approximation for the given system.

dimen-ion	ratio of matrices	mean values of square root for sum of			cosine	error ratio
		signal ①	signal by CLS ②	error ③	$\cos\theta$	③/①
$a_{1,2,3}$	0.05	1.32	1.33	0.04	0.999	0.03
$a_{1,2,3,5}$	0	1.32	1.32	0	1	0

Example 7.26. Let the signals be the input response map of the following 5-dimensional affine dynamical system: $\sigma = ((\boldsymbol{R}^5, F), g, h, h^0)$, where

$$F(u_1) = \begin{bmatrix} -1 & -0.4 & -0.6 & -1 & -1 \\ 0 & -0.2 & 0.5 & 0.4 & -0.2 \\ 0 & -0.2 & 0 & 0.8 & 0 \\ 1 & 0.2 & 0.1 & 0.7 & 1 \\ 0 & 0.2 & 0.7 & 0.3 & 0 \end{bmatrix}, F(u_2) = \begin{bmatrix} 0 & -0.4 & -0.4 & 0 & 0 \\ -1 & 0.21 & -0.2 & -0.5 & 0 \\ 0 & 0.2 & 0.6 & 0.4 & 0 \\ 0 & -0.2 & 0.2 & 0.5 & 0 \\ 1 & 0.7 & 0.2 & 0.5 & 0.9 \end{bmatrix},$$

$$F(u_3) = \begin{bmatrix} 0 & -0.6 & -0.31 & -0.5 & -0.7 \\ 0 & 0.7 & 0.1 & 0.2 & 0.7 \\ -1 & 0 & -0.1 & 0.4 & -0.3 \\ 0 & 0.5 & 0.5 & 0.8 & 0.6 \\ 0 & 0.2 & -0.1 & -0.2 & 0.2 \end{bmatrix}, g(u_1) = \mathbf{e}_1, g(u_2) = \mathbf{e}_2,$$

$g(u_3) = \mathbf{e}_3$, $h = [12, -1, -2, 1, 7]$, $h^0 = 1$.

Then the approximate realization problem is solved as follows:

covariance matrix	eigenvalues							
	1	2	3	4	5	6	$\cdots$	12
$H_{\underline{a}\,(3,40)}^T(1,2,3)H_{\underline{a}\,(3,40)}(1,2,3)$	1412	1148	242					
$H_{\underline{a}\,(4,40)}^T(1,\cdots,4)H_{\underline{a}\,(4,40)}(1,\cdots,4)$	3577	1202	243	151				
$H_{\underline{a}\,(5,40)}^T(1,\cdots,5)H_{\underline{a}\,(5,40)}(1,\cdots,5)$	4943	1219	433	151	1.8			
$H_{\underline{a}\,(6,40)}^T(1,\cdots,6)H_{\underline{a}\,(6,40)}(1,\cdots,6)$	4943	1219	433	151	1.6	0		
$H_{\underline{a}\,(12,40)}^T(1,\cdots,12)H_{\underline{a}\,(12,40)}(1,\cdots,12)$	16485	3728	1243	170	1.9	0	$\cdots$	0

covariance matrix	square root of eigenvalues							
$H_{\underline{a}\,(3,40)}^T(1,2,3)H_{\underline{a}\,(3,40)}(1,2,3)$	37.6	33.9	15.6					
$H_{\underline{a}\,(4,40)}^T(1,\cdots,4)H_{\underline{a}\,(4,40)}(1,\cdots,4)$	59.8	34.7	15.6	12.3				
$H_{\underline{a}\,(5,40)}^T(1,\cdots,5)H_{\underline{a}\,(5,40)}(1,\cdots,5)$	70.3	34.9	20.8	12.3	1.3			
$H_{\underline{a}\,(6,40)}^T(1,\cdots,6)H_{\underline{a}\,(6,40)}(1,\cdots,6)$	70.3	34.9	20.8	12.3	1.3	0		
$H_{\underline{a}\,(12,40)}^T(1,\cdots,12)H_{\underline{a}\,(12,40)}(1,\cdots,12)$	128	61	35.3	13	1.4	0	$\cdots$	0

1) Since the ratio $\frac{1.3}{70.3} = 0.02$ obtained by the square root of $H_{\underline{a}\,(5,40)}^T(1,\cdots,5) \times H_{\underline{a}\,(5,40)}(1,\cdots,5)$ and the ratio $\frac{1.4}{128} = 0.01$ obtained by the square root of $H_{\underline{a}\,(12,40)}^T(1,\cdots,12)H_{\underline{a}\,(12,40)}(1,\cdots,12)$ are small, the approximate 4-dimensional affine dynamical system obtained by the CLS method may be good.

2) After determining the independent vectors $\underline{S}_l(u_1)\underline{a} - \underline{a}$, $\underline{S}_l(u_2)\underline{a} - \underline{a}$ and $\underline{S}_l(u_3)\underline{a} - \underline{a}$ whose numerical value of input are 1, 2, 3 and 4, we will continue the approximate realization algorithm by the CLS method.

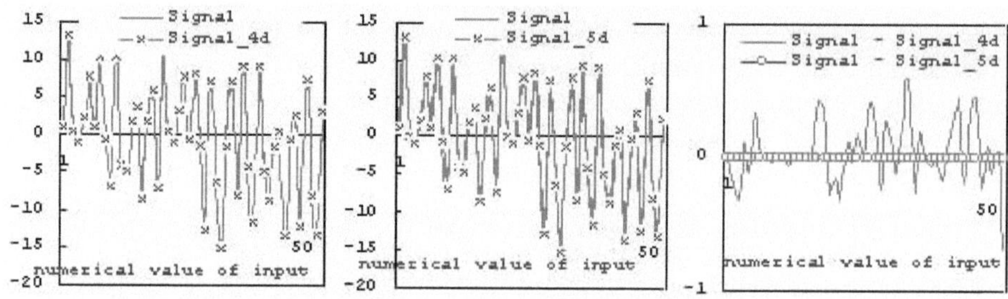

Fig. 7.5. The left is the original input response map and the behavior of a 4-dimensional affine dynamical system obtained by the CLS method. The middle is the original input response map and the behavior of a 5-dimensional affine dynamical system obtained by the CLS method. The right is the difference between the original one and the behavior of the 4-dimensional affine dynamical system obtained by the CLS method or the 5-dimensional affine dynamical system obtained by the CLS method in Example (7.26).

Therefore, an approximate 4-dimensional affine dynamical system $\sigma_1 = ((\boldsymbol{R}^4, F_1), g_1, h_1, h^0)$ obtained by the CLS method is constructed as follows:

$$F_1(u_1) = \begin{bmatrix} -1 & -0.3 & -0.25 & -0.85 \\ 0 & 0 & 1.2 & 0.7 \\ 0 & -0.3 & -0.26 & 0.7 \\ 1 & 0.3 & 0.35 & 0.8 \end{bmatrix}, \quad F_1(u_2) = \begin{bmatrix} 0.5 & 0.05 & -0.3 & 0.25 \\ 0 & 0.9 & 0 & 0 \\ -0.37 & 0.06 & 0.5 & 0.22 \\ 0.36 & 0.05 & 0.27 & 0.7 \end{bmatrix},$$

$$F_1(u_3) = \begin{bmatrix} 0 & -0.5 & -0.36 & -0.6 \\ 0 & 0.94 & 0 & 0 \\ -1 & 0.07 & 0.06 & 0.47 \\ 0 & 0.6 & 0.46 & 0.73 \end{bmatrix}, \quad g_1(u_1) = \mathbf{e}_1, \quad g_1(u_2) = \mathbf{e}_2,$$

$g_1(u_3) = \mathbf{e}_3$, $h_1 = [12.1, \ 0.68, \ -2.1, \ 1.1]$, $h^0 = 1$.

For reference, a 5-dimensional affine dynamical system $\sigma_2 = ((\boldsymbol{R}^4, F_2), g_2, h_2, h^0)$ obtained by the CLS method can be expressed as follows:

$$F_2(u_1) = \begin{bmatrix} -1 & -0.4 & -0.6 & -1 & -1 \\ 0 & -0.2 & 0.5 & 0.4 & -0.2 \\ 0 & -0.2 & 0 & 0.8 & 0 \\ 1 & 0.2 & 0.14 & 0.7 & 1 \\ 0 & 0.2 & 0.7 & 0.3 & 0 \end{bmatrix}, \quad F_2(u_2) = \begin{bmatrix} 0 & -0.4 & -0.4 & 0 & 0 \\ -1 & 0.2 & -0.2 & -0.5 & 0 \\ 0 & 0.2 & 0.6 & 0.4 & 0 \\ 0 & -0.2 & 0.2 & 0.5 & 0 \\ 1 & 0.7 & 0.2 & 0.5 & 0.9 \end{bmatrix},$$

$$F_2(u_3) = \begin{bmatrix} 0 & -0.6 & -0.31 & -0.5 & -0.7 \\ 0 & 0.7 & 0.1 & 0.2 & 0.7 \\ -1 & 0 & -0.1 & 0.4 & -0.3 \\ 0 & 0.2 & -0.1 & -0.2 & 0.2 \end{bmatrix}, \quad g_2(u_1) = \mathbf{e}_1, \quad g_2(u_2) = \mathbf{e}_2,$$

$g_2(u_3) = \mathbf{e}_3$, $h_2 = [12, \ -1, \ -2, \ 1, \ 7]$, $h^0 = 1$.

In this example, the original signals are considered as the input response map of a 5-dimensional affine dynamical system and the desirable input response map is obtained by the CLS method within our expectations. The model obtained by the CLS method is a 4-dimensional affine dynamical system.

For reference, a 5-dimensional affine dynamical system is also given by the CLS method. The system completely reconstructs the original system.

Just as we expected, the following table and Fig. 7.5 truly indicate that the 3-dimensional affine dynamical system obtained by the CLS method is a somewhat good approximation. For reference, the input response map of the same dimensional affine dynamical system as the original system is shown. Hence, there exists a good approximation for the given system.

dimenion	ratio of matrices	mean values of square root for sum of			cosine	error ratio
		signal	signal by CLS	error	$\cos\theta$	
		①	②	③		③/①
$a_{1,2,3,4}$	0.02	1.05	1.04	0.03	0.999	0.03
$a_{1,2,3,4,5}$	0	1.05	1.05	0	1	0

Example 7.27. Let the signals be the input response map of the following 5-dimensional affine dynamical system: $\sigma = ((\boldsymbol{R}^5, F), g, h, h^0)$, where

$$F(u_1) = \begin{bmatrix} -1 & -0.2 & -0.7 & -0.2 & 0.2 \\ 0 & 0.2 & -0.4 & 0.2 & 0.4 \\ 0 & 0 & -0.7 & 0 & 0 \\ 1 & 0 & 0 & 0 & 0.1 \\ 0 & -0.2 & -0.4 & -0.2 & -0.2 \end{bmatrix}, F(u_2) = \begin{bmatrix} 0 & 0.5 & 0.8 & 0.2 & 0.7 \\ -1 & 0.1 & -0.2 & -0.4 & 0.2 \\ 0 & 0 & 0.6 & 0 & 0 \\ 0 & 0 & 0.1 & 0 & 0.1 \\ 1 & 0.9 & 0.2 & 0.4 & 0.7 \end{bmatrix},$$

$$F(u_3) = \begin{bmatrix} 0 & -0.4 & 0 & 0 & -0.3 \\ 0 & 0 & 0.2 & 0 & 0 \\ 0 & 0 & 0.1 & 0 & 0 \\ 0 & 0 & 0 & 0 & 0 \\ 0 & 0.1 & 0 & 0 & 0.1 \end{bmatrix}, \ g(u_1) = \mathbf{e}_1, \ g(u_2) = \mathbf{e}_2, \ g(u_3) = \mathbf{e}_3,$$

$h = [6, \ -3, \ -7, \ 0.6, \ -6], \ h^0 = 1.$

Then the approximate realization problem is solved as follows:

covariance matrix	eigenvalues							
	1	2	3	4	5	6	$\cdots$	12
$H^T_{\underline{a}\,(4,40)}(1,\cdots,4)H_{\underline{a}\,(4,40)}(1,\cdots,4)$	234	110	52	0.7				
$H^T_{\underline{a}\,(5,40)}(1,\cdots,5)H_{\underline{a}\,(5,40)}(1,\cdots,5)$	297	113	88	1.7	0.5			
$H^T_{\underline{a}\,(6,40)}(1,\cdots,6)H_{\underline{a}\,(6,40)}(1,\cdots,6)$	433	149	92	1.7	0.5	0		
$H^T_{\underline{a}\,(12,40)}(1,\cdots,12)H_{\underline{a}\,(12,40)}(1,\cdots,12)$	1280	425	252	1.7	0.5	0	$\cdots$	0

covariance matrix	square root of eigenvalues							
$H^T_{\underline{a}\,(4,40)}(1,\cdots,4)H_{\underline{a}\,(4,40)}(1,\cdots,4)$	15.3	10.5	7.2	0.8				
$H^T_{\underline{a}\,(5,40)}(1,\cdots,5)H_{\underline{a}\,(5,40)}(1,\cdots,5)$	17.2	10.6	9.4	1.3	0.7			
$H^T_{\underline{a}\,(6,40)}(1,\cdots,6)H_{\underline{a}\,(6,40)}(1,\cdots,6)$	20.8	12.2	9.6	1.3	0.7	0		
$H^T_{\underline{a}\,(12,40)}(1,\cdots,12)H_{\underline{a}\,(12,40)}(1,\cdots,12)$	35.8	20.6	15.9	1.3	0.7	0	$\cdots$	0

1) Since the ratio $\frac{0.7}{17.2} = 0.04$ obtained by the square root of $H^T_{\underline{a}\,(5,40)}(1,\cdots,5) \times H_{\underline{a}\,(5,40)}(1,\cdots,5)$ and the ratio $\frac{0.7}{35.8} = 0.02$ obtained by the square root of $H^T_{\underline{a}\,(12,40)}(1,\cdots,12)H_{\underline{a}\,(12,40)}(1,\cdots,12)$ are somewhat small, the approximate 4-dimensional affine dynamical system obtained by the CLS method may be good. 2) After determining the independent vectors $\underline{S}_l(u_1)\underline{a}-\underline{a}$, $\underline{S}_l(u_2)\underline{a}-\underline{a}$, $\underline{S}_l(u_3)\underline{a}-\underline{a}$ and $\underline{S}_l(u_1|u_1)\underline{a} - \underline{a}$ whose numerical value of input are 1, 2, 3 and 4, we will continue the approximate realization algorithm by the CLS method.

Therefore, an approximate 4-dimensional affine dynamical system $\sigma_1 = ((\boldsymbol{R}^4, F_1), g_1, h_1, h^0)$ obtained by the CLS method is constructed as follows:

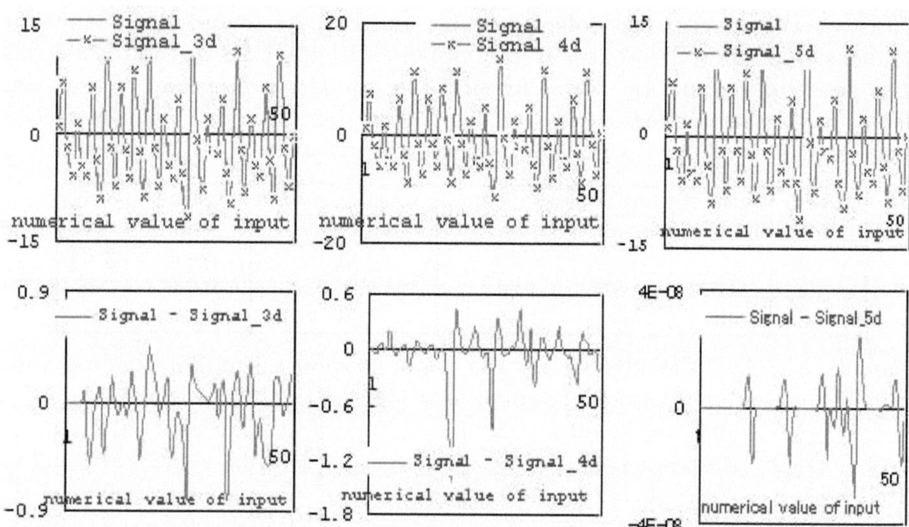

Fig. 7.6. The left are the original input response map and the behavior of a 3-dimensional affine dynamical system obtained by the CLS method. The middle are the original input response map and the behavior of a 4-dimensional affine dynamical system obtained by the CLS method. The right are the difference between the original one and the behavior of a 5-dimensional affine dynamical system obtained by the CLS method in Example (7.27).

$$F_1(u_1) = \begin{bmatrix} -1 & -0.08 & -0.46 & -0.08 \\ 0 & 0.01 & -0.8 & 0.01 \\ 0 & 0.01 & -0.68 & 0.01 \\ 1 & -0.2 & -0.48 & -0.2 \end{bmatrix}, F_1(u_2) = \begin{bmatrix} -0.8 & 0 & 0.69 & -0.04 \\ -0.03 & 0.9 & -0.01 & -0.01 \\ -0.07 & -0.05 & 0.6 & -0.02 \\ 3.2 & 1.3 & 0.3 & 0.5 \end{bmatrix},$$

$$F_1(u_3) = \begin{bmatrix} 0 & -0.46 & 0 & 0 \\ 0 & 0.1 & 0.2 & 0 \\ 0 & -0.01 & 0.1 & 0 \\ 0 & 0.1 & 0 & 0 \end{bmatrix}, \ g_1(u_1) = \mathbf{e}_1, \ g_1(u_2) = \mathbf{e}_2, \ g_1(u_3) = \mathbf{e}_3,$$

$h = [6, \ -3, \ -7, \ 0.4], \ h^0 = 1.$

For reference, a 5-dimensional affine dynamical system σ_2
$= ((\mathbf{R}^4, \ F_2), \ g_2, \ h_2, \ h^0)$ obtained by the CLS method can be expressed as follows:

$$F_2(u_1) = \begin{bmatrix} -1 & -0.2 & -0.7 & -0.2 & 0.2 \\ 0 & 0.2 & -0.4 & 0.2 & 0.4 \\ 0 & 0 & -0.7 & 0 & 0 \\ 1 & 0 & 0 & 0 & 0.1 \\ 0 & -0.2 & -0.4 & -0.2 & -0.2 \end{bmatrix}, \ F_2(u_2) = \begin{bmatrix} 0 & 0.5 & 0.8 & 0.2 & 0.7 \\ -1 & 0.1 & -0.2 & -0.4 & 0.2 \\ 0 & 0 & 0.6 & 0 & 0 \\ 0 & 0 & 0.1 & 0 & 0.1 \\ 1 & 0.8 & 0.2 & 0.4 & 0.7 \end{bmatrix},$$

$$F_2(u_3) = \begin{bmatrix} 0 & -0.4 & 0 & 0 & -0.3 \\ 0 & 0 & 0.2 & 0 & 0 \\ 0 & 0 & 0.1 & 0 & 0 \\ 0 & 0 & 0 & 0 & 0 \\ 0 & 0.1 & 0 & 0 & 0.1 \end{bmatrix}, \ g_2(u_1) = \mathbf{e}_1, \ g_2(u_2) = \mathbf{e}_2,$$

$g_2(u_3) = \mathbf{e}_3, \ h_2 = [6, \ -3, \ -7, \ 0.8, \ -6], \ h^0 = 1.$

In this example, the original signals are considered as the input response map of a 5-dimensional affine dynamical system and the desirable input response map is obtained by the CLS method within our expectations. The model obtained by the CLS method is a 4-dimensional affine dynamical system.

For reference, a 5-dimensional affine dynamical system is also given by the CLS method. The system completely reconstructs the original system.

Just as we thought, the following table and Fig. 7.6 truly indicate that the 4-dimensional affine dynamical system obtained by the CLS method is a somewhat good approximation. For reference, the input response map of the same dimensional affine dynamical system as the original system is shown. Hence, there exists a good approximation for the given system.

dimen-ion	ratio of matrices	mean values of square root for sum of			cosine ① and ②	error ratio
		signal ①	signal by CLS ②	error ③	$\cos\theta$	③/①
$a_{1,2,3}$	0.07	0.925	0.917	0.04	0.9990	0.04
$a_{1,2,3,4}$	0.03	0.925	0.92	0.03	0.9995	0.03
$a_{1,2,3,4,5}$	0	0.925	0.925	0	1	0

7.5 Noisy Realization of Affine Dynamical Systems

In this section, we discuss noisy realization problems of affine dynamical systems.

For noise $\{\bar{\gamma}(t) : t \in N\}$ added to an unknown affine dynamical system a, we will obtain the observed data $\{\hat{\gamma}(|\omega|) + \bar{\gamma}(|\omega|) : \omega \in U^*\}$.

For any given $\{\hat{\gamma}(|\omega|) + \bar{\gamma}(|\omega|) : \omega \in U^*\}$, σ which satisfies $a_\sigma(\omega) \approx \hat{\gamma}(|\omega|) : \omega \in U^*$ is called a noisy realization of a.

Roughly speaking, we can propose the following noisy realization problem: For any given $\{\hat{\gamma}(|\omega|) + \bar{\gamma}(|\omega|) : \omega \in U^*\}$, find an affine dynamical system σ which satisfies $a_\sigma(\omega) \approx \hat{\gamma}(|\omega|)$ for any $\omega \in U^*$.

In order to make our discussion simple, we assume that the set Y of outout is the set R of real numbers, namely 1-output.

A situation for noisy realization problem 7.28
Let the observed object be an affine dynamical system and noise be added to the output. Then we will obtain the data $\{\gamma(t) = \hat{\gamma}(t) + \bar{\gamma}(t) : 0 \leq t \leq \underline{N}\}$ for some integer $\underline{N} \in N$, where $\hat{\gamma}(t)$ is the exact signal which comes from the observed affine dynamical system and $\bar{\gamma}(t)$ is the noise added at the time of observation.

Problem statement of noisy realization for affine dynamical systems 7.29
Let $H_{\underline{a}\ (p,\bar{p})}$ be the measured finite-sized Input/output matrix. Then find the cleaned-up Input/output matrix $\hat{H}_{\underline{a}\ (p,\bar{p})}$ such that $H_{\underline{a}\ (p,\bar{p})} = \hat{H}_{\underline{a}\ (p,\bar{p})} + \bar{H}_{\underline{a}\ (p,\bar{p})}$ holds.

Namely, find a minimal dimensional affine dynamical system $\sigma = ((R^n, F_r), g_r, h_r, h^0))$ which realizes $\hat{H}_{\underline{a}\ (p,\bar{p})}$.

Theorem 7.30. *Algorithm of noisy realization for Affine Danamical systems*
Let an input response map $\underline{a}$ be a considered object which is an affine dynamical
system. Then an approximate realization $\sigma = ((\mathbf{R}^n, F_s), g_s, h_s, h^0)$ of $\underline{a}$ is given
by the following algorithm:
1) Based on the square root of eigenvalues for a matrix
  $H_{\underline{a}\ (n,\bar{p})}(|||\omega_1|||, \cdots, |||\omega_n|||) H_{\underline{a}\ (n,\bar{p})}(|||\omega_1|||, \cdots, |||\omega_n|||)^T$, determine the value n
  of rank for the matrix, where $|||\omega_1|||$, $|||\omega_2|||$, $\cdots$ and $|||\omega_n|||$ are suitably
  selected in order of numerical value of input and
  $\{\underline{S_l}(\omega_i)\underline{a} - \underline{a}; 1 \leq i \leq n,\ \omega_i \in U^*\}$ is a set of independent vectors.
  Namely, determine the value n of rank for the matrix $H_{\underline{a}\ (n,\bar{p})}(|||\omega_1|||, \cdots, |||\omega_n|||)$
  such that a set of the square root of eigenvalues for the covariance matrix
  composed of relatively small and equally-sized numbers is excluded, where
  the signal part effected by the set may be the noisy part of the observed
  data.
2) In order to determine g_s, the CLS method is used as follows:
  ① In particular, set $g_s(\omega_i) := \mathbf{e}_i$ for $\omega_i \in U$. Namely, $g_s(\omega_1) := \mathbf{e}_1$,
  $g_s(\omega_2) := \mathbf{e}_2, \cdots, g_s(\omega_k) := \mathbf{e}_k$ for some $k \in N$.
  For $u \in U$ such that $u \notin \{\omega_i; 1 \leq i \leq n\}$ and $\omega_r < u$,
  $g_s(u) = \sum_{j=1}^{r} b_{u,j}(\underline{S_l}(\omega_j)\underline{a} - \underline{a})$ is obtained as follows:
  Let a matrix $A_u \in \mathbf{R}^{1 \times (r+1)}$ be $A_u := [b_{u,1}, b_{u,2}, \cdots, b_{u,r}, -1]$.
  Choose the coefficients $\{b_{u,j} : 1 \leq j \leq r\}$ such that
  $\sum_{j=1}^{r}(\underline{S_l}(\omega_j)\bar{\underline{a}} - \bar{\underline{a}}) \cdot (\underline{S_l}(\omega_j)\bar{\underline{a}} - \bar{\underline{a}})) + (\underline{S_l}(u)\bar{\underline{a}} - \bar{\underline{a}}) \cdot (\underline{S_l}(u)\bar{\underline{a}} - \bar{\underline{a}})$ take
  a minimum value, where $\{\underline{S_l}(\omega_j)\bar{\underline{a}} - \bar{\underline{a}} \in \mathbf{R}^{L \times 1} : 0 \leq j \leq r\}$ are given by
  the equation $[\underline{S_l}(\omega_1)\bar{\underline{a}} - \bar{\underline{a}}, \underline{S_l}(\omega_2)\bar{\underline{a}} - \bar{\underline{a}}, \cdots, \underline{S_l}(\omega_r)\bar{\underline{a}} - \bar{\underline{a}}, \underline{S_l}(u)\bar{\underline{a}} - \bar{\underline{a}}]^T :=$
  $A_u^T[A_u A_u^T]^{-1} A_u H_{\underline{a}\ (r+1,L)}^T(|||\omega_1|||, |||\omega_2|||, \cdots, |||\omega_r|||, |||u|||)$
  and $H_{\underline{a}\ (|||u|||+1,L)}^T(|||\omega_1|||, |||\omega_2|||, \cdots, |||\omega_r|||, |||u|||) :=$
  $[\underline{S_l}(\omega_1)\underline{a} - \underline{a}, \cdots, \underline{S_l}(\omega_r)\underline{a} - \underline{a}, \underline{S_l}(u)\underline{a} - \underline{a}]$. And $\cdot$ denotes the inner product
  of two vectors.
3) In order to obtain F_s, the CLS method is used as follows:
  ① Let $_i\mathbf{f}_j \in \mathbf{R}^n$ in $F_s(u_i) := [_i\mathbf{f}_1\ _i\mathbf{f}_2\ \cdots\ _i\mathbf{f}_n]$ be $_i\mathbf{f}_j := [_if_{j,1}\ _if_{j,2}\ \cdots\ _if_{j,n}]^T$
  for $1 \leq i \leq n$, where $_if_{j,k}$ is given by the following:
  $\underline{S_l}(u_i)(\underline{S_l}(\omega_j)\underline{a} - \underline{a}) = \sum_{k=1}^{n}\ _if_{j,k}(\underline{S_l}(\omega_k)\underline{a} - \underline{a})$, $_if_{j,k} \in \mathbf{R}$ in the sense of
  $F(U_{L-p}^*, Y)$.
  We cannot directly obtain the coefficients $\{_i\mathbf{f}_j; 1 \leq i \leq 3,\ 1 \leq j \leq n\}$.
  Firstly, we will determine the coefficients $\{_i\bar{f}_{j,k}; 1 \leq k \leq n\}$ from the
  equation $\underline{S_l}(u_i)\underline{S_l}(\omega_j)\underline{a} - \underline{a} = \sum_{k=1}^{n}\ _i\bar{f}_{j,k}(\underline{S_l}(\omega_k)\underline{a} - \underline{a})$ by using the CLS
  method.
  ② For $i\ (1 \leq i \leq 3)$, $j\ (1 \leq j \leq n)$ and for the maximum number
  $r\ (1 \leq r \leq n)$ such that $\omega_r, \omega_j \in \{\omega_j; 1 \leq j \leq n\}$ and $|||\omega_r||| < |||u_i|\omega_j|||$,
  let a matrix $_iA_j \in \mathbf{R}^{1 \times (r+1)}$ be $_iA_j := [_i\bar{f}_{j,1}, _i\bar{f}_{j,2}, \cdots, _i\bar{f}_{j,r}, -1]$.
  Choose the coefficients $_i\bar{f}_{j,k} = 0$ for $k\ (r + 1 \leq k \leq n)$ and $\{_i\bar{f}_{j,k} : 1 \leq k \leq r\}$
  such that $\sum_{i=1}^{r}(\underline{S_l}(\omega_i)\bar{\underline{a}} - \bar{\underline{a}}) \cdot (\underline{S_l}(\omega_i)\bar{\underline{a}} - \bar{\underline{a}}) + (\underline{S_l}(u_i|\omega_j)\bar{\underline{a}} - \bar{\underline{a}}) \cdot (\underline{S_l}(u_i|\omega_j)\bar{\underline{a}} - \bar{\underline{a}})$
  takes a minimum value, where $\{\underline{S_l}(\omega_i)\bar{\underline{a}} - \bar{\underline{a}} \in \mathbf{R}^{L \times 1} : 0 \leq i \leq n\}$
  and $\underline{S_l}(u_i|\omega_j)\bar{\underline{a}} - \bar{\underline{a}} \in \mathbf{R}^{L \times 1}$ are given by the equation
  $[\underline{S_l}(\omega_1)\bar{\underline{a}} - \bar{\underline{a}}, \underline{S_l}(\omega_2)\bar{\underline{a}} - \bar{\underline{a}}, \cdots, \underline{S_l}(\omega_r)\bar{\underline{a}} - \bar{\underline{a}}, \underline{S_l}(u_i|\omega_j)\bar{\underline{a}} - \bar{\underline{a}}]^T :=$

$_iA_j^T[_iA_j\ _iA_j^T]^{-1}\ _iA_j\ H_{\underline{a}\ (r+1,L)}^T(\||\omega_1\||, \||\omega_2\||, \cdots, \||\omega_r\||, \||u_i|\omega_j\||)$ and
$H_{\underline{a}\ (r+1,L)}^T(\||\omega_1\||, \cdots, \||\omega_r\||, \||u_i|\omega_j\||) :=$
$[\underline{S_l}(\omega_1)\underline{a} - \underline{a}, \cdots, \underline{S_l}(\omega_r)\underline{a} - \underline{a}, \underline{S_l}(u_i|\omega_j)\underline{a} - \underline{a}].$
And $\bullet$ denotes the inner product of two vectors.

③ Next, using the equations
$\underline{S_l}(u_i)(\underline{S_l}(\omega_j)\underline{a} - \underline{a})$
$= \underline{S_l}(u_i)\underline{S_l}(\omega_j)\underline{a} - \underline{a} - \underline{S_l}(u_i)\underline{a} + \underline{a}$
$= \sum_{k=1}^n {}_i\bar{f}_{j,k}(\underline{S_l}(\omega_k)\underline{a} - \underline{a}) - \sum_{j=1}^{r_{u_i}} b_{u_i j}(\underline{S_l}(u_j)\underline{a} - \underline{a}),$
we obtain $\underline{S_l}(u_i)(\underline{S_l}(\omega_j)\underline{a} - \underline{a})$
$= \sum_{k=1}^n {}_i\bar{f}_{j,k}(\underline{S_l}(\omega_k)\underline{a} - \underline{a}) - \sum_{j=1}^{r_{u_i}} b_{u_i,i}(\underline{S_l}(u_j)\underline{a} - \underline{a})$
$= \sum_{j=1}^{r_{u_i}}(_i\bar{f}_{j,k} - b_{u_i,j})(\underline{S_l}(\omega_j)\underline{a} - \underline{a}) + \sum_{j=r_{u_i}+1}^n {}_i\bar{f}_{j,k}(\underline{S_l}(\omega_j)\underline{a} - \underline{a}).$
On the other hand, the equation
$\underline{S_l}(u_i)(\underline{S_l}(\omega_j)\underline{a} - \underline{a}) = \sum_{k=1}^n {}_if_{j,k}(\underline{S_l}(\omega_k)\underline{a} - \underline{a})$ holds. Therefore, we obtain
the following equation:
$\sum_{k=1}^n {}_if_{j,k}(\underline{S_l}(\omega_k)\underline{a} - \underline{a})$
$= \sum_{j=1}^{r_{u_i}}(_i\bar{f}_{j,k} - b_{u_i,j})\underline{S_l}(\omega_j)\underline{a} - \underline{a} + \sum_{j=r_{u_i}+1}^n {}_i\bar{f}_{j,k}(\underline{S_l}(\omega_j)\underline{a} - \underline{a}).$
Comparing the coefficients, we obtain the following:
$\sum_{j=1}^{r_{u_i}}(_if_{j,k} - {}_i\bar{f}_{j,k} + b_{u_i,j})(\underline{S_l}(\omega_j)\underline{a} - \underline{a}) + \sum_{j=r_{u_i}+1}^n(_if_{j,k} - {}_i\bar{f}_{j,k})(\underline{S_l}(\omega_j)\underline{a} - \underline{a})$
$= 0.$
Finally, coefficients $_if_{j,k}$ of $_i\mathbf{f}_j$ in F_s are obtained as follows:
$_if_{j,k} = {}_i\bar{f}_{j,k} - b_{u_i,j}$ for $1 \leq j \leq r_{u_i},$
$_if_{j,k} = {}_i\bar{f}_{j,k}$ for $r_{u_i} + 1 \leq j \leq n.$

4) In order to determine $h_s \in \mathbf{R}^{1 \times n}$, the CLS method is used as follows:
① For the first $\omega_{r_1+1}, \omega_{r_1} \in \{\omega_i; 1 \leq i \leq n\}$ such that $\omega_{r_1+1} > \omega_{r_1}$ and
$\||\omega_{r_1+1}\||$ - $\||\omega_{r_1}\|| > 1$ when starting out from $\omega_1,$
set $\underline{S_l}(\lambda_1)\underline{a} - \underline{a} := \sum_{i=1}^{r_1} b_{\lambda_1,i}\mathbf{e}_i$ for the obtained equation $\underline{S_l}(\lambda_1)\underline{a} - \underline{a} =$
$\sum_{i=1}^{r_1} b_{\lambda_1,i}(\underline{S_l}(\omega_i)\underline{a} - \underline{a})$ for λ_1 such that $\||\lambda_1\|| = \||\omega_{r_1}\|| + 1.$
Let a matrix $A_{\lambda_1} \in \mathbf{R}^{1 \times (r_1+1)}$ be $A_{\lambda_1} := [b_{\lambda_1,1}, b_{\lambda_1,2}, \cdots, b_{\lambda_1,r_1}, -1].$
Choose the coefficients $\{b_{\lambda_1,i} : 1 \leq i \leq r_1\}$ such that
$\sum_{i=1}^{r_1}(\underline{S_l}(\omega_i)\bar{\underline{a}} - \bar{\underline{a}}) \bullet (\underline{S_l}(\omega_i)\bar{\underline{a}} - \bar{\underline{a}}) + (\underline{S_l}(\lambda_1)\bar{\underline{a}} - \bar{\underline{a}}) \bullet (\underline{S_l}(\lambda_1)\bar{\underline{a}} - \bar{\underline{a}})$ take
a minimum value, where $\{\underline{S_l}(\omega_i)\bar{\underline{a}} - \bar{\underline{a}} \in \mathbf{R}^{L \times 1} : 0 \leq i \leq r_1\}$ are given by
the equation $[\underline{S_l}(\omega_1)\bar{\underline{a}} - \bar{\underline{a}}, \underline{S_l}(\omega_2)\bar{\underline{a}} - \bar{\underline{a}}, \cdots, \underline{S_l}(\omega_{r_1})\bar{\underline{a}} - \bar{\underline{a}}, \underline{S_l}(\lambda_1)\bar{\underline{a}} - \bar{\underline{a}}]^T :=$
$A_{\lambda_1}^T[A_{\lambda_1} A_{\lambda_1}^T]^{-1}A_{\lambda_1} H_{\underline{a}\ (r_1+1,L)}^T(\||\omega_1\||, \||\omega_2\||, \cdots, \||\omega_{r_1}\||, \||\lambda_1\||)$
and $H_{\underline{a}\ (\||\lambda_1\||+1,L)}^T(\||\omega_1\||, \||\omega_2\||, \cdots, \||\omega_{r_1}\||, \||\lambda_1\||) :=$
$[\underline{S_l}(\omega_1)\underline{a} - \underline{a}, \cdots, \underline{S_l}(\omega_{r_1})\underline{a} - \underline{a}, \underline{S_l}(\lambda_1)\underline{a} - \underline{a}].$ And $\bullet$ denotes the inner product
of two vectors.
Then let $h_{\lambda_1 s} \in \mathbf{R}^{1 \times r_1}$ be
$h_{\lambda_1 s} := [\underline{a}(\omega_1) - \underline{a}(1) - (\bar{\underline{a}}(\omega_1) - \bar{\underline{a}}(1)), \underline{a}(\omega_2) - \underline{a}(1) - (\bar{\underline{a}}(\omega_2) - \bar{\underline{a}}(1)), \cdots,$
$\underline{a}(\omega_{r_1}) - \underline{a}(1) - (\bar{\underline{a}}(\omega_{r_1}) - \bar{\underline{a}}(1))].$

② For the first $\omega_{r_2+1}, \omega_{r_2} \in \{\omega_i; 1 \leq i \leq n\}$ such that $\omega_{r_2+1} > \omega_{r_2}$ and
$\||\omega_{r_2+1}\||$ - $\||\omega_{r_2}\|| > 1$ when starting out from $\omega_{r_1+1},$
set $\underline{S_l}(\lambda_2)\underline{a} - \underline{a} := \sum_{i=1}^{r_2} b_{\lambda_2,i}\mathbf{e}_i$ for the obtained equation $\underline{S_l}(\lambda_2)\underline{a} - \underline{a} =$
$\sum_{i=1}^{r_2} b_{\lambda_2,i}(\underline{S_l}(\omega_i)\underline{a} - \underline{a})$ for λ_2 such that $\||\lambda_2\|| = \||\omega_{r_2}\|| + 1.$
Let a matrix $A_{\lambda_2} \in \mathbf{R}^{1 \times (r_2+1)}$ be $A_{\lambda_2} := [b_{\lambda_2,1}, b_{\lambda_2,2}, \cdots, b_{\lambda_2,r_2}, -1].$

Choose the coefficients $\{b_{\lambda_2,i} : 1 \leq i \leq r_2\}$ such that
$\sum_{i=1}^{r_2}(\underline{S_l}(\omega_i)\bar{a} - \bar{a}) \cdot (\underline{S_l}(\omega_i)\bar{a} - \bar{a}) + (\underline{S_l}(\lambda_2)\bar{a} - \bar{a}) \cdot (\underline{S_l}(\lambda_2)\bar{a} - \bar{a})$ *take*
a minimum value, where $\{\underline{S_l}(\omega_i)\bar{a} - \bar{a} \in \mathbf{R}^{L \times 1} : 0 \leq i \leq r_2\}$ are given by
the equation $[\underline{S_l}(\omega_1)\bar{a} - \bar{a}, \underline{S_l}(\omega_2)\bar{a} - \bar{a}, \cdots, \underline{S_l}(\omega_{r_2})\bar{a} - \bar{a}, \underline{S_l}(\lambda_2)\bar{a} - \bar{a}]^T :=$
$A_{\lambda_2}^T [A_{\lambda_2} A_{\lambda_2}^T]^{-1} A_{\lambda_2} H_{\underline{a}\ (r_2+1,L)}^T (|||\omega_1|||, |||\omega_2|||, \cdots, |||\omega_{r_2}|||, |||\lambda_2|||)$
and $H_{\underline{a}\ (|||\lambda_2|||+1,L)}^T (|||\omega_1|||, |||\omega_2|||, \cdots, |||\omega_{r_2}|||, |||\lambda_2|||) :=$
$[\underline{S_l}(\omega_1)\underline{a} - \underline{a}, \cdots, \underline{S_l}(\omega_{r_2})\underline{a} - \underline{a}, \underline{S_l}(\lambda_2)\underline{a} - \underline{a}]$. *And* ▪ *denotes the inner product*
of two vectors.
Then let $h_{\lambda_2 s} \in \mathbf{R}^{1 \times r_2}$ be
$h_{\lambda_2 s} := [\underline{a}(\omega_{r_1+1}) - \underline{a}(1) - (\bar{a}(\omega_{r_1+1}) - \bar{a}(1)),$
$\underline{a}(\omega_{r_1+2}) - \underline{a}(1) - (\bar{a}(\omega_{r_1+2}) - \bar{a}(1)), \cdots, \underline{a}(\omega_{r_2}) - \underline{a}(1) - (\bar{a}(\omega_{r_2}) - \bar{a}(1))].$

$\vdots$

$\textcircled{t}$ *For $\omega \in U^*$ such that $|||\omega||| = |||\omega_n||| + 1$, let a matrix $A_\omega \in \mathbf{R}^{1 \times (n+1)}$*
be $A_\omega := [b_{\omega,1}, b_{\omega,2}, \cdots, b_{\omega,n}, -1].$
Choose the coefficients $\{b_{\omega,i} : 1 \leq i \leq n\}$ such that
$\sum_{i=1}^{n}(\underline{S_l}(\omega_i)\bar{a} - \bar{a}) \cdot (\underline{S_l}(\omega_i)\bar{a} - \bar{a}) + (\underline{S_l}(\omega)\bar{a} - \bar{a}) \cdot (\underline{S_l}(\omega)\bar{a} - \bar{a})$ *takes*
a minimum value, where $\{\underline{S_l}(\omega_i)\bar{a} - \bar{a} : 0 \leq i \leq n\}$ and $\underline{S_l}(\omega)\bar{a} - \bar{a}$ are
given by the following equation:
$[\underline{S_l}(\omega_1)\bar{a} - \bar{a}, \underline{S_l}(\omega_2)\bar{a} - \bar{a}, \cdots, \underline{S_l}(\omega_n)\bar{a} - \bar{a}, \underline{S_l}(\omega)\bar{a} - \bar{a}]^T :=$
$A_\omega^T [A_\omega A_\omega^T]^{-1} A_\omega H_{\underline{a}\ (n+1,L)}^T (|||\omega_1|||, |||\omega_2|||, \cdots, |||\omega_n|||, |||\omega|||)$
and $H_{\underline{a}\ (n+1,L)}^T (|||\omega_1|||, |||\omega_2|||, \cdots, |||u|||) :=$
$[\underline{S_l}(\omega_1)\underline{a} - \underline{a}, \cdots, \underline{S_l}(\omega_n)\underline{a} - \underline{a}, \underline{S_l}(\omega)\underline{a} - \underline{a}]$, *and* ▪ *denotes the inner product*
of two vectors.
Then let $h_{\omega s}$ be
$h_{\omega s} := [\underline{a}(\omega_{r_t+1}) - \underline{a}(1) - (\bar{a}(\omega_{r_t+1}) - \bar{a}(1)),$
$\underline{a}(\omega_{r_t+2}) - \underline{a}(1) - (\bar{a}(\omega_{r_t+2}) - \bar{a}(1)), \cdots, \underline{a}(\omega_n) - \underline{a}(1) - (\bar{a}(\omega_n) - \bar{a}(1))].$
Finally, let $h_s \in \mathbf{R}^{1 \times n}$ be
$h_s := [h_{\lambda_1 s}, h_{\lambda_2 s}, \cdots, h_{\omega s}].$

[proof]. In 1), the number of dimensions is determined by checking what parts are noisy parts and by using the ratio of the Hankel matrix norm, which implies the noise to signal ratio. According to Theorem (7.20), an affine dynamical system $\sigma = ((\mathbf{R}^n, F_s), g_s, h_s, h^0)$ is obtained as follows. In 2), g_s is obtained directly or by using the CLS method for A_u corresponding to the matrix A in Proposition (2.14). In 3), F_s is obtained by using the CLS method for $_iA_j$ corresponding to the matrix A in Proposition (2.14). In 4), h_s is obtained by using the CLS method for A_{λ_1}, A_{λ_2}, A_{λ_ω} corresponding to the matrix A in Proposition (2.14). Remark : Let S and N be the norm of a signal and noise. Then the selected ratio of matrices in the algorithm may be considered as $\frac{N}{S+N}$.

In the figures of this chapter, we use a notation *Signal_n d* as the input response map obtained by a n-dimensional affine dynamical system.

In the examples of this chapter, a notation $H_{\underline{a}\ (r,40)}^T (1, \cdots, r)$ is used in place of $H_{\underline{a}\ (r,40)}^T (1, 2, \cdots, r-1, r)$.

Example 7.31. Let the signals be the input response map of the following 3-dimensional affine dynamical system: $\sigma = ((\boldsymbol{R}^3, F), g, h, h^0)$, where $F(u_1) =$
$$\begin{bmatrix} -1 & -0.3 & -0.5 \\ 0 & 0 & -0.8 \\ 0 & -0.3 & -0.7 \end{bmatrix}, \quad F(u_2) = \begin{bmatrix} -0.6 & 0 & 0.7 \\ 0 & 1 & 0 \\ -0.7 & 0 & 0.6 \end{bmatrix}, \quad F(u_3) = \begin{bmatrix} 0 & -0.5 & 0 \\ 0 & 0.1 & 0.3 \\ 0 & 0 & 0.1 \end{bmatrix},$$
$g(u_1) = \mathbf{e}_1$, $g(u_2) = \mathbf{e}_2$, $g(u_3) = \mathbf{e}_3$, $h = [6, \ -3, \ -6]$, $h^0 = 1$.
Then the noisy realization problem is solved as follows:

covariance matrix	eigenvalues					
	1	2	3	4	5	6
$H^T_{\underline{a}\ (2,40)}(1,2)H_{\underline{a}\ (2,40)}(1,2)$	182	117				
$H^T_{\underline{a}\ (3,40)}(1,2,3)H_{\underline{a}\ (3,40)}(1,2,3)$	244	138	69			
$H^T_{\underline{a}\ (4,40)}(1,\cdots,4)H_{\underline{a}\ (4,40)}(1,\cdots,4)$	245	139	69	3.6		
$H^T_{\underline{a}\ (5,40)}(1,\cdots,5)H_{\underline{a}\ (5,40)}(1,\cdots,5)$	286	183	139	4.7	1.9	
$H^T_{\underline{a}\ (6,40)}(1,\cdots,6)H_{\underline{a}\ (6,40)}(1,\cdots,6)$	378	240	146	5.0	1.9	1.8
covariance matrix	square root of eigenvalues					
$H^T_{\underline{a}\ (2,40)}(1,2)H_{\underline{a}\ (2,40)}(1,2)$	13.5	10.8				
$H^T_{\underline{a}\ (3,40)}(1,2,3)H_{\underline{a}\ (3,40)}(1,2,3)$	15.6	11.7	8.3			
$H^T_{\underline{a}\ (4,40)}(1,\cdots,4)H_{\underline{a}\ (4,40)}(1,\cdots,4)$	15.7	11.8	8.3	1.9		
$H^T_{\underline{a}\ (5,40)}(1,\cdots,5)H_{\underline{a}\ (5,40)}(1,\cdots,5)$	16.9	13.5	11.8	2.2	1.4	
$H^T_{\underline{a}\ (6,40)}(1,\cdots,6)H_{\underline{a}\ (6,40)}(1,\cdots,6)$	19.4	15.5	12.1	2.2	1.4	1.3

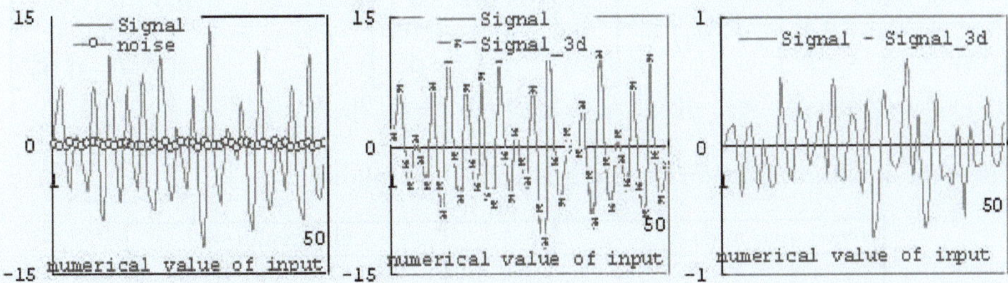

Fig. 7.7. The left is the original input response map and noise added to the original 3-dimensional affine dynamical system. The middle is the original input response map and the behavior of a 3-dimensional affine dynamical system obtained by the CLS method. The right is the difference between the original input response map and the behavior of the 3-dimensional affine dynamical system obtained by the CLS method in Example (7.31).

1) A set $\{2.2, 1.4, 1.3\}$ is composed of relatively small and equally-sized numbers in the square root of eigenvalues for $H^T_{\underline{a}\ (6,40)}(1,\cdots,6)H_{\underline{a}\ (6,40)}(1,\cdots,6)$.
2) After determining the independent vectors $\underline{S}_l(u_1)\underline{a} - \underline{a}$, $\underline{S}_l(u_2)\underline{a} - \underline{a}$ and $\underline{S}_l(u_3)\underline{a} - \underline{a}$ whose numerical value of input are 1, 2 and 3, we will continue the noisy realization algorithm by the CLS method.

Therefore, a noisy 3-dimensional affine dynamical system $\sigma_1 = ((\boldsymbol{R}^3, F_1), g_1, h_1, h^0)$ obtained by the CLS method is constructed as follows:

$$F_1(u_1) = \begin{bmatrix} -0.94 & -0.25 & -0.42 \\ 0.05 & 0 & -0.8 \\ 0.06 & -0.3 & -0.6 \end{bmatrix}, \ F_1(u_2) = \begin{bmatrix} -0.6 & 0.01 & 0.7 \\ 0.1 & 1 & -0.06 \\ -0.7 & -0.1 & 0.6 \end{bmatrix},$$

$$F_1(u_3) = \begin{bmatrix} 0.01 & -0.5 & -0.04 \\ -0.02 & 0.14 & 0.3 \\ 0.03 & -0.03 & 0.08 \end{bmatrix}, \ g_1(u_1) = \mathbf{e}_1, \ g_1(u_2) = \mathbf{e}_2, \ g_1(u_3) = \mathbf{e}_3, \ h_1 =$$

$[5.9, \ -3.2, \ -5.6], \ h^0 = 1.$

In this example, the original signals are considered as the input response map of a 3-dimensional affine dynamical system and the desirable input response map is obtained by the CLS method. The model obtained by the CLS method is a 3-dimensional affine dynamical system.

Just as we expected, the following table and Fig. 7.7 truly indicate that the 3-dimensional affine dynamical system obtained by the CLS method is a good noisy realization.

dimen-ion	ratio of matrices	mean values of square root for sum of			cosine	error ratio
		signal ①	signal by CLS ②	error ③	① and ② $\cos\theta$	③/①
$a_{1,2,3}$	0.11	0.883	0.851	0.04	0.999	0.05

Example 7.32. Let the signals be the input response map of the following 3-dimensional affine dynamical system: $\sigma = ((\mathbf{R}^3, F), g, h, h^0)$, where $F(u_1) =$
$$\begin{bmatrix} -0.6 & -0.3 & 0 \\ 1 & 0.1 & 0.1 \\ 0 & 1 & 0 \end{bmatrix}, \ F(u_2) = \begin{bmatrix} -0.9 & -0.1 & 0 \\ 0.5 & 0.3 & 0.05 \\ 0 & 0.5 & 0 \end{bmatrix}, \ F(u_3) = \begin{bmatrix} 0.5 & 0 & 0 \\ 1 & 0.3 & 0 \\ 1 & 0 & 0.05 \end{bmatrix},$$
$g(u_1) = \mathbf{e}_1, \ g(u_2) = \mathbf{e}_2, \ g(u_3) = \mathbf{e}_3, \ h = [6, \ -4, \ 2], \ h^0 = 1.$

Then the noisy realization problem is solved as follows:

covariance matrix	eigenvalues					
	1	2	3	4	5	6
$H_{\underline{a} \ (2,40)}^T(1,2) H_{\underline{a} \ (2,40)}(1,2)$	544	48				
$H_{\underline{a} \ (3,40)}^T(1,2,3) H_{\underline{a} \ (3,40)}(1,2,3)$	545	52	10			
$H_{\underline{a} \ (4,40)}^T(1,\cdots,4) H_{\underline{a} \ (4,40)}(1,\cdots,4)$	675	81	11	4.2		
$H_{\underline{a} \ (5,40)}^T(1,\cdots,5) H_{\underline{a} \ (5,40)}(1,\cdots,5)$	1010	164	11	4.5	3	
$H_{\underline{a} \ (6,40)}^T(1,\cdots,6) H_{\underline{a} \ (6,40)}(1,\cdots,6)$	1226	183	12.8	4.7	4.2	2.8
covariance matrix	square root of eigenvalues					
$H_{\underline{a} \ (2,40)}^T(1,2) H_{\underline{a} \ (2,40)}(1,2)$	23.3	6.9				
$H_{\underline{a} \ (3,40)}^T(1,2,3) H_{\underline{a} \ (3,40)}(1,2,3)$	23.3	7.2	3.2			
$H_{\underline{a} \ (4,40)}^T(1,\cdots,4) H_{\underline{a} \ (4,40)}(1,\cdots,4)$	26	9	3.3	2		
$H_{\underline{a} \ (5,40)}^T(1,\cdots,5) H_{\underline{a} \ (5,40)}(1,\cdots,5)$	31.8	12.8	3.3	2.1	1.7	
$H_{\underline{a} \ (6,40)}^T(1,\cdots,6) H_{\underline{a} \ (6,40)}(1,\cdots,6)$	35	13.5	3.6	2.2	2	1.7

1) A set $\{2.2, \ 2, \ 1.7\}$ is composed of relatively small and equally-sized numbers in the square root of eigenvalues for $H_{\underline{a} \ (6,40)}^T(1,\cdots,6) H_{\underline{a} \ (6,40)}(1,\cdots,6)$.

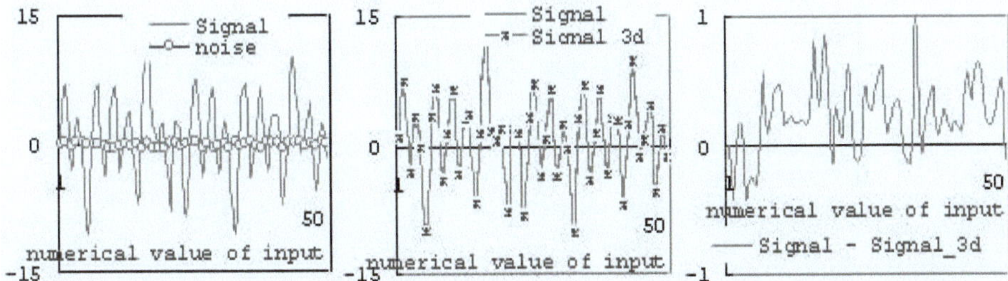

Fig. 7.8. The left is the original input response map and noise added to original 3-dimensional affine dynamical system. The middle is the original input response map and the behavior of a 3-dimensional affine dynamical system obtained by the CLS method. The right is the difference between the original input response map and the behavior of the 3-dimensional affine dynamical system obtained by the CLS method in Example (7.32).

2) After determining the independent vectors $\underline{S}_l(u_1)\underline{a} - \underline{a}$, $\underline{S}_l(u_2)\underline{a} - \underline{a}$ and $\underline{S}_l(u_3)\underline{a} - \underline{a}$ whose numerical value of input are 1, 2 and 3, we will continue the noisy realization algorithm by the CLS method.

Therefore, a noisy 3-dimensional affine dynamical system $\sigma_1 = ((\boldsymbol{R}^3,\ F_1)$, $g_1,\ h_1,\ h^0)$ obtained by the CLS method is constructed as follows:

$$F_1(u_1) = \begin{bmatrix} -0.6 & -0.27 & 0 \\ 0.97 & 0 & 0.25 \\ 0.05 & 0.41 & 0.04 \end{bmatrix},\ F_1(u_2) = \begin{bmatrix} -0.9 & -0.09 & -0.02 \\ 0.38 & 0.22 & 0.02 \\ 0.18 & 0.19 & -0.02 \end{bmatrix},$$

$$F_1(u_3) = \begin{bmatrix} 0.6 & 0.01 & 0.01 \\ 0.6 & 0.3 & 0.02 \\ -0.45 & -0.3 & -0.2 \end{bmatrix},\ g_1(u_1) = \mathbf{e}_1,\ g_1(u_2) = \mathbf{e}_2,$$

$g_1(u_3) = \mathbf{e}_3, h_1 = [6.4,\ -4.2,\ 2.4],\ h^0 = 1.$

In this example, the original signals are considered as the input response map of a 3-dimensional affine dynamical system and the desirable input response map is obtained by the CLS method. The model obtained by the CLS method is a 3-dimensional affine dynamical system.

Just as we expected, the following table and Fig. 7.8 truly indicate that the 3-dimensional affine dynamical system obtained by the CLS method is a somewhat good noisy realization.

dimen-ion	ratio of matrices	mean values of square root for sum of			cosine ① and ②	error ratio
		signal	signal by CLS	error	$\cos\theta$	③/①
		①	②	③		
$a_{1,2,3}$	0.06	0.733	0.721	0.05	0.997	0.07

Example 7.33. Let the signals be the input response map of the following 4-dimensional affine dynamical system: $\sigma = ((\boldsymbol{R}^4, F), g, h, h^0)$, where

$$F(u_1) = \begin{bmatrix} 0 & -0.5 & -0.4 & -0.4 \\ 0 & 0.6 & 0 & -0.03 \\ -0.6 & 0 & -0.1 & 0.4 \\ 0 & -0.5 & 0.5 & -0.6 \end{bmatrix}, \ F(u_2) = \begin{bmatrix} 0.6 & 0 & -0.3 & 0.3 \\ 0 & 0.8 & 0 & 0 \\ -0.3 & 0 & 0.6 & 0.3 \\ 0.3 & 0 & 0.3 & 0.7 \end{bmatrix},$$

$$F(u_3) = \begin{bmatrix} -1 & -0.3 & 0 & -0.7 \\ 0 & 0 & 1 & 0.8 \\ 0 & -0.3 & -0.3 & 0.7 \\ 0.8 & 0.3 & 0.4 & 0.7 \end{bmatrix}, \ g(u_1) = \mathbf{e}_1, \ g(u_2) = \mathbf{e}_2, \ g(u_3) = \mathbf{e}_3,$$

$$h = [12, \ -8, \ 0, \ 1], \ h^0 = 1.$$

Then the noisy realization problem is solved as follows:

covariance matrix	eigenvalues					
	1	2	3	4	5	6
$H_{\underline{a}\,(3,40)}^{T}(1,2,3)H_{\underline{a}\,(3,40)}(1,2,3)$	1657	1152	709			
$H_{\underline{a}\,(4,40)}^{T}(1,\cdots,4)H_{\underline{a}\,(4,40)}(1,\cdots,4)$	1820	1554	1121	3.8		
$H_{\underline{a}\,(5,40)}^{T}(1,\cdots,5)H_{\underline{a}\,(5,40)}(1,\cdots,5)$	4152	1614	1308	21	3.8	
$H_{\underline{a}\,(6,40)}^{T}(1,\cdots,6)H_{\underline{a}\,(6,40)}(1,\cdots,6)$	5876	3153	1448	125	6	2.8
covariance matrix	square root of eigenvalues					
$H_{\underline{a}\,(3,40)}^{T}(1,2,3)H_{\underline{a}\,(3,40)}(1,2,3)$	40.7	34	27			
$H_{\underline{a}\,(4,40)}^{T}(1,\cdots,4)H_{\underline{a}\,(4,40)}(1,\cdots,4)$	42.7	39.4	33.5	1.9		
$H_{\underline{a}\,(5,40)}^{T}(1,\cdots,5)H_{\underline{a}\,(5,40)}(1,\cdots,5)$	64	40	36	4.6	1.9	
$H_{\underline{a}\,(6,40)}^{T}(1,\cdots,6)H_{\underline{a}\,(6,40)}(1,\cdots,6)$	76	56	38	11	2.4	1.7

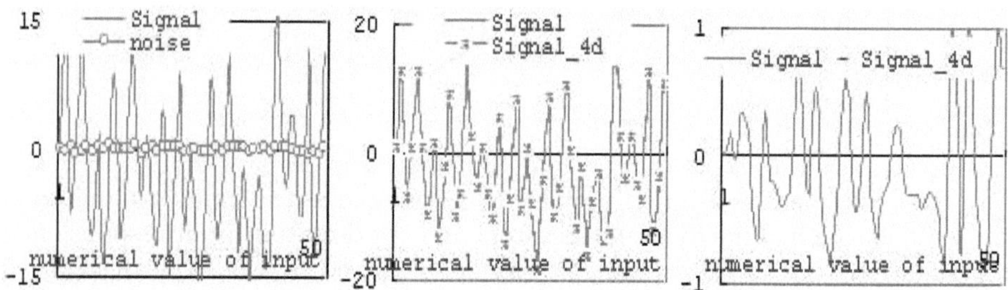

Fig. 7.9. The left is the original input response map and noise added to the original 4-dimensional affine dynamical system. The middle is the original input response map and the behavior of a 4-dimensional affine dynamical system obtained by the CLS method. The right is the difference between the original input response map and the behavior of the 4-dimensional affine dynamical system obtained by the CLS method in Example (7.33).

1) A set $\{2.4, 1.7\}$ is composed of relatively small and equally-sized numbers in the square root of eigenvalues for $H_{\underline{a}\,(6,40)}^{T}(1,\cdots,6)H_{\underline{a}\,(6,40)}(1,\cdots,6)$.

2) After determining the independent vectors $\underline{S}_l(u_1)\underline{a}-\underline{a}$, $\underline{S}_l(u_2)\underline{a}-\underline{a}$, $\underline{S}_l(u_3)\underline{a}-\underline{a}$ and $\underline{S}_l(u_2|u_1)\underline{a}-\underline{a}$ whose numerical value of input are 1, 2, 3 and 5 we will continue the noisy realization algorithm by the CLS method.

Therefore, a noisy 4-dimensional affine dynamical system $\sigma_1 = ((\mathbf{R}^4, F_1), g_1, h_1, h^0)$ obtained by the CLS method is constructed as follows:

$$F_1(u_1) = \begin{bmatrix} 0 & 0.5 & -1.4 & 1.2 \\ 0.03 & 2.3 & -1.7 & 3.4 \\ -0.6 & -0.5 & 0.39 & -1.04 \\ 0 & -1.7 & 1.7 & -2.8 \end{bmatrix}, \ F_1(u_2) = \begin{bmatrix} 0 & 0.01 & -0.94 & -0.1 \\ -1 & 0.8 & -1.07 & -0.28 \\ 0 & -0.02 & 0.9 & 0.03 \\ 1 & 0 & 1 & 1.07 \end{bmatrix},$$

$$F_1(u_3) = \begin{bmatrix} -2.7 & -0.9 & -0.9 & -2.9 \\ -2.7 & -1 & -0.4 & -3 \\ 0.8 & 0.01 & 0.01 & 0.9 \\ 2.7 & 1 & 1.4 & 2.9 \end{bmatrix}, \ g_1(u_1) = \mathbf{e}_1, \ g_1(u_2) = \mathbf{e}_2,$$

$g_1(u_3) = \mathbf{e}_3, \ h_1 = [11.8, \ -8, \ -0.3, \ -0.2], \ h^0 = 1.$

In this example, the original signals are considered as the input response map of a 4-dimensional affine dynamical system and the desirable input response map is obtained by the CLS method. The model obtained by the CLS method is a 4-dimensional affine dynamical system.

Just as we expected, the following table and Fig. 7.9 truly indicate that the 4-dimensional affine dynamical system obtained by the CLS method is a somewhat good noisy realization.

dimen-ion	ratio of matrices	mean values of the square root for sum of			cosine ① and ②	error ratio
		signal	signal by CLS	error		
		①	②	③	$\cos\theta$	③/①
$a_{1,2,3,5}$	0.03	1.3	1.27	0.07	0.999	0.05

Example 7.34. Let the signals be the input response map of the following 5-dimensional affine dynamical system: $\sigma = ((\mathbf{R}^5, F), g, h, h^0)$, where

$$F(u_1) = \begin{bmatrix} -1 & -0.4 & -0.6 & -1 & -1 \\ 0 & -0.2 & 0.5 & 0.4 & -0.2 \\ 0 & -0.2 & 0 & 0.8 & 0 \\ 1 & 0.2 & 0.1 & 0.7 & 1 \\ 0 & 0.2 & 0.7 & 0.3 & 0 \end{bmatrix}, \ F(u_2) = \begin{bmatrix} 0 & -0.4 & -0.4 & 0 & 0 \\ -1 & 0.21 & -0.2 & -0.5 & 0 \\ 0 & 0.2 & 0.6 & 0.4 & 0 \\ 0 & -0.2 & 0.2 & 0.5 & 0 \\ 1 & 0.7 & 0.2 & 0.5 & 0.9 \end{bmatrix},$$

$$F(u_3) = \begin{bmatrix} 0 & -0.6 & -0.31 & -0.5 & -0.7 \\ 0 & 0.7 & 0.1 & 0.2 & 0.7 \\ -1 & 0 & -0.1 & 0.4 & -0.3 \\ 0 & 0.5 & 0.5 & 0.8 & 0.6 \\ 0 & 0.2 & -0.1 & -0.2 & 0.2 \end{bmatrix}, \ g(u_1) = \mathbf{e}_1, \ g(u_2) = \mathbf{e}_2,$$

$g(u_3) = \mathbf{e}_3, \ h = [12, \ -1, \ -2, \ 1, \ 7], \ h^0 = 1.$

Then the noisy realization problem is solved as follows:

covariance matrix	eigenvalues							
	1	2	3	4	5	6	7	8
$H_{\underline{a}\,(3,40)}^T(1,2,3)H_{\underline{a}\,(3,40)}(1,2,3)$	1377	1180	251					
$H_{\underline{a}\,(4,40)}^T(1,\cdots,4)H_{\underline{a}\,(4,40)}(1,\cdots,4)$	3524	1215	256	144				
$H_{\underline{a}\,(5,40)}^T(1,\cdots,5)H_{\underline{a}\,(5,40)}(1,\cdots,5)$	4874	1228	432	144	6			
$H_{\underline{a}\,(6,40)}^T(1,\cdots,6)H_{\underline{a}\,(6,40)}(1,\cdots,6)$	4874	1228	432	144	8.5	4.8		
$H_{\underline{a}\,(7,40)}^T(1,\cdots,7)H_{\underline{a}\,(7,40)}(1,\cdots,7)$	5381	1607	442	145	8.9	4.9	3.1	
$H_{\underline{a}\,(8,40)}^T(1,\cdots,8)H_{\underline{a}\,(8,40)}(1,\cdots,8)$	7003	2962	536	153	9.6	5	3.1	2
covariance matrix	square root of eigenvalues							
$H_{\underline{a}\,(3,40)}^T(1,2,3)H_{\underline{a}\,(3,40)}(1,2,3)$	37	34	16					
$H_{\underline{a}\,(4,40)}^T(1,\cdots,4)H_{\underline{a}\,(4,40)}(1,\cdots,4)$	59	35	16	12				
$H_{\underline{a}\,(5,40)}^T(1,\cdots,5)H_{\underline{a}\,(5,40)}(1,\cdots,5)$	70	35	21	12	2.5			
$H_{\underline{a}\,(6,40)}^T(1,\cdots,6)H_{\underline{a}\,(6,40)}(1,\cdots,6)$	70	35	21	12	3	2.2		
$H_{\underline{a}\,(7,40)}^T(1,\cdots,7)H_{\underline{a}\,(7,40)}(1,\cdots,7)$	73	40	21	12	3	2.2	1.8	
$H_{\underline{a}\,(8,40)}^T(1,\cdots,8)H_{\underline{a}\,(8,40)}(1,\cdots,8)$	84	54	23	12	3.1	2.2	1.8	1.4

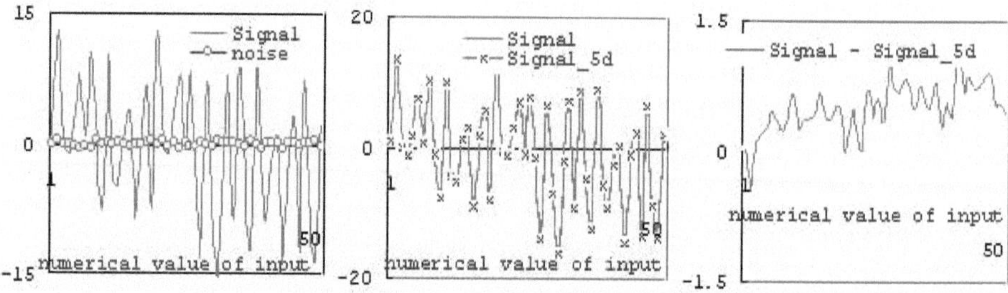

Fig. 7.10. The left is the original input response map and noise added to the original 5-dimensional affine dynamical system. The middle is the original input response map and the behavior of a 5-dimensional affine dynamical system obtained by the CLS method. The right is the difference between the original input response map and the behavior of the 5-dimensional affine dynamical system obtained by the CLS method in Example (7.34).

1) A set $\{2.2,\ 1.8,\ 1.4\}$ is composed of relatively small and equally-sized numbers in the square root of eigenvalues for $H_{\underline{a}\,(9,40)}^T(1,\cdots,8)H_{\underline{a}\,(8,40)}(1,\cdots,8)$.
2) After determining the independent vectors $\underline{S}_l(u_1)\underline{a}-\underline{a}$, $\underline{S}_l(u_2)\underline{a}-\underline{a}$, $\underline{S}_l(u_3)\underline{a}-\underline{a}$, $\underline{S}_l(u_1|u_1)\underline{a}-\underline{a}$ and $\underline{S}_l(u_2|u_1)\underline{a}-\underline{a}$ whose numerical value of input are 1, 2, 3, 4 and 5 we will continue the noisy realization algorithm by the CLS method.

Therefore, a noisy 5-dimensional affine dynamical system $\sigma_1 = ((\boldsymbol{R}^5,\ F_1),\ g_1,\ h_1,\ h^0)$ obtained by the CLS method is constructed as follows:

$$F_1(u_1) = \begin{bmatrix} -1 & -0.32 & -0.7 & -1.2 & -1 \\ 0 & -0.01 & 0.18 & -0.08 & -0.2 \\ 0 & -0.2 & 0.06 & 0.9 & -0.08 \\ 1 & 0.28 & 0.06 & 0.6 & 1 \\ 0 & 0 & 0.97 & 0.74 & 0 \end{bmatrix},$$

$$F_1(u_2) = \begin{bmatrix} 0 & -0.5 & -0.4 & -0.2 & 0.07 \\ -1 & 0.05 & -0.3 & -0.85 & 0.1 \\ 0 & 0.3 & 0.6 & 0.5 & -0.06 \\ 0 & -0.3 & 0.14 & 0.43 & 0.06 \\ 1 & 1 & 0.33 & 0.82 & 0.75 \end{bmatrix},$$

$$F_1(u_3) = \begin{bmatrix} 0.45 & -0.77 & -0.39 & -0.65 & -0.66 \\ 0.9 & 0.36 & -0.08 & -0.14 & 0.76 \\ -1.3 & 0.1 & -0.01 & 0.5 & -0.35 \\ 0.3 & 0.4 & 0.4 & 0.7 & 0.6 \\ -0.9 & 0.5 & 0.07 & 0.1 & 0.14 \end{bmatrix}, \; g_1(u_1) = \mathbf{e}_1,$$

$$g_1(u_2) = \mathbf{e}_2, \; g_1(u_3) = \mathbf{e}_3, \; h_1 = [12.5, \; -1, \; -2.2, \; 0.72, \; 6.6], \; h^0 = 1.$$

In this example, the original signals are considered as the input response map of a 5-dimensional affine dynamical system and the desirable input response map is obtained by the CLS method. The model obtained by the CLS method is a 5-dimensional affine dynamical system.

Just as we expected, the following table and Fig. 7.10 truly indicate that the 5-dimensional affine dynamical system obtained by the CLS method is a good noisy realization.

dimen-ion	ratio of matrices	mean values of square root for sum of			cosine ① and ②	error ratio
		signal ①	signal by CLS ②	error ③	$\cos\theta$	③/①
$a_{1,2,3,4}$	0.04	1.053	1.08	0.11	0.994	0.10
$a_{1,2,3,4,5}$	0.03	1.053	1.08	0.08	0.997	0.07

7.6 Historical Notes and Concluding Remarks

According to realization theory and many facts regarding affine dynamical systems in the reference [Matsuo and Hasegawa, 2003], we obtained a partial realization algorithm of the systems needed for approximation and noisy realization problems of the systems.

In regards to important facts regarding the systems, there are a representation of their behaviors and a partial realization algorithm. The representation of their behaviors means that any affine dynamical system can be completely characterized by the input response map itself.

As for the general non-linear systems, approximation and noisy realization problems were proposed for the first time. The approximate realization problem was attempted to be solved by presenting an approximate realization algorithm. The algorithm is made up of the ratio of Hankel matrix norm and the CLS method, namely, the constrained least square method. Using the ratio of Hankel matrix norm which is a square norm, we decrease the number of dimensions of state space while considering information loss in mind. By using the CLS method, we make a full effort to characterize a relation of a linear combination.

By applying this algorithm to several examples of affine dynamical systems, we have shown that this algorithm is practical and useful. In the case that the ratio of Hankel matrix norm is within some percent, we have shown this approximate realization algorithm produces good results. Our several examples show that the changing relations among the ratio of matrices and the error to signal ratio are proportional relations, and the ratio of Hankel matrices is 0.01 while the error to signal ratio ranges from 0.005 to 0.02.

This approximate realization algorithm appears to be very promising.

We treated a noisy realization problem by attempting to solve it by presenting a noisy realization algorithm. The algorithm is made up by making a set of singular values of a matrix and the CLS method, namely the constrained least square method. By making a set of singular values of a matrix, we determine the number of dimensions of state space by drawing a distinction between a noiseless part and a noisy part in the given signal. By using the CLS method, we make a full effort to characterize a relation of a linear combination in the noiseless part.

By applying this algorithm to several examples of affine dynamical systems, we have shown that this algorithm is practical and useful. In the case that we can make the set composed of relatively small and equally-sized numbers, we have shown this noisy realization algorithm produces good results. Our several examples show that the changing relations among the ratio of matrices and the error to signal ratio are proportional relations, and the ratio of Hankel matrices is 0.01 while the error to signal ratio ranges from 0.005 to 0.025.

This noisy realization algorithm also appears to be very promising.

As we have mentioned before, the concrete discussions of approximate and noisy realization for non linear systems are very new.

7.7   Appendix

7.7.1   Partial Realization

(7-A.1) Lemma

Let an affine dynamical system $\sigma = ((X, F), g, h, h^0)$ be p-quasi-reachable. Then $G_p := G \cdot J_p : V(U^+)_p \rightarrow X$ is surjective, where $J_p : V(U^+)_p \rightarrow V(U^+)$ is a canonical injection and G is a linear map $G : V(U^+) \rightarrow X$ such that $G(e_\omega) = \sum_{j=1}^{|\omega|} F(\omega(|\omega|))F(\omega(|\omega| - 1)) \cdots F(\omega(|\omega| - j))g(\omega(j))$ holds for any $\omega \in U^*$.

[proof]. This is easily obtained.

(7-A.2) Proposition

If a linear sub-space S of $V(U^+)_p$ satisfies the next two conditions, then there uniquely exists an ideal $\underline{S} \subseteq V(U^+)$ such that $\underline{S} \cap V(U^+)_{p+1} = S$ and $V(U^+)_{p+1}/S$ is isomorphic to $V(U^+)/\underline{S}$. Moreover, an affine dynamical system $((V(U^+)/\underline{S}, \bar{S}_r), \bar{e}, \bar{a}_l, a(1))$ is p-quasi-reachable, where $\bar{e}$ is given by $\bar{e}(u) = \mathbf{e_u} + \underline{S}$, $\bar{S}_r$ is given by $\bar{S}_r(\lambda + \underline{S}) = S_r\lambda + \underline{S}$ and $\bar{a}_l$ is given by $\bar{a}_l(e_\omega + \underline{S}) = a_l(e_\omega)$ in Example (7.4).

condition 1: $\lambda \in V(U^+)_p \cap S$ implies $S_r(u)\lambda \in S$ for any $u \in U$.
condition 2: There exist coefficients $\lambda(\omega_i) \in K$ such that $e_\omega - \sum_{\omega_i \in U_p^*} \lambda(\omega_i) e_{\omega_i}$
$\in S$ for any $\omega \in U^*$, $|\omega| = p + 1$.

[proof]. Let $J_{(p,p+1)} : V(U^+)_p \to V(U^+)_{p+1}$ be the canonical injection and
$\pi_S : V(U^+)_{p+1} \to V(U^+)_{p+1}/S$ be the canonical surjection. Then condition 2
implies that a composition map $\pi_S \cdot J_{(p,p+1)}$ is surjective. And condition 1 implies that $S_r(u)\lambda \in S$ holds for any $u \in U$ and $\lambda \in S$. Therefore, by setting
$\tilde{S}_r(u)(\lambda + S) = S_r\lambda + S$ for any $\lambda \in V(U^+)_{p+1}$, we can uniquely define a map
$\tilde{S}_r : U \to V(U^+)_{p+1}/S$. And $((V(U^+)_{p+1}/S, \tilde{S}_r), \tilde{e}, \tilde{a}_l, a(1))$ is an affine dynamical system and p-quasi-reachable, where where $\tilde{e}$ is given by $\tilde{e}(u) = \mathbf{e_u} + S$, $\tilde{S}_r$
is given by $\tilde{S}_r(\lambda + S) = S_r\lambda + S$ and $\tilde{a}_l$ is given by $\tilde{a}_l(e_\omega + S) = a_l(e_\omega)$. Then
there uniquely exists a linear input map $G : V(U^+) \to V(U^+)_{p+1}/S$ which
satisfies $GS_r(u) = \tilde{S}_r G(u)$, $u \in U$. Setting $G_{p+1} := G \cdot J_{p+1}$, $\ker G_{p+1} = S$
holds and $\underline{S} := \ker G$ satisfies $\underline{S} \cap V(U^+)_{p+1} = S$. Since G is a linear map
with the property, $\underline{S}$ is an invariant sub-space under $S_r(u)$ for any $u \in U$.
Moreover, the surjection of G implies that $((V(U^+)_{p+1}/S, \tilde{S}_r), \tilde{e}, \tilde{a}_l, a(1))$ is isomorphic to $((V(U^+)/\underline{S}, \bar{S}_r), \bar{e}, \bar{a}_l, a(1))$. Therefore, $((V(U^+)/\underline{S}, \bar{S}_r), \bar{e}, \bar{a}_l, a(1))$
is a p-quasi-reachable affine dynamical system. The uniqueness of $\underline{S}$ is obtained
by the uniqueness of $\tilde{S}_r$ and G.

Set $F(U_q^*, Y) := \{$a function $: U_q^* \to Y\}$, let P_q be the canonical surjection:
$F(U^*, Y) \to F(U_q^*, Y); a \mapsto [; \omega \mapsto a(\omega)]$, and define $\underline{S}_l$ by setting $\underline{S}_l(\omega) :$
$F(U_q^*, Y) \to F(U_{q-|\omega|}^*, Y); a \mapsto \underline{S}_l(\omega)a[; \bar{\omega} \mapsto a(\bar{\omega}|\omega)]$.

(7-A.3) Definition
Let an affine dynamical system $\sigma = ((X, F), g, h, h^0)$ be q-quasi-reachable. Then
a linear map $H_q := P_q \cdot H$ is injective, where H is a linear map $H : X \to F(U, Y)$
which satisfies $Hx(\omega) = hF(\omega(|\omega|))F(\omega(|\omega| - 1)) \cdots F(\omega(1)x$.

[proof]. This is easily obtained.

(7-A.4) Proposition
If a sub-space Z of $F(U_{q+1}^*, Y)$ satisfies the next two conditions, then there
uniquely exists a linear space X which satisfies $S_l(u)x \in X$ for any $x \in X$ and
$u \in U$ such that a map $P_{q|X} : X \to Z$ is isomorphic, where $P_{q|X}$ is a restriction
of the canonical surjection $P_q : F(U^*, Y) \to F(U_q^*, Y)$ to X.

And an affine dynamical system $((X, S_l), \xi, 1, a(1))$ is q-distinguishable,
where a map $\xi : U \to F(U^*, Y)$ is given by $u \mapsto \xi(u)[; \omega \mapsto a(\omega|u) - a(\omega)]$, $a \in$
$F(U^*, Y)$ and 1 is a map $: Z \to Y; a \mapsto a(1)$ in Example (7.5).

condition 3: A composition map $\pi \cdot j : Z \xrightarrow{j} F(U_{q+1}^*, Y) \xrightarrow{\pi} F(U_q^*, Y)$ is injective.
condition 4: im $(\underline{S}_l(u) \cdot j) \subseteq$ im $(j \cdot \pi)$ holds in the sense of $F(U_q^*, Y)$,
where π is the canonical surjection.

[proof]. By conditions 3 and 4, we can define $F(u)z = (\pi \cdot j)^{-1} \underline{S}_l(u) \cdot jz$ for
any $u \in U$ and $z \in Z$. Then F is a map $: U \to L(Z)$. Injection of $\pi \cdot j$ implies
that $((Z, F), \xi, 1, a(1))$ is q-distinguishable affine dynamical system. It follows
that the linear observation map H corresponding to $((Z, F), 1)$ is injective by

[matsuo and Hasegawa, 2003]. Set $X := \operatorname{im} H$, a map $H^{-1} : X \to Z$ is clearly the restriction of the map $P_q : F(U^*, Y) \to F(U_q^*, Y)$ to X.

An equation $1 = 1 \cdot H$ implies that $((X, S_l), \xi, 1, a(1))$ is isomorphic to $((Z, F), \xi, 1, a(1))$ in the sense of an affine dynamical system. Therefore, $((X, S_l), \xi, 1, a(1))$ is a q-distinguishable affine dynamical system. A uniqueness of X is obtained by the uniqueness of F and H.

We can consider a partial linear input/output map $A_{(p,\underline{N}-p)} : V(U^+)_p \to F(U^*_{\underline{N}-p}, Y)$ for $\underline{a} \in F(U^*_{\underline{N}}, Y)$ the same as the linear input/output map $A : V(U^+) \to F(U^*, Y)$ which satisfies $S_l(u)A = A\psi(u)$ and $A(e_\omega)(\bar{\omega}) = a(\bar{\omega}|\omega) - a(\bar{\omega})$ holds for any $u \in U$, ω, $\bar{\omega} \in U^*$ and any $a \in F(U^*, Y)$.

(7-A.5) Lemma

Let $A_{(p,\underline{N}-p)}$ be the partial linear input/output map corresponding to $\underline{a} \in F(U^*_{\underline{N}}, Y)$. Then the following diagrams commute:

1)

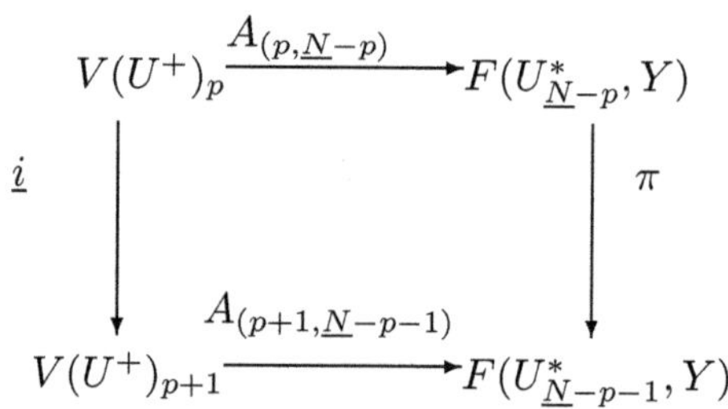

where $\underline{i}$ is the canonical injection and π is the canonical surjection.

2)

$$
\begin{array}{ccc}
V(U^+)_p & \xrightarrow{\ A_{(p,\underline{N}-p)}\ } & F(U^*_{\underline{N}-p}, Y) \\
{\scriptstyle \psi(u)}\downarrow & & \downarrow{\scriptstyle \underline{S_l}} \\
V(U^+)_{p+1} & \xrightarrow{\ A_{(p+1,\underline{N}-p-1)}\ } & F(U^*_{\underline{N}-p-1}, Y)
\end{array}
$$

[proof]. These can be obtained by direct calculation.

(7-A.6) Proposition

Let $A_{(p_1,\underline{N}-p_1)}$ be the partial linear input/output map corresponding to $\underline{a} \in F(U^*_{\underline{N}}, Y)$ and p_2 be any integers such that $0 \leq p_2 \leq p_1 < \underline{N}$.

If im $A_{(p_2+1,\underline{N}-p_2-1)} = $ im $A_{(p_2,\underline{N}-p_2-1)}$, then im $A_{(p_1,\underline{N}-p_1)} = $ im $A_{(p_2,\underline{N}-p_1)}$ holds.

[proof]. Note that this proposition holds if and only if im $A_{(p_2+1,\underline{N}-p_2-1)} = $ im $A_{(p_2,\underline{N}-p_2-1)}$ implies im $A_{(p_2+1+n,\underline{N}-p_2-1-n)} = $ im $A_{(p_2,\underline{N}-p_2-1-n)}$ holds for any non-negative integer n. Therefore, we prove the latter by the inductive method. When $n = 0$, it holds by assumption. Let's assume it holds for $n = k$, i.e., assume that im $A_{(p_2+1+k,\underline{N}-p_2-1-k)} = $ im $A_{(p_2,\underline{N}-p_2-1-k)}$. Then for any $\bar{\omega} \in U^*, |\bar{\omega}| = p_2 + 1 + k + 1$ given by $\bar{\omega} = u|\omega_1$. By assumption, there exist $\omega_j \in U^*_{p_2}, \alpha_j \in \boldsymbol{R}$ and $m \in N(1 \le j \le m)$ such that $\underline{S_l}(\omega_1)\underline{a} - \underline{a} = \sum_{j=1}^{m} \alpha_j(\underline{S_l}(\omega_j)\underline{a} - \underline{a})$ in the sense of $F(U^*_{\underline{N}-p_2-1-k}, Y)$. Therefore, $\underline{S_l}(\bar{\omega})\underline{a} - \underline{a} = \underline{S_l}(u)\underline{S_l}(\omega_1) - \underline{a} = \sum_{j=1}^{m} \alpha_i(\underline{S_l}(u)\underline{S_l}(\omega_j)\underline{a} - \underline{a}) = \sum_{j=1}^{m} \alpha_i(\underline{S_l}(u|\omega_j)\underline{a} - \underline{a})$ hold.

Therefore, im $A_{(p_2+1+k+1,\underline{N}-p_2-1-k-1)} = $ im $A_{(p_2+1,\underline{N}-p_2-1-k-1)}$ holds. On the other hand, im $A_{(p_2+1,\underline{N}-p_2-1)} = $ im $A_{(p_2,\underline{N}-p_2-1)}$ is equivalent to im $A_{(p_2+1,j)} = $ im $A_{(p_2,j)}$ for any $j \le \underline{N} - p_2 - 1$.

Therefore, im $A_{(p_2+1+k+1,\underline{N}-p_2-1-k-1)} = $ im $A_{(p_2,\underline{N}-p_2-1-k-1)}$ holds. The condition equation holds for $n = k + 1$.

(7-A.7) Proposition

Let $A_{(\ ,\ )}$ be the partial linear input/output map corresponding to $\underline{a} \in F(U^*_{\underline{N}}, Y)$. For p_1 and p_2 be any integers such that $0 \le p_2 < p_1 < \underline{N}$.

If ker $A_{(p_1,\underline{N}-p_1)} = $ ker $A_{(p_1,\underline{N}-p_1-1)}$ hold, then ker $A_{(p_2,\underline{N}-p_1-1)}$ $= $ ker $A_{(p_2,\underline{N}-p_2)}$ holds.

[proof]. Note that this proposition holds if and only if ker $A_{(p_1,\underline{N}-p_1)} = $ ker $A_{(p_1,\underline{N}-p_1-1)}$ implies ker $A_{(p_1-n,\underline{N}-p_1-1)} = $ ker $A_{(p_1-n,\underline{N}-p_1+n)}$ for any n in $0 \le n \le p_1$. Therefore, we prove the latter by the inductive method. When $n = 0$, it holds by assumption. Let's assume that it holds for $n = k$, i.e., assume that ker $A_{(p_1-k,\underline{N}-p_1-1)} = $ ker $A_{(p_1-k,\underline{N}-p_1+k)}$. Then, for any $u \in U$ and $\bar{\omega} \in U^*, |\bar{\omega}| = \underline{N} - p_1 + k + 1$, let $\bar{\omega} = \bar{\omega}_1|u$. By assumption, there exist $\omega_j \in U^*_{\underline{N}-p_1-1}, \alpha_j \in \boldsymbol{R}$ and $m \in N(1 \le j \le m)$ such that $\underline{a}(\bar{\omega}|\omega) - \underline{a}(\bar{\omega}) = \underline{a}(\bar{\omega}_1|u|\omega) - \underline{a}(\bar{\omega}_1|u) = \sum_{j=1}^{m} \alpha_j(\underline{a}(\omega_j|u|\omega) - \underline{a}(\omega_j|u))$. Since $\omega_j|u \in U^*_{\underline{N}-p_1}$, ker $A_{(p_1-k-1,\underline{N}-p_1)} = $ ker $A_{(p_1-k-1,\underline{N}-p_1+k+1)}$ holds. On the other hand, if we note that ker $A_{(p_1,\underline{N}-p_1)} = $ ker $A_{(p_1,\underline{N}-p_1-1)}$ is equivalent to ker $A_{(i,\underline{N}-p_1)} = $ ker $A_{(i,\underline{N}-p_1-1)}$ for any i in $0 \le i \le p_1$, ker $A_{(p_1-k-1,\underline{N}-p_1-1)} = $ ker $A_{(p_1-k-1,\underline{N}-p_1+k+1)}$ holds. Therefore, the condition's equation holds for $n = k + 1$.

(7-A.8) Lemma

For a partial linear input/output map $A_{(\ ,\ )}$ corresponding to $\underline{a} \in F(U^*_{\underline{N}}, Y)$ and an affine dynamical system $\sigma = ((X, F), g, h, h^0)$, the next matters hold, where $G_p := G \cdot J_p, H_q := P_q \cdot H$ for the linear input map G corresponding to x^0 and the linear output map H corresponding to h. $A_{p,q} := H_q \cdot J_p$.

1) σ is a partial realization of $\underline{a}$ if and only if the following figure commutes for any p such that $0 \le p < \underline{N}$.

2) σ is a natural partial realization of $\underline{a}$ if and only if the following figure commutes, G_p is surjective and $H_{\underline{N}-p-1}$ is injective for some p such that $0 \leq p < \underline{N}$.

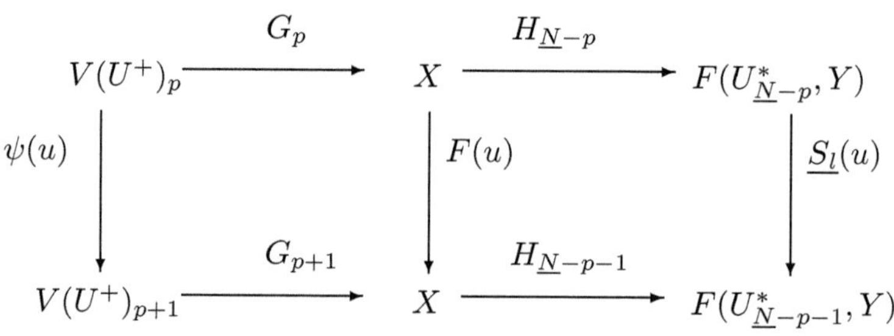

[proof]. These can be obtained by definition of the partial and natural partial realization.

(7-A.9) Proof of Theorem (7.17)

We prove the theorem by rewriting the conditions of a partial Hankel matrix in Theorem (7.18) to a partial linear input/output map $A_{(\ ,\ )}$ corresponding to $\underline{a} \in F(U_{\underline{N}}^*, Y)$. By using Proposition (7-A.6) and (7-A.7), the conditions of Hankel matrix can be equivalently changed to the following equations (1) and (2):

(1) im $A_{(p,\underline{N}-p-1)} = $ im $A_{(p,\underline{N}-p-1)}$
(2) ker $A_{(p,\underline{N}-p)} = $ ker $A_{(p,\underline{N}-p-1)}$

Therefore, we will prove the theorem by using (1) and (2).

First, we show that the above equations (1) and (2) are necessary. Let $\sigma = ((X,F), g, h, h^0)$ be a natural partial realization of $\underline{a} \in F(U_{\underline{N}}^*, Y)$, then σ is p-quasi-reachable, and q-distinguishable for some p and q such that $p+q < N$. Let G satisfy $G(e_\omega) = \sum_{j=1}^{|\omega|} F(\omega(|\omega|))F(\omega(|\omega|-1))\cdots F(\omega(|\omega|-j))g(\omega(j))$ for any $\omega \in U^*$. Let H satisfy $Hx(\omega) = hF(\omega(|\omega|))F(\omega(|\omega|-1))\cdots F(\omega(1))x$ for any $x \in X$, $\omega \in U^*$. Let $p \leq p'$ and $q \leq q'$, then $G_{p'} := G \cdot J_{p'}$ is onto, $H_{q'} := P_{q'} \cdot H$ is one-to-one. Therefore, $A_{(p',q')} := H_{q'} \cdot J_{p'}$ satisfies equations (1) and (2).

Next, we show that the equations (1) and (2) are sufficient. Set $S := $ ker $A_{(p+1,\underline{N}-p-1)}$ and $Z := $ im $A_{(p,\underline{N}-p)}$. Then equation (2) implies that a composition map $\pi \cdot j : Z \xrightarrow{j} F(U_{\underline{N}-p}^*, Y) \xrightarrow{\pi} F(U_{\underline{N}-p-1}^*, Y)$ is injective, where π and j are the same as in Proposition (7-A.4). Therefore, Z satisfies condition 3 in Proposition (7-A.4). Equation (1) implies that there exist $\lambda(\omega_i)e_{\omega_i}$ such that $A_{(p+1,\underline{N}-p-1)}(e_\omega) = A_{(p,\underline{N}-p-1)}(\sum_i \lambda(\omega_i)e_{\omega_i}$ for any $\omega \in U^*$ and $|\omega| = p+1$.

By Lemma (7-A.5), we obtain that $A_{(p+1,\underline{N}-p-1)}(e_\omega - \sum_i \lambda(\omega_i)e_{\omega_i}) = 0$, and $e_\omega - \sum_i \lambda(\omega_i)e_{\omega_i} \in S$ holds. This implies that S satisfies condition 2 in Proposition (7-A.2).

Let $\underline{j}$ be the canonical injection $:A_{(p,\underline{N}-p-1)} \to F(U_{\underline{N}-p-1}^*, Y)$ and π is the same as in Proposition (7-A.4), $B := (\underline{j})^{-1} \cdot \pi \cdot j : Z \to $ im $A_{(p,\underline{N}-p-1)}$ is a bijective linear map by (2) in Proposition (7-A.2). When we consider the bijective linear map $A^b := A^b_{(p+1,\underline{N}-p-1)} : V(U^+)_{p+1}/S \to $ im $A_{(p+1,\underline{N}-p-1)}$ associated with $A_{(p+1,\underline{N}-p-1)} : V(U^+)_{p+1} \to F(U^*_{\underline{N}-p-1},Y)$, equation (2) implies that a linear map $B^{-1} \cdot A^b$ is a bijective linear map $:V(U^+)_{p+1}/S \to Z$. For any $\lambda \in V(U^+)_p \cap S$, $A_{(p,\underline{N}-p)}(\lambda) = 0$ holds by injection of $B^{-1} \cdot A^b$. Therefore, $A_{(p+1,\underline{N}-p-1)}(S_r(u)\lambda) = \underline{S_l}(u)A_{(p,\underline{N}-p)}(\lambda) = 0$ holds by using 2) in Lemma (7-A.5) for any $u \in U$. This implies that $S_r(u)\lambda \in S$. Therefore, S satisfies condition 1 in Proposition (7-A.2). Then Proposition (7-A.2) implies that an affine dynamical system $((V(U^+)_{p+1})/S, \tilde{S}_r), \tilde{e}, \tilde{a}_l, a(1))$ is p-quasi-reachable. Here, equation (1) implies that there exists $x \in \text{im } A_{(p,\underline{N}-p-1)}$ such that $\underline{j}(x) = \underline{S_l} \cdot j(z)$ for any $z \in Z$ and $u \in U$. Moreover, by surjection of B, there exists $z' \in Z$ such that $B(z') = x$. Therefore, $\underline{S_l} \cdot j(z) = j(x) = \underline{j} \cdot B(z') = \pi \cdot j(z')$, which implies that im $(\underline{S_l}(u) \cdot j) \subseteq \text{im } (\pi \cdot j)$. It follows that Z satisfies condition 4 in Proposition (7-A.4) and $((Z,F), \psi, 1, a(1))$ is $(\underline{N} - p - 1)$-distinguishable. We can also show that $B^{-1} \cdot A^b$ is a linear map $B^{-1} \cdot A^b : V(U^+)_{p+1}/S \to Z$ which satisfies $B^{-1} \cdot A^b \tilde{S}_r(u) = F(u)B^{-1} \cdot A^b$ for any $u \in U$, and we can show that an affine dynamical system $\sigma_1 = ((V(U^+)_{p;1}/S, \tilde{S}_r), \tilde{e}, \tilde{a}_l, a(1))$ is isomorphic to an affine dynamical system $\sigma_2 = ((Z,F), \xi, 1, a(1))$. It follows that σ_1 and σ_2 are the natural partial realizations of $\underline{a} \in F(U^*_{\underline{N}},Y)$. Therefore, there exist the natural partial realizations of $\underline{a} \in F(U^*_{\underline{N}},Y)$.

(7-A.10) Lemma
Two canonical affine dynamical systems are isomorphic if and only if their behavior is the same.

[proof]. This can be obtained from Theorem (7.7) and Corollary (7.3).

(7-A.11) Proof of Theorem (7.18)
Let $A_{(\ ,\ )}$ be the partial linear input/output map corresponding to $\underline{a} \in F(U^*_{\underline{N}},Y)$. In order to prove necessity, we assume existence of the natural partial realization of $\underline{a}$. Let Theorem (7.18) hold for integers p and p' that are different. Namely,
(1) im $A_{(p,\underline{N}-p-1)} = \text{im } A_{(p+1,\underline{N}-p-1)}$
(2) ker $A_{(p,\underline{N}-p)} = \text{ker } A_{(p,\underline{N}-p-1)}$
(3) im $A_{(p',\underline{N}-p'-1)} = \text{im } A_{(p'+1,\underline{N}-p'-1)}$
(4) ker $A_{(p',\underline{N}-p')} = \text{ker } A_{(p',\underline{N}-p'-1)}$
Then Propositions (7-A.6) and (7-A.7) imply that the dimension of $Z = \text{im } A_{(p,\underline{N}-p-1)}$ is equal to one of $Z' = \text{im } A_{(p',\underline{N}-p'-1)}$. Let σ and σ' be the natural partial realizations of $\underline{a}$ whose state spaces are Z and Z' respectively and which can be obtained by the same procedure as in (7-A.9). Then σ is clearly isomorphic to σ' and the behavior of σ is equal to one of σ' by Lemma (7-A.10). This implies that the behavior of the natural partial realization is always the same regardless of different integers p and p'. Therefore, the natural partial realization of $\underline{a}$ is a unique modulo isomorphism by Lemma (7-A.10).

Next, we show sufficiency by the contrapositive. We assume that there does not exist a natural partial realization of $\underline{a} \in F(U^*_{\underline{N}},Y)$. Then minimum dimensional partial realization σ of $\underline{a}$ is p-quasi-reachable and q-distinguishable for $p + q \geq \underline{N}$. It cannot be quasi-reachable within $p - 1$ and not be distinguishable within $q - 1$. Then, there exists a state x in σ such that x can be first reachable by a input ω with length p. The remaining data of $F(U^*_{\underline{N}-p-1}, Y)$ can't determine a new state $F(u)x$ for any $u \in U$, because of $\underline{N}-p-1 < q$. Therefore, we can't determine the transition matrix $F(u)$ uniquely by q-distinguishability. This implies that the minimum dimensional realization of $\underline{a}$ is not unique.

(7-A.12) Proof of Theorem (7.19)

Let's consider the natural partial realization $\sigma_2 = ((Z, F), \xi, 1, a(1))$ of $\underline{a} \in F(U^*_{\underline{N}}, Y)$ given in (7-A.9). Then we can obtain the quasi-reachable standard system $\sigma_s = ((\boldsymbol{R}^n, F_s), \mathbf{e}_1, h_s)$ from σ_2 in the same manner as theorem for a realization procedure (7.15).

8 Approximate and Noisy Realization of Linear Representation Systems

Let the set of output's values Y be a linear space over the field $\boldsymbol{R}$. In the reference [Matsuo and Hasegawa, 2003], linear representation systems were presented with the following main theorem. The main theorem says that for any causal input/output map, i.e., any input response map, there exist at least two canonical, namely quasi-reachable and distinguishable linear representation systems which realize, equivalently, faithfully describe it, and any two canonical linear representation systems with the same behavior are isomorphic.

For self contained, the results obtained in the reference are stated.

Firstly, the realization theory is listed.

Secondly, the results of finite dimensional linear representation systems are stated. They consist of a criterion for a canonical finite dimensional linear representation systems, a representation theorem of isomorphic classes for canonical linear representation systems, a criterions for the behavior of finite dimensional linear representation systems, and a procedure to obtain a canonical linear representation system.

Thirdly, their partial realization is remarked on according to the above results. The existence of minimum partial realization is listed. It rarely happens for minimum partial realizations to be unique up to an isomorphism. To solve the uniqueness problem, we define a notion of natural partial realizations and state the following main results for this partial realization:

1) A necessary and sufficient condition for the existence of natural partial realizations is given by the rank condition of a finite-sized Hankel matrix.

2) The existence condition of natural partial realizations is equivalent to the uniqueness condition of minimum partial realizations.

3) An algorithm to obtain a natural partial realization from a given partial input response map is given.

We can easily understand that the above results of our systems are the same as the ones obtained in linear system theory.

8.1 Basic Facts about Linear Representation Systems

Definition 8.1. *Linear Representation System*
1) A system given by the following system equation is written as a collection $\sigma = ((X, F), x^0, h)$ and it is said to be a linear representation system.

Y. Hasegawa: Approxi. & Noisy Reali. of Discrete-Time Dyn. Sys., LNCIS 376, pp. 205–237, 2008.
springerlink.com

$$\begin{cases} x(t+1) = F(\omega(t+1))x(t) \\ x(0) \quad = x^0 \\ \gamma(t) \quad = hx(t) \end{cases}$$

for any $t \in N$, $x(t) \in X$, $\gamma(t) \in Y$,

where X is a linear space over the field $\mathbf{R}$, F is a linear operator on X, an initial state $x^0 \in X$ and $h : X \to Y$ is a linear operator.

2) *The input response map $a_\sigma : U^* \to Y; \omega \mapsto h\phi_F(\omega)x^0$ is said to be the behavior of σ. For an input response map $a \in F(U^*, Y)$, σ which satisfies $a_\sigma = a$ is called a realization of a,*
where $\phi_F(\omega) := F(\omega(|\omega|))F(\omega(|\omega| - 1)) \cdots F(\omega(1))$.

3) *A linear representation system σ is said to be quasi-reachable if the linear hull of the reachable set $\{\phi_F(\omega)x^0; \omega \in U^*\}$ is equal to X and a linear representation system σ is called distinguishable if $h\phi_F(\omega)x_1 = h\phi_F(\omega)x_2$ for any $\omega \in U^*$ implies $x_1 = x_2$.*

4) *A linear representation system σ is called canonical if σ is quasi-reachable and distinguishable.*

Remark 1: The $x(t)$ in the system equation of σ is the state that produces output values of a_σ at the time t, namely the state $x(t)$ and linear operator $h : X \to Y$ generate the output value $a_\sigma(t)$ at the time t.

Remark 2: It is meant for σ to be a faithful model for the input response map a that σ realizes a.

Remark 3: Notice that a canonical linear representation system $\sigma = ((X, F), x^0, h)$ is a system that has the most reduced state space X among systems that have the behavior a_σ.

Example 8.2. $A(U^*) := \{\lambda = \sum_\omega \lambda(\omega)e_\omega$ (finite sum) $\}$, where $\omega = \bar{\omega}$ implies $e_\omega(\bar{\omega}) = 1$, and $\omega \neq \bar{\omega}$ implies $e_\omega(\bar{\omega}) = 0$. Let S_r be a map : $U \to L(A(U^*)); u \mapsto S_r(u)[\lambda \mapsto \sum_\omega \lambda(\omega)e_{u|\omega}$, an initial state be e_1 and a linear output map be $a : A(U^*) \to Y; \lambda \mapsto a(\lambda) = \sum \omega\lambda(\omega)a(\omega)$. Then a collection $((A(U^*), S_r), e_1, a)$ is a quasi-reachable linear representation system that realizes a.

Let $F(U^*, Y)$ be a set of any input response maps, let $S_l : U \to L(A(U^*)); u \mapsto S_l(u)[a \mapsto [\omega \mapsto a(\omega|u)]]$. Let a linear output map be $1 : F(U^*, Y) \to Y; a \mapsto a(1)$. Then a collection $(F(U^*, Y), S_l), a, 1)$ is a distinguishable linear representation system that realizes a.

Remark: Note that the linear output map $a : A(U^*) \to Y$ is introduced by the fact $F(U^*, Y) = L(A(U^*), Y)$.

Theorem 8.3. *The following two linear representation systems are canonical realizations of any input response map $a \in F(U^*, Y)$.*

1) $((A(U^*)/_a, \hat{S}_r), [e_1], \hat{a})$,

where $A(U^*)/_a$ is a quotient space obtained by equivalence relation

$\sum_{\omega} \lambda(\omega)e_\omega = \sum_{\bar{\omega}} \lambda(\bar{\omega})e_{\bar{\omega}} \iff \sum_{\omega} \lambda(\omega)a(\omega) = \sum_{\bar{\omega}} \lambda(\bar{\omega})a(\bar{\omega})$. $\hat{S}_r$ *is given by a map* :
$U \to L(A(U^*)/_a); u \mapsto \hat{S}_r(u)[\lambda \mapsto \sum_{\omega} \lambda(\omega)[e_{u|\omega}]$, *and* $\hat{a}$ *is given by* $\hat{a} : A(U^*)/_a \to$
$Y; [\lambda] \mapsto \hat{a}([\lambda]) = \sum_{\omega} \lambda(\omega)a(\omega)]$.

2) $((\ll S_l(U^*)a \gg), S_l), a, 1)$,
where $\ll S_l(U^*)a \gg$ *is the smallest linear space which contains* $S_l(U^*)a :=$
$\{S_l(\omega)a; \omega \in U^*\}$.

Definition 8.4. *Let* $\sigma_1 = ((X_1, F_1), x_1^0, h_1)$ *and* $\sigma_2 = ((X_2, F_2), x_2^0, h_2)$ *be linear representation systems, then a linear operator* $T : X_1 \to X_2$ *is said to be a linear representation system morphism* $T : \sigma_1 \to \sigma_2$ *if* T *satisfies* $TF_1(u) = F_2(u)T$ *for any* $u \in U$, $Tx_1^0 = x_2^0$ *and* $h_1 = h_2 T$. *If* $T : X_1 \to X_2$ *is bijective, then* $T : \sigma_1 \to \sigma_2$ *is said to be an isomorphism.*

Theorem 8.5. *Realization Theorem of linear representation systems*
Existence part: For any input response map $a \in F(U^*, Y)$, *there exist at least*
  two canonical linear representation systems which realize a.
Uniqueness part: Let σ_1 *and* σ_2 *be any two canonical linear representation*
  systems that realize $a \in F(U^*, Y)$, *then there exists an isomorphism*
  $T : \sigma_1 \to \sigma_2$.

8.2 Finite Dimensional Linear Representation Systems

Based on the realization theory (8.5), we again state structures of finite-dimensional linear representation systems in this section that have been previously described. To obtain concrete and meaningful results, we assume that the set U of input values is finite; i.e., $U := \{u_i; 1 \leq i \leq m$ for some $m \in N\}$. This assumption implies that the difference morphism F of a linear representation system $\sigma = ((X, F, x^0, h)$ is completely determined by the finite matrices $\{F(u_i); 1 \leq i \leq m\}$. But it will be presented that the assumption is not so special. The main results can be stated in the following four steps:

Firstly, we present conditions when finite dimensional linear representation system is canonical.

Secondly, we obtain a representation theorem for finite dimensional canonical linear representation systems, namely, we show two standard systems as a representative in their equivalence classes. One is a quasi-reachable standard system, and the other is a distinguishable standard system.

Thirdly, we give two criteria for the behavior of finite dimensional linear representation systems. One is the rank condition of infinite Hankel matrix, and the other is the application of Kleene's Theorem obtained in automata theory.

Lastly, we give a procedure to obtain the quasi-reachable standard system which realizes a given input response map.

Corollary 8.6. *Let* T *be a linear representation system morphism* $T : \sigma_1 \to \sigma_2$, *then* $a_{\sigma_1} = a_{\sigma_2}$ *holds.*

There is a fact about finite dimensional linear spaces that a n-dimensional linear space over the field $\boldsymbol{R}$ is isomorphic to $\boldsymbol{R}^n$ and $L(\boldsymbol{R}^n, \boldsymbol{R}^m)$ is isomorphic to $\boldsymbol{R}^{m \times n}$ (See Halmos [1958]). Therefore, without loss of generality, we can consider a n-dimensional linear representation system as $\sigma = ((\boldsymbol{R}^n, F), x^0, h)$, where F is a map $: U \to \boldsymbol{R}^{n \times n}$, $x^0 \in \boldsymbol{R}^n$ and $h \in \boldsymbol{R}^{p \times n}$.

Now we will show that the assumption of finiteness for input value's set U is not so special.

$U = \{u_1, u_2\}$ **8.7.**
In this case, a linear representation system $\sigma = ((\boldsymbol{R}^n, F), x^0, h)$ can be completely determined by $\{F(u_i); u_i \in U$ for $i = 1, 2\}$.

If on-off inputs are applied to a black-box, any non-linear system can be treated in this case.

Moreover, if an optimal solution is a bang-bang control, when a controlled object is in the optimal controlled condition, then it can be treated in this case.

Cases where U is a convex set in R^m 8.8
Let the set U be a convex set in R^m and a set V of the extreme points be a finite set $\{u_j; 1 \leq j \leq m\}$. Let F in $\sigma = ((\boldsymbol{R}^n, F), x^0, h)$ be a linear operator: $U \to \boldsymbol{R}^{n \times n}$, i.e. $F(\sum_{i=1}^{m} \alpha_i \mathbf{e_i}) = \sum_{i=1}^{m} \alpha_i F(u_i)$, $\sum_{i=1}^{m} \alpha_i = 1$. Then the linear representation system $\sigma = ((\boldsymbol{R}^n, F), x^0, h)$ can be rewritten as a linear representation system $\tilde{\sigma} = ((\boldsymbol{R}^n, \tilde{F}), x^0, h)$,
where $\tilde{F} : V \to \boldsymbol{R}^{n \times n}$ is given by $\tilde{F}(u_i) = F(u_i)$ for any $u_i \in V$.

Note that the quasi-reachability of σ is equivalent to the quasi-reachability of $\tilde{\sigma}$.

$U = R^m$ **8.9.**
Let $\boldsymbol{R} = R$ and $V = \{0, \mathbf{e_1}, \mathbf{e_2}, \cdots, \mathbf{e_m}\}$ for basis $\mathbf{e_i}$ in $R^m (1 \leq i \leq m)$. Let F in $\sigma = ((R^n, F), x^0, h)$ be an affine operator $: U \to R^{n \times n}$, i.e. $F(\sum_{i=1}^{m} \alpha_i \mathbf{e_i}) = A + (\sum_{i=1}^{m} \alpha_i \tilde{N}_i)$, A, $\tilde{N}_i \in R^{n \times n}$, $i \in N$. Then the linear representation system $\sigma = ((R^n, F), x^0, h)$ can be rewritten as a linear representation system $\tilde{\sigma} = ((R^n, \tilde{F}), x^0, h)$, where $\tilde{F} : V \to \boldsymbol{R}^{n \times n}$ is given by $F(0) = A$, $F(\mathbf{e_i}) = A + \tilde{N}_i$ for any $i(1 \leq j \leq m)$. Note that this $\tilde{\sigma}$ is a homogeneous bilinear system investigated by Tarn & Nonoyama [1976].

Note that the quasi-reachability of σ is equivalent to the quasi-reachability of $\tilde{\sigma}$.

Theorem 8.10. *A linear representation system $\sigma = ((K^n, F), x^0, h)$ is canonical if and only if the following conditions 1) and 2) hold.*
1) rank $[x^0, F(u_1)x^0, \cdots, F(u_m)x^0, \cdots, F(u_1)^2 x^0, F(u_1)F(u_2)x^0,$
$\cdots, F(u_1)F(u_m)x^0, \cdots, F(u_m)^2 x^0, \cdots, F(u_1)^{n-1} x^0, F(u_2)F(u_1)^{n-2} x^0, \cdots,$
$F(u_m)^{n-1} x^0] = n.$
2) rank $[h^T, (hF(u_1))^T, \cdots, (hF(u_m))^T, (hF(u_1)^2)^T, \cdots, (hF(u_1)F(u_m))^T,$
$\cdots, (hF(u_1)^{n-1})^T, (hF(u_1)^{n-2}F(u_m))^T, \cdots, (hF(u_m)^{n-1})^T] = n.$

Definition 8.11. *Let the input value's set U be $U := \{u_i; 1 \leq i \leq m\}$ and let a map $\| \ \| : U \to N$ be $u_i \mapsto \|u_i\| = i$. And let a numerical value $\||\omega\||$ of an input $\omega \in U^*$ be $\||\omega\|| = \|\omega(|\omega|)\| + \|\omega(|\omega| - 1)\| \times m + \cdots + \|\omega(1)\| \times m^{|\omega|-1}$ and $\||1\|| = 0.$*

Then, we can define totally ordered relation by this numerical value in U^*. Namely, $\omega_1 \leq \omega_2 \iff |||\omega_1||| \leq |||\omega_2|||$.

Definition 8.12. *A canonical linear representation system $\sigma = ((R^n, F_s), \mathbf{e_1}, h_s)$ is said to be a quasi-reachable standard system if input sequences $\{\omega_i; 1 \leq i \leq n\}$ given by $\mathbf{e_i} = \phi_{F_s}(\omega_i)\mathbf{e_1}$ satisfy the following conditions:*
1) $1 = \omega_1 < \omega_2 < \cdots < \omega_n$ and $|\omega_i| \leq i - 1$ for $i(1 \leq i \leq n)$ hold.
2) $\phi_{F_s}(\omega)\mathbf{e_1} = \sum_{i=1}^{j} \mathbf{e_j}$ holds for any input sequence such that $\omega_j < \omega < \omega_{j+1}(1 \leq i \leq n - 1)$, $\omega \in U^*$.

Theorem 8.13. *Representation Theorem for equivalence classes*
For any finite dimensional canonical linear representation system, there exists a uniquely determined isomorphic quasi-reachable standard system.

Definition 8.14. *Let Y be a field R for convenience. A canonical linear representation system $\sigma_d = ((R^n, F_d), x_d^0, h_d)$ is said to be a distinguishable standard system if input sequences $\{\omega_i; 1 \leq i \leq n\}$ given by $\mathbf{e_1}^T = h_d\mathbf{e_1}^T \phi_{F_d}(\omega_i)$ satisfy the following conditions:*
1) $1 = \omega_1 < \omega_2 < \cdots < \omega_n$ and $|\omega_i| \leq i - 1$ for $i(1 \leq i \leq n)$ hold.
2) $\mathbf{e_1}^T \phi_{F_d}(\omega) = \sum_{i=1}^{j} \alpha_i \mathbf{e_1}^T$ holds for any input sequence ω such that $\omega_j < \omega < \omega_{j+1}(1 \leq i \leq n - 1)$.

Theorem 8.15. *Representation Theorem for equivalence classes*
For any finite dimensional canonical linear representation system, there exists a uniquely determined isomorphic distinguishable standard system.

Definition 8.16. *For any input response map $a \in F(U^*, Y)$, the corresponding linear input/output map $A : (A(U^*), S_r) \to (F(U^*, Y), S_l)$ satisfies $A(\mathbf{e_\omega})(\bar{\omega}) = a(\bar{\omega}|\omega)$ for $\omega, \bar{\omega} \in U^*$.*
*   Hence, A can be represented by the next infinite matrix H_a^L. This H_a^L is said to be a Hankel matrix of a.*

$$
H_a^L = \begin{array}{c} \\ \bar{\omega} \end{array} \overset{\omega}{\left(\begin{array}{ccc} \vdots & & \\ \vdots & & \\ \vdots & & \\ \cdots & \cdots & a(\bar{\omega}|\omega) \end{array} \right)}
$$

Theorem 8.17. *Theorem for existence criterion*
For an input response map $a \in F(U^*, Y)$, the following conditions are equivalent:
1) The input response map $a \in F(U^*, Y)$ has the behavior of a n-dimensional canonical linear representation system.
2) There exist n linearly independent vectors and no more than n linearly independent vectors in a set $\{S_l(\omega)a; |\omega| \leq n - 1$ for $\omega \in U^*\}$.
3) The rank of the Hankel matrix H_a^L of a is n.

Remark: Fliess [1974] has introduced the Hankel matrix of non-commutative formal power series and shown that the recognizability of the formal power series is equal to the finite rank of its Hankel matrix.

Let $A(U^*)$ have the following operation $\times$.
$\times : A(U^*) \times A(U^*) \to A(U^*); (\sum_\omega \lambda(\omega)\mathbf{e}_\omega, \sum_{\bar{\omega}} \lambda(\bar{\omega})\mathbf{e}_{\bar{\omega}}) \mapsto (\sum_\omega \lambda(\omega)\mathbf{e}_\omega) \times (\sum_{\bar{\omega}} \lambda(\bar{\omega})\mathbf{e}_{\bar{\omega}}) = \sum_{\omega'} (\sum_{\omega|\bar{\omega}=\omega'} \lambda(\omega)\lambda(\bar{\omega})\mathbf{e}_{\omega'})$
Then $A(U^*)$ is an algebra over $\mathbf{R}$, and is a free algebra over $\mathbf{R}$.

As $a \in F(U^*, \mathbf{R})$ can be expressed as a formal power series $a = \sum_\omega a(\omega)\mathbf{e}_\omega$, it can be considered that $F(U^*, \mathbf{R})$ contains $A(U^*)$. Here, we introduce an operation $* : F(U^*, \mathbf{R}) \times F(U^*, \mathbf{R}); a \mapsto a^* = \sum_{j=1}^{\infty} (a - a(1))^j$.

Theorem 8.18. *Theorem for a realization procedure*
Let an input response $a \in F(U^*, Y)$ satisfy the condition of Theorem (8.18), then the quasi-reachable standard system $\sigma_s = ((\mathbf{R}^n, F_s), \mathbf{e}_1, h_s)$ which realizes the input response map a can be obtained by the following procedure:
1) Select the linearly independent vectors $\{S_l(\omega_i)a; 1 \leq i \leq n\}$ of the set $\{S_l(\omega)a; |\omega| \leq n - 1, \omega \in U^*\}$ in order of their numerical value.
2) Let the state space be $\mathbf{R}^n$, the initial state be $\mathbf{e}_1$.
3) Let the output map $h_s = [a(\omega_1), a(\omega_2), a(\omega_3), \cdots, a(\omega_n)]$.
4) Let $_if_j \in \mathbf{R}^n$ in $F_s(u_i) := [_if_1 \; _if_2 \; \cdots \; _if_n]$ be $_if_j := [_if_{j,1} \; _if_{j,2} \; \cdots \; _if_{j,n}]^T$, where $S_l(u_i)S_l(\omega_j)a = \sum_{k=1}^{j} _if_{1,k} \, S_l(\omega_k)a$, $_if_{j,k} \in \mathbf{R}$ and $\mathbf{e}_1 = [1, 0, 0, \cdots, 0, 0]^T \in \mathbf{R}^n$.

8.3 Partial Realization Theory of Linear Representation Systems

Here we consider a partial realization problem by multi-experiment. Let $\underline{a}$ be an $\underline{N}$ sized input response map($\in F(U_{\underline{N}}^*, Y)$), where $\underline{N} \in N$ and $U_{\underline{N}}^* := \{\omega \in U^*; |\omega| \leq \underline{N}\}$. The $\underline{a}$ is said to be a partial input response map. A finite dimensional linear representation system $\sigma = ((X, F), x^0, h)$ is called a partial realization of $\underline{a}$ if $h\phi_F(\omega)x^0 = \underline{a}(\omega)$ holds for any $\omega \in U_{\underline{N}}^*$.

A partial realization problem of linear representation systems can be stated as follows:
< For any given $\underline{a} \in F(U_{\underline{N}}^*, Y)$, find a partial realization σ of $\underline{a}$ such that the dimensions of state space X of σ is minimal, where the σ is said to be a minimal partial realization of $\underline{a}$. Moreover, show when the minimal realizations are isomorphic.>

Proposition 8.19. *For any given $\underline{a} \in F(U_{\underline{N}}^*, Y)$, there always exists a minimal partial realization of it.*

[proof]. For any $\omega \notin U_{\underline{N}}^*$, set $\underline{a}(\omega) = 0$. Then $\underline{a} \in F(U^*, Y)$, and Theorem (8.18) implies that there exists a finite dimensional partial realization of $\underline{a}$. Therefore, there exists a minimal partial realization.

Minimal partial realizations are, in general, not unique modulo isomorphism. Therefore, we introduce a natural partial realization, and we show that natural partial realizations exist if and only if they are isomorphic.

Definition 8.20. *For a linear representation system $\sigma = ((X, F), x^0, h)$ and some $p \in N$, if $X =\ll \{\phi_F(\omega)x^0; \omega \in U_p^*\} \gg$, then σ is said to be p-quasi-reachable,*
where $\ll S \gg$ denotes the smallest linear space which contains a set S.

Let q be some integer. If $h\phi_F(\omega)x = 0$ implies $x=0$ for any $\omega \in U_q^*$, then σ is said to be q-distinguishable.

For a given $\underline{a} \in F(U_L^*, Y)$, if there exist p and $q \in N$ such that $p + q < L$ and σ is p-quasi-reachable and q-distinguishable then σ is said to be a natural partial realization of $\underline{a}$.

For a partial input response map $\underline{a} \in F(U_L^*, Y)$, the following matrix $H_{\underline{a}\,(p,L-p)}^L$ is said to be a finite-sized Hankel matrix of $\underline{a}$.

$$
H_{\underline{a}\,(p,L-p)}^L = \quad
\begin{matrix} & & \omega \\ & & \vdots \\ & & \vdots \\ & & \vdots \\ \overline{\omega} & & \vdots \end{matrix}
\begin{pmatrix} & & & \\ & & \vdots & \\ & & \vdots & \\ & & \vdots & \\ \cdots & \cdots & a(\overline{\omega}\,|\omega) & \end{pmatrix}
$$

where $\overline{\omega} \in U_p^*$ and $\omega \in U_{L-p}^*$.

Theorem 8.21. *Let $H_{\underline{a}\,(p,L-p)}^L$ be the finite Hankel-matrix of $\underline{a} \in F(U_L^*, Y)$. Then there exists a natural partial realization of $\underline{a}$ if and only if the following conditions hold:*
rank $H_{\underline{a}\,(p,L-p)}^L =$ rank $H_{\underline{a}\,(p,L-p-1)}^L =$ rank $H_{\underline{a}\,(p+1,L-p-1)}^L$ for some $p \in N$.

Theorem 8.22. *There exists a natural partial realization of a given partial input response map $\underline{a} \in F(U_L^*, Y)$ if and only if the minimal partial realization of $\underline{a}$ are unique modulo isomorphisms.*

Theorem 8.23. *Let a partial input response $\underline{a} \in F(U_L^*, Y)$ satisfy the condition of Theorem (8.26), then the quasi-reachable standard system $\sigma_s = ((X, F_s), \mathbf{e_1}, h_s)$ which realizes $\underline{a}$ can be obtained by the following algorithm.*

Set $n := $ rank $H_{\underline{a}\,(p,L-p)}^L$, where $H_{\underline{a}\,(p,L-p)}^L$ is the finite Hankel matrix of $\underline{a} \in F(U_L^*, Y)$.

1) Select the linearly independent vectors $\{S_l(\omega_i)\underline{a} \in F(U_{L-p}^*, Y); 1 \leq i \leq n\}$ from $H_{\underline{a}\,(p,L-p)}^L$ in order of their numerical value.

2) Let the state space be $\mathbf{R}^n$, the initial state be $\mathbf{e_1} = [100, \cdots, 0]^T$.

3) Let the output map $h_s = [\underline{a}(1)\underline{a}(\omega_2)\underline{a}(\omega_3) \cdots \underline{a}(\omega_n)]$.

4) Let $_if_j$ in $F_s(u_i) := [_if_1 \; _if_2 \; \cdots \; _if_n \;]$ be $_if_j := [_if_{j,1} \; _{i}f_{j,2} \; \cdots \; _if_{j,n}]$ for $1 \leq i \leq n$, where $_if_j$ is given by the following.
$\underline{S_l}(u_i)\underline{S_l}(\omega_j)\underline{a} = \sum_{k=1}^{j} {_if_{j,k}}\underline{S_l}(\omega_i)\underline{a}, \; _if_{j,k} \in \mathbf{R}$ in the sense of $F(U_{L-p}^*, Y)$ and
$\underline{S_l}(\omega) : F(U_s^*, Y) \to F(U_{s-|\omega|}^*, Y) \; ; \; \underline{a} \mapsto \underline{S_l}(\omega)\underline{a}[; \overline{\omega} \mapsto \underline{a}(\overline{\omega}|\omega)]$.

8.4 Approximate Realization of Linear Representation Systems

In this section, we discuss an approximate realization problem of linear representation systems.

We will discuss the approximate realization problem under the assumption the set U of input values is a finite set $U = \{u_j : 1 \leq j \leq m\}$ for an finite integer $m \in N$. The reference [Matsuo and Hasegawa, 2003] showed that this assumption is not so special. However, for simplicity of our discussion, we assume that the set U of input values is $U = \{u_1, u_2\}$ or $U = \{u_1,\ u_2,\ u_3\}$.

Roughly speaking, the approximate realization of linear representation systems can be stated as follows:

< For any given partial data of a linear representation system, find a linear representation system which approximates the given data. >

In order to make our discussion simple, we assume that the set Y of outout is the set R of real numbers, namely 1-output.

Theorem 8.24. *Algorithm for approximate realization*
Let an input response map $\underline{a}$ be a considered object which is a linear representation system and $U := \{u_j; 1 \leq j \leq m, m \in N\}$. Then an approximate realization $\sigma = ((R^n, F_s), x_s^0, h_s)$ of $\underline{a}$ is given by the following algorithm:
1) Based on the ratio of the square root of eigenvalues for a matrix
$H_{\underline{a}\ (n,\bar{p})}(|||\omega_1|||, \cdots, |||\omega_n|||) H_{\underline{a}\ (n,\bar{p})}(|||\omega_1|||, \cdots, |||\omega_n|||)^T$, determine the value n
of rank for the matrix, where $|||\omega_1|||, |||\omega_2|||, \cdots$ and $|||\omega_n|||$ are selected in
order of numerical value of input and $\{\underline{S_l}(\omega_i)\underline{a}; 1 \leq i \leq n,\ \omega_i \in U^*\}$
is a set of independent vectors.
Namely, determine the value n of rank for the matrix
$H_{\underline{a}\ (n,\bar{p})}(|||\omega_1|||, \cdots, |||\omega_n|||)$ such that the ratio of the square root of
eigenvalues for the covariance matrix becomes very small.
The small ratio means the nearness of approximation degree.
2) In order to determine g_s, the CLS method is used as follows:
① In particular, set $g_s(\omega_i) := \mathbf{e}_i$ for $\omega_i \in U$. Namely, $g_s(\omega_1) := \mathbf{e}_1$,
$g_s(\omega_2) := \mathbf{e}_2,\ \cdots, g_s(\omega_k) := \mathbf{e}_k$ for some $k \in N$.
For $u \in U$ such that $u \notin \{\omega_i; 1 \leq i \leq n\}$ and $\omega_r < u$,
$g_s(u) = \sum_{j=1}^r b_{u,j} \underline{S_l}(\omega_j)\underline{a}$ is obtained as follows:
Let a matrix $A_u \in R^{1 \times (r+1)}$ be $A_u := [b_{u,1}, b_{u,2}, \cdots, b_{u,r}, -1]$.
Choose the coefficients $\{b_{u,j} : 1 \leq j \leq r\}$ such that
$\sum_{j=1}^r (\underline{S_l}(\omega_j)\underline{\bar{a}}) \cdot (\underline{S_l}(\omega_j)\underline{\bar{a}}) + (\underline{S_l}(u)\underline{\bar{a}}) \cdot (\underline{S_l}(u)\underline{\bar{a}})$ take a minimum value,
where $\{\underline{S_l}(\omega_j)\underline{\bar{a}} \in R^{L \times 1} : 0 \leq j \leq r\}$ are given by
the equation $[\underline{S_l}(\omega_1)\underline{\bar{a}}, \underline{S_l}(\omega_2)\underline{\bar{a}}, \cdots, \underline{S_l}(\omega_r)\underline{\bar{a}}, \underline{S_l}(u)\underline{\bar{a}}]^T :=$
$A_u^T [A_u A_u^T]^{-1} A_u H_{\underline{a}\ (r+1,L)}^T (|||\omega_1|||, |||\omega_2|||, \cdots, |||\omega_r|||, |||u|||)$
and $H_{\underline{a}\ (|||u|||+1,L)}^T (|||\omega_1|||, |||\omega_2|||, \cdots, |||\omega_r|||, |||u|||) :=$
$[\underline{S_l}(\omega_1)\underline{a}, \cdots, \underline{S_l}(\omega_r)\underline{a}, \underline{S_l}(u)\underline{a}]$. And $\cdot$ denotes the inner product
of two vectors.
3) In order to obtain F_s, the CLS method is used as follows:
① Let $_i\mathbf{f}_j \in R^n$ in $F_s(u_i) := [_i\mathbf{f}_1\ _i\mathbf{f}_2\ \cdots\ _i\mathbf{f}_n\ ]$ be $_i\mathbf{f}_j := [_if_{j,1}\ _{ii}f_{j,2}\ \cdots\ _if_{j,n}]^T$

for $1 \leq i \leq m$, where $_if_{j,k}$ is given by the following:
$\underline{S_l}(u_i)\underline{S_l}(\omega_j)\underline{a} = \sum_{k=1}^{n} {}_if_{j,k}\underline{S_l}(\omega_k)\underline{a}$, $_if_{j,k} \in \mathbf{R}$ in the sense of
$F(U_{L-p}^*, Y)$.

② For i $(1 \leq i \leq m)$, j $(1 \leq j \leq n)$ and for the maximum number
r $(1 \leq r \leq n)$ such that $\omega_r, \omega_j \in \{\omega_j; 1 \leq j \leq n\}$ and $|||\omega_r||| < |||u_i|\omega_j|||$,
let a matrix $_iA_j \in \mathbf{R}^{1 \times (r+1)}$ be $_iA_j := [_if_{j,1}, _if_{j,2}, \cdots, _if_{j,r}, -1]$.
Choose the coefficients $_if_{j,k} = 0$ for k $(r+1 \leq k \leq n)$ and $\{_if_{j,k} : 1 \leq k \leq r\}$
such that $\sum_{i=1}^{r}(\underline{S_l}(\omega_i)\underline{\bar{a}}) \cdot (\underline{S_l}(\omega_i)\underline{\bar{a}}) + (\underline{S_l}(u_i|\omega_j)\underline{\bar{a}}) \cdot (\underline{S_l}(u_i|\omega_j)\underline{\bar{a}})$ takes a
minimum value, where $\{\underline{S_l}(\omega_i)\underline{\bar{a}} \in \mathbf{R}^{L \times 1} : 0 \leq i \leq r\}$ and $\underline{S_l}(u_i|\omega_j)\underline{\bar{a}}$ are
given by the following equation:
$[\underline{S_l}(\omega_1)\underline{\bar{a}}, \underline{S_l}(\omega_2)\underline{\bar{a}}, \cdots, \underline{S_l}(\omega_r)\underline{\bar{a}}, \underline{S_l}(u_i|\omega_j)\underline{\bar{a}}]^T :=$
$_iA_j^T [_iA_j \; _iA_j^T]^{-1} \; _iA_j \; H_{\underline{a}}^T{}_{(n+1,L)}(|||\omega_1|||, |||\omega_2|||, \cdots, |||\omega_r|||, |||u_i|\omega_j|||)$ and
$H_{\underline{a}}^T{}_{(r+1,L)}(|||\omega_1|||, |||\omega_2|||, \cdots, |||\omega_r|||, |||u_i|\omega_j|||) :=$
$[\underline{S_l}(\omega_1)\underline{a}, \underline{S_l}(\omega_2)\underline{a}, \cdots, \underline{S_l}(\omega_r)\underline{a}, \underline{S_l}(u_i|\omega_j)\underline{a}]$, and $\cdot$ denotes the inner product
of two vectors.

4) In order to determine h_s, the CLS method is used as follows:
① For the first ω_{r_1+1}, $\omega_{r_1} \in \{\omega_i; 1 \leq i \leq n\}$ such that $\omega_{r_1+1} > \omega_{r_1}$ and
$|||\omega_{r_1+1}||| - |||\omega_{r_1}||| > 1$ when starting out from ω_1,
set $\underline{S_l}(\lambda_1)\underline{a} := \sum_{i=1}^{r_1} b_{\lambda_1,i}\mathbf{e}_i$ for the obtained equation $\underline{S_l}(\lambda_1)\underline{a} =$
$\sum_{i=1}^{r_1} b_{\lambda_1,i}(\underline{S_l}(\omega_i)\underline{a})$ for λ_1 such that $|||\lambda_1||| = |||\omega_{r_1}||| + 1$.
Let a matrix $A_{\lambda_1} \in \mathbf{R}^{1 \times (r_1+1)}$ be $A_{\lambda_1} := [b_{\lambda_1,1}, b_{\lambda_1,2}, \cdots, b_{\lambda_1,r_1}, -1]$.
Choose the coefficients $\{b_{\lambda_1,i} : 1 \leq i \leq r_1\}$ such that
$\sum_{i=1}^{r_1}(\underline{S_l}(\omega_i)\underline{\bar{a}}) \cdot (\underline{S_l}(\omega_i)\underline{\bar{a}}) + (\underline{S_l}(\lambda_1)\underline{\bar{a}}) \cdot (\underline{S_l}(\lambda_1)\underline{\bar{a}})$ take a minimum value,
where $\{\underline{S_l}(\omega_i)\underline{\bar{a}} \in \mathbf{R}^{L \times 1} : 0 \leq i \leq r_1\}$ are given by
the equation $[\underline{S_l}(\omega_1)\underline{\bar{a}}, \underline{S_l}(\omega_2)\underline{\bar{a}}, \cdots, \underline{S_l}(\omega_{r_1})\underline{\bar{a}}, \underline{S_l}(\lambda_1)\underline{\bar{a}}]^T :=$
$A_{\lambda_1}^T [A_{\lambda_1} A_{\lambda_1}^T]^{-1} A_{\lambda_1} H_{\underline{a}}^T{}_{(r_1+1,L)}(|||\omega_1|||, |||\omega_2|||, \cdots, |||\omega_{r_1}|||, |||\lambda_1|||)$
and $H_{\underline{a}}^T{}_{(|||\lambda_1|||+1,L)}(|||\omega_1|||, |||\omega_2|||, \cdots, |||\omega_{r_1}|||, |||\lambda_1|||) :=$
$[\underline{S_l}(\omega_1)\underline{a}, \cdots, \underline{S_l}(\omega_{r_1})\underline{a}, \underline{S_l}(\lambda_1)\underline{a}]$. And $\cdot$ denotes the inner product of two
vectors.
Then let $h_{\lambda_1 s} \in \mathbf{R}^{1 \times r_1}$ be
$h_{\lambda_1 s} := [\underline{a}(\omega_1) - (\underline{\bar{a}}(\omega_1), \underline{a}(\omega_2) - (\underline{\bar{a}}(\omega_2), \cdots, \underline{a}(\omega_{r_1}) - (\underline{\bar{a}}(\omega_{r_1})]$.

② For the first ω_{r_2+1}, $\omega_{r_2} \in \{\omega_i; 1 \leq i \leq n\}$ such that $\omega_{r_2+1} > \omega_{r_2}$ and
$|||\omega_{r_2+1}||| - |||\omega_{r_2}||| > 1$ when starting out from ω_{r_1+1},
set $\underline{S_l}(\lambda_2)\underline{a} := \sum_{i=1}^{r_2} b_{\lambda_2,i}\mathbf{e}_i$ for the obtained equation $\underline{S_l}(\lambda_2)\underline{a} =$
$\sum_{i=1}^{r_2} b_{\lambda_2,i}(\underline{S_l}(\omega_i)\underline{a})$ for λ_2 such that $|||\lambda_2||| = |||\omega_{r_2}||| + 1$.
Let a matrix $A_{\lambda_2} \in \mathbf{R}^{1 \times (r_2+1)}$ be $A_{\lambda_2} := [b_{\lambda_2,1}, b_{\lambda_2,2}, \cdots, b_{\lambda_2,r_2}, -1]$.
Choose the coefficients $\{b_{\lambda_2,i} : 1 \leq i \leq r_2\}$ such that
$\sum_{i=1}^{r_2}(\underline{S_l}(\omega_i)\underline{\bar{a}}) \cdot (\underline{S_l}(\omega_i)\underline{\bar{a}}) + (\underline{S_l}(\lambda_2)\underline{\bar{a}}) \cdot (\underline{S_l}(\lambda_2)\underline{\bar{a}})$ take a minimum value,
where $\{\underline{S_l}(\omega_i)\underline{\bar{a}} \in \mathbf{R}^{L \times 1} : 0 \leq i \leq r_2\}$ are given by
the equation $[\underline{S_l}(\omega_1)\underline{\bar{a}}, \underline{S_l}(\omega_2)\underline{\bar{a}}, \cdots, \underline{S_l}(\omega_{r_2})\underline{\bar{a}}, \underline{S_l}(\lambda_2)\underline{\bar{a}}]^T :=$
$A_{\lambda_2}^T [A_{\lambda_2} A_{\lambda_2}^T]^{-1} A_{\lambda_2} H_{\underline{a}}^T{}_{(r_2+1,L)}(|||\omega_1|||, |||\omega_2|||, \cdots, |||\omega_{r_2}|||, |||\lambda_2|||)$
and $H_{\underline{a}}^T{}_{(|||\lambda_2|||+1,L)}(|||\omega_1|||, |||\omega_2|||, \cdots, |||\omega_{r_2}|||, |||\lambda_2|||) :=$
$[\underline{S_l}(\omega_1)\underline{a}, \cdots, \underline{S_l}(\omega_{r_2})\underline{a}, \underline{S_l}(\lambda_2)\underline{a}]$. And $\cdot$ denotes the inner product of two
vectors.

Then let $h_{\lambda_2 s} \in \mathbf{R}^{1 \times r_2}$ be
$$h_{\lambda_2 s} := [\underline{a}(\omega_{r_1+1}) - (\underline{\bar{a}}(\omega_{r_1+1}), \ \underline{a}(\omega_{r_1+2}) - (\underline{\bar{a}}(\omega_{r_1+2}), \cdots, \ \underline{a}(\omega_{r_2}) - (\underline{\bar{a}}(\omega_{r_2})].$$

$\vdots$

$\textcircled{t}$ *For $\omega \in U^*$ such that $|||\omega||| = |||\omega_n||| + 1$, let a matrix $A_\omega \in \mathbf{R}^{1 \times (n+1)}$*
be $A_\omega := [b_{\omega,1}, b_{\omega,2}, \cdots, b_{\omega,n}, -1]$.
Choose the coefficients $\{b_{\omega,i} : 1 \leq i \leq n\}$ such that
$\sum_{i=1}^{n}(\underline{S_l}(\omega_i)\underline{\bar{a}}) \cdot (\underline{S_l}(\omega_i)\underline{\bar{a}}) + (\underline{S_l}(\omega)\underline{\bar{a}}) \cdot (\underline{S_l}(\omega)\underline{\bar{a}})$ *takes a minimum value,*
where $\{\underline{S_l}(\omega_i)\underline{\bar{a}} : 1 \leq i \leq n\}$ and $\underline{S_l}(\omega)\underline{\bar{a}}$ are
given by the following equation:
$[\underline{S_l}(\omega_1)\underline{\bar{a}}, \underline{S_l}(\omega_2)\underline{\bar{a}}, \cdots, \underline{S_l}(\omega_n)\underline{\bar{a}}, \underline{S_l}(\omega)\underline{\bar{a}}]^T :=$
$A_\omega^T [A_\omega A_\omega^T]^{-1} A_\omega H_{\underline{a}}^T{}_{(n+1,L)}(|||\omega_1|||, |||\omega_2|||, \cdots, |||\omega_n|||, |||\omega|||)$
and $H_{\underline{a}}^T{}_{(n+1,L)}(|||\omega_1|||, |||\omega_2|||, \cdots, |||\omega_n|||, |||\omega|||) :=$
$[\underline{S_l}(\omega_1)\underline{a}, \cdots, \underline{S_l}(\omega_n)\underline{a}, \underline{S_l}(\omega)\underline{a}]$, *and $\cdot$ denotes the inner product of two*
vectors.
Then let $h_{\omega s}$ be
$$h_{\omega s} := [\underline{a}(\omega_{r_t+1}) - (\underline{\bar{a}}(\omega_{r_t+1}), \ \underline{a}(\omega_{r_t+2}) - \underline{\bar{a}}(\omega_{r_t+2}), \cdots, \ \underline{a}(\omega_n) - \underline{\bar{a}}(\omega_n)].$$
Finally, let $h_s \in \mathbf{R}^{1 \times n}$ be
$$h_s := [h_{\lambda_1 s}, h_{\lambda_2 s}, \cdots, h_{\omega s}].$$

[proof]. In 1), the number of dimensions is determined by considering the ratio of Hankel matrix norm, which means a degree of information loss. According to Theorem (8.23), a linear representation system $\sigma = ((\mathbf{R}^n, F_s), g_s, h_s)$ is obtained as follows. In 2), g_s is obtained directly or by using the CLS method for A_u corresponding to the matrix A in Proposition (2.14). In 3), F_s is obtained by using the CLS method for ${}_i A_j$ corresponding to the matrix A in Proposition (2.14). In 4), h_s is obtained by using the CLS method for A_{λ_1}, A_{λ_2}, A_{λ_ω} corresponding to the matrix A in Proposition (2.14).

In the figures of this chapter, we use a notation $Signal\_n\ d$ as an input response map obtained by a n-dimensional linear representation system.

In examples of this chapter, a notation $H_{\underline{a}}^T{}_{(r+1,40)}(0, \cdots, r)$ is used in place of $H_{\underline{a}}^T{}_{(r+1,40)}(0, 1, 2, 3, \cdots, r-1, r)$.

Example 8.25. Let the signals be the behavior of the following 3-dimensional linear representation system: $\sigma = ((\mathbf{R}^3, F), x^0, h)$, where $F(u_1) = \begin{bmatrix} 0 & 0.8 & -1.2 \\ 1 & 0 & 1.5 \\ 0 & 0 & -0.1 \end{bmatrix}$,

$F(u_2) = \begin{bmatrix} -1 & 0 & -1.5 \\ 0 & 0 & 1.5 \\ 0 & 1 & -0.8 \end{bmatrix}$, $F(u_3) = \begin{bmatrix} 0 & 0.3 & -0.7 \\ 0 & -0.2 & 0.1 \\ 0 & -0.4 & 0.4 \end{bmatrix}$, $x^0 = \mathbf{e}_1$, $h = [10, 2, 0.1]$.

Then the approximate realization problem is solved as follows:

covariance matrix	eigenvalues							
	1	2	3	4	5	6	$\cdots$	13
$H_{\underline{a}\ (3,40)}^{T}(0,1,2)H_{\underline{a}\ (3,40)}(0,1,2)$	2492	1241	0					
$H_{\underline{a}\ (4,40)}^{T}(0,\cdots,3)H_{\underline{a}\ (4,40)}(0,\cdots,3)$	2492	1241	0	0				
$H_{\underline{a}\ (5,40)}^{T}(0,\cdots,4)H_{\underline{a}\ (5,40)}(0,\cdots,4)$	2912	1401	0	0	0			
$H_{\underline{a}\ (6,40)}^{T}(0,\cdots,5)H_{\underline{a}\ (6,40)}(0,\cdots,5)$	10591	1572	104	0	0	0		
$H_{\underline{a}\ (13,40)}^{T}(0,\cdots,12)H_{\underline{a}\ (13,40)}(0,\cdots,12)$	13019	2388	187	0	0	0	$\cdots$	0
covariance matrix	square root of eigenvalues							
$H_{\underline{a}\ (3,40)}^{T}(0,1,2)H_{\underline{a}\ (3,40)}(0,1,2)$	50	35.2	0					
$H_{\underline{a}\ (4,40)}^{T}(0,\cdots,3)H_{\underline{a}\ (4,40)}(0,\cdots,3)$	50	35.2	0	0				
$H_{\underline{a}\ (5,40)}^{T}(0,\cdots,4)H_{\underline{a}\ (5,40)}(0,\cdots,4)$	54	37.4	0	0	0			
$H_{\underline{a}\ (6,40)}^{T}(0,\cdots,5)H_{\underline{a}\ (6,40)}(0,\cdots,5)$	103	39.6	10.2	0	0	0		
$H_{\underline{a}\ (13,40)}^{T}(0,\cdots,12)H_{\underline{a}\ (13,40)}(0,\cdots,12)$	114	49	13.7	0	0	0	$\cdots$	0

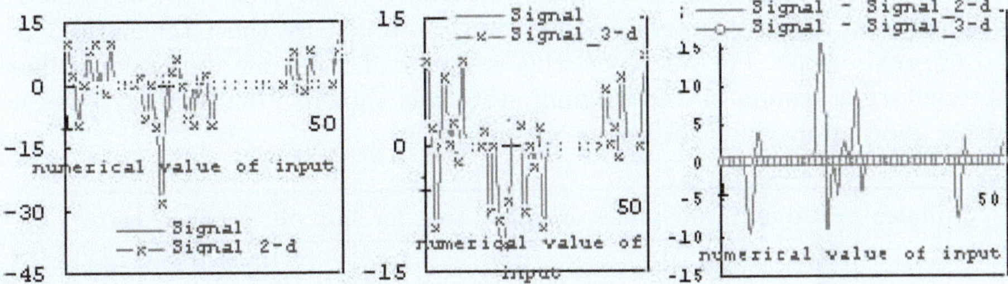

Fig. 8.1. The left is the original input response map and the behavior of a 2-dimensional linear representation system obtained by the CLS method. The middle is the original input response map and the behavior of a 3-dimensional linear representation system obtained by the CLS method. The right is the difference between the original one and the behavior of the 2-dimensional linear representation system obtained by the CLS method or the 3-dimensional linear representation system obtained by the CLS method in Example (8.25).

1) Since the ratio $\frac{10.2}{103} = 0.10$ obtained by the square root of $H_{\underline{a}\ (6,40)}^{T}(0,\cdots,5)H_{\underline{a}\ (6,40)}(0,\cdots,5)$ is a little large, the approximate 2-dimensional linear representation system obtained by the CLS method may not be good.
2) After determining the independent vectors $\underline{a}$ and $S_l(u_1)\underline{a}$ whose numerical value of input are 0 and 1, we will continue the approximate realization algorithm by the CLS method.

Therefore, an approximate 2-dimensional linear representation system $\sigma_1 = ((\mathbf{R}^2, F_1), x_1^0, h_1)$ obtained by the CLS method is constructed as follows:

$$F_1(u_1) = \begin{bmatrix} 0 & 0.8 \\ 1 & 0 \end{bmatrix} , \; F_1(u_2) = \begin{bmatrix} -1 & 1.3 \\ 0 & -1.6 \end{bmatrix} , \; F_1(u_3) = \begin{bmatrix} 0 & -0.18 \\ 0 & 0.43 \end{bmatrix} ,$$

$x_1^0 = e_1, \; h_1 = [10, \; 2]$.

For reference, a 3-dimensional linear representation system $\sigma_2 = ((\mathbf{R}^3, F_2), x_2^0, h_2)$ obtained by the CLS method can be expressed as follows:

$$F_2(u_1) = \begin{bmatrix} 0 & 0.8 & -1.2 \\ 1 & 0 & 1.5 \\ 0 & 0 & -0.1 \end{bmatrix} , \; F_2(u_2) = \begin{bmatrix} -1 & 0 & -1.5 \\ 0 & 0 & 1.5 \\ 0 & 1 & 0.8 \end{bmatrix} , \; F_2(u_3) = \begin{bmatrix} 0 & 0.3 & -0.7 \\ 0 & -0.2 & 0.1 \\ 0 & -0.4 & 0.4 \end{bmatrix} ,$$

$x_2^0 = e_1, \; h_2 = [10, \; 2, \; 0.1]$.

In this example, the original signals are considered as the behavior of a 3-dimensional linear representation system and the desirable input response map is obtained by the CLS method with our bad feeling. The model obtained by the CLS method is a 2-dimensional linear representation system.

For reference, a 3-dimensional linear representation system is also obtained by the CLS method. The system completely reconstructs the original system.

Just as we thought, the following table and Fig. 8.1 truly indicate that the 2-dimensional linear representation system obtained by the CLS method is a bad approximation. For reference, the behavior of the same dimensional linear representation system as the original system is shown. Finally, there does not exist a good approximation for the given system.

dimen-ion	ratio of matrices	mean values of square root for sum of			cosine ① and ②	error ratio
		signal ①	signal by CLS ②	error ③	$\cos\theta$	③/①
$a_{0,1}$	0.1	0.672	0.891	0.51	0.82	0.76
$a_{0,1,5}$	0	0.672	0.672	0	1	0

Example 8.26. Let the signals be the behavior of the following 3-dimensional linear representation system: $\sigma = ((\mathbf{R}^3, F), x^0, h)$, where $F(u_1)$

$$= \begin{bmatrix} 0 & -1.4 & -1.3 \\ 1 & -1.5 & -0.3 \\ 0 & 0 & 1 \end{bmatrix} , \; F(u_2) = \begin{bmatrix} 0 & 0.2 & 0.1 \\ 0 & 0.3 & 0.3 \\ 1 & -0.6 & 0.6 \end{bmatrix} , \; F(u_3) = \begin{bmatrix} -0.6 & -0.1 & -0.7 \\ 0.5 & -0.5 & 0.1 \\ 0 & 0.5 & 0.5 \end{bmatrix} ,$$

$x^0 = e_1, \; h = [6, \; -7.5, \; 1]$.

Then the approximate realization problem is solved as follows:

covariance matrix	eigenvalues							
	1	2	3	4	5	6	$\cdots$	13
$H^T_{\underline{a}\,(3,40)}(0,1,2)H_{\underline{a}\,(3,40)}(0,1,2)$	815	84	1.1					
$H^T_{\underline{a}\,(4,40)}(0,\cdots,3)H_{\underline{a}\,(4,40)}(0,\cdots,3)$	1180	92	1.1	0				
$H^T_{\underline{a}\,(5,40)}(0,\cdots,4)H_{\underline{a}\,(5,40)}(0,\cdots,4)$	1181	120	5.2	0	0			
$H^T_{\underline{a}\,(6,40)}(0,\cdots,5)H_{\underline{a}\,(6,40)}(0,\cdots,5)$	1229	123	5.9	0	0	0		
$H^T_{\underline{a}\,(13,40)}(0,\cdots,12)H_{\underline{a}\,(13,40)}(0,\cdots,12)$	2082	241	13	0	0	0	$\cdots$	0
covariance matrix	square root of eigenvalues							
$H^T_{\underline{a}\,(3,40)}(0,1,2)H_{\underline{a}\,(3,40)}(0,1,2)$	28.5	9.2	1					
$H^T_{\underline{a}\,(4,40)}(0,\cdots,3)H_{\underline{a}\,(4,40)}(0,\cdots,3)$	34.5	9.6	1	0				
$H^T_{\underline{a}\,(5,40)}(0,\cdots,4)H_{\underline{a}\,(5,40)}(0,\cdots,4)$	34.3	11	2.3	0	0			
$H^T_{\underline{a}\,(6,40)}(0,\cdots,5)H_{\underline{a}\,(6,40)}(0,\cdots,5)$	35	11	2.4	0	0	0		
$H^T_{\underline{a}\,(13,40)}(0,\cdots,12)H_{\underline{a}\,(13,40)}(0,\cdots,12)$	46	15.5	3.6	0	0	0	$\cdots$	0

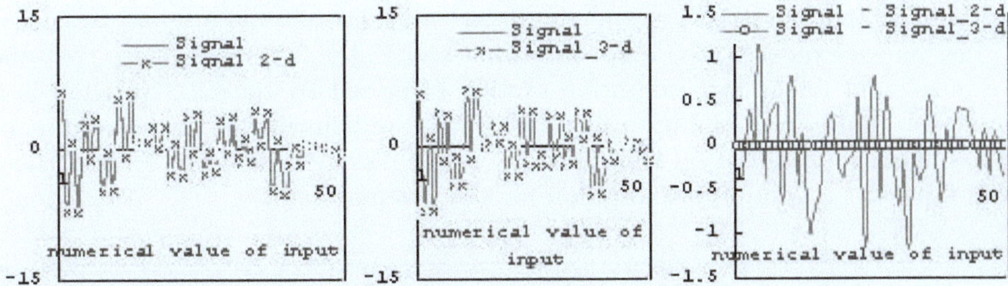

Fig. 8.2. The left is the original input response map and the behavior of a 2-dimensional linear representation system obtained by the CLS method. The middle is the original input response map and the behavior of a 3-dimensional linear representation system obtained by the CLS method. The right is the difference between the original one and the behavior of the 2-dimensional linear representation system obtained by the CLS method or the 3-dimensional linear representation system obtained by the CLS method in Example (8.26).

1) Since the ratio $\frac{1}{28.5} = 0.04$ obtained by the square root of $H^T_{\underline{a}\,(3,40)}(0,1.2)H_{\underline{a}\,(3,40)}(0,1,2)$ is a little small, an approximate 2-dimensional linear representation system obtained by the CLS method may be somewhat good.
2) After determining the independent vectors $\underline{a}$ and $\underline{S}_l(u_1)\underline{a}$ whose numerical value of input are 0 and 1, we will continue the approximate realization algorithm by the CLS method.

Therefore, an approximate 2-dimensional linear representation system $\sigma_1 = ((\boldsymbol{R}^2,\ F_1),\ x^0_1,\ h_1)$ obtained by the CLS method is constructed as follows:

$$F_1(u_1) = \begin{bmatrix} 0 & -0.4 \\ 1 & -0.8 \end{bmatrix}, \ F_1(u_2) = \begin{bmatrix} 1 & -0.4 \\ 0.8 & -0.2 \end{bmatrix}, \ F_1(u_3) = \begin{bmatrix} -0.6 & 0.4 \\ 0.5 & -0.1 \end{bmatrix}, \ x_1^0 = \mathbf{e}_1,$$

$h_1 = [6.4, \ -7.2]$.

For reference, a 3-dimensional linear representation system $\sigma_2 = ((\mathbf{R}^3, \ F_2), \ x_2^0, \ h_2)$ obtained by the CLS method can be expressed as follows:

$$F_2(u_1) = \begin{bmatrix} 0 & -1.4 & -1.3 \\ 1 & -1.5 & -0.3 \\ 0 & 1 & 1 \end{bmatrix}, \ F_2(u_2) = \begin{bmatrix} 0 & 0.2 & 0.1 \\ 0 & 0.3 & 0.3 \\ 1 & -0.6 & 0.6 \end{bmatrix},$$

$$F_2(u_3) = \begin{bmatrix} -0.6 & -0.1 & -0.7 \\ 0.5 & -0.5 & 0.1 \\ 0 & 0.5 & 0.5 \end{bmatrix}, \ x_2^0 = \mathbf{e}_1, \ h_2 = [6, \ -7.5, \ 1].$$

In this example, the original signals are considered as the behavior of a 3-dimensional linear representation system and the desirable input response map is obtained by the CLS method. The model obtained by the CLS method is a 2-dimensional linear representation system.

For reference, a 3-dimensional linear representation system is also obtained by the CLS method. The system completely reconstructs the original system.

Just as we expected, the following table and Fig. 8.2 truly indicate that the 2-dimensional linear representation system obtained by the CLS method is a somewhat good approximation. For reference, the behavior of the same dimensional linear representation system as the original system is shown. Hence, there does not exist a good approximation for the given system.

dimension	ratio of matrices	mean values of square root for sum of			cosine ① and ②	error ratio
		signal ①	signal by CLS ②	error ③	$\cos\theta$	③/①
$a_{0,1}$	0.04	0.455	0.442	0.07	0.988	0.15
$a_{0,1,5}$	0	0.455	0.455	0	1	0

Example 8.27. Let the signals be the behavior of the following 4-dimensional linear representation system: $\sigma = ((\mathbf{R}^4, F), x^0, h)$, where $F(u_1) =$

$$\begin{bmatrix} 0 & 1.2 & -0.1 & 0.7 \\ 1 & 0.4 & 0 & 0 \\ 0 & -2 & 1.2 & 0 \\ 0 & 0.2 & 0.3 & 0.8 \end{bmatrix}, \ F(u_2) = \begin{bmatrix} 0 & -1 & -0.1 & -0.8 \\ 0 & 0.5 & 0 & 0.3 \\ 1 & 0 & -0.5 & 0.4 \\ 0 & 0 & 0 & -0.6 \end{bmatrix}, \ F(u_3) = \begin{bmatrix} 0 & 0.4 & 0 & -0.2 \\ 0 & 0.5 & 0.1 & 1.3 \\ 0 & -0.9 & -0.3 & -0.4 \\ 1 & 0.3 & 0.2 & -0.5 \end{bmatrix},$$

$x^0 = \mathbf{e}_1, \ h = [12, \ 8, \ 1, \ 1]$.

Then the approximate realization problem is solved as follows:

covariance matrix	eigenvalues							
	1	2	3	4	5	6	$\cdots$	13
$H_{\underline{a}\,(3,40)}^{T}(0,1,2)\,H_{\underline{a}\,(3,40)}(0,1,2)$	4420	456	133					
$H_{\underline{a}\,(4,40)}^{T}(0,\cdots,3)\,H_{\underline{a}\,(4,40)}(0,\cdots,3)$	6619	982	250	35				
$H_{\underline{a}\,(5,40)}^{T}(0,\cdots,4)\,H_{\underline{a}\,(5,40)}(0,\cdots,4)$	9161	1854	284	79	0			
$H_{\underline{a}\,(6,40)}^{T}(0,\cdots,5)\,H_{\underline{a}\,(6,40)}(0,\cdots,5)$	9434	2179	491	78	0	0		
$H_{\underline{a}\,(13,40)}^{T}(0,\cdots,12)\,H_{\underline{a}\,(13,40)}(0,\cdots,12)$	18732	2826	1196	143	0	0	$\cdots$	0
covariance matrix	square root of eigenvalues							
$H_{\underline{a}\,(3,40)}^{T}(0,1,2)\,H_{\underline{a}\,(3,40)}(0,1,2)$	66.4	21.3	11.5					
$H_{\underline{a}\,(4,40)}^{T}(0,\cdots,3)\,H_{\underline{a}\,(4,40)}(0,\cdots,3)$	81.4	31.3	15.8	5.9				
$H_{\underline{a}\,(5,40)}^{T}(0,\cdots,4)\,H_{\underline{a}\,(5,40)}(0,\cdots,4)$	95.7	43	16.9	8.9	0			
$H_{\underline{a}\,(6,40)}^{T}(0,\cdots,5)\,H_{\underline{a}\,(6,40)}(0,\cdots,5)$	97	46.7	22.2	8.8	0	0		
$H_{\underline{a}\,(13,40)}^{T}(0,\cdots,12)\,H_{\underline{a}\,(13,40)}(0,\cdots,12)$	137	53	34.6	11.9	0	0	$\cdots$	0

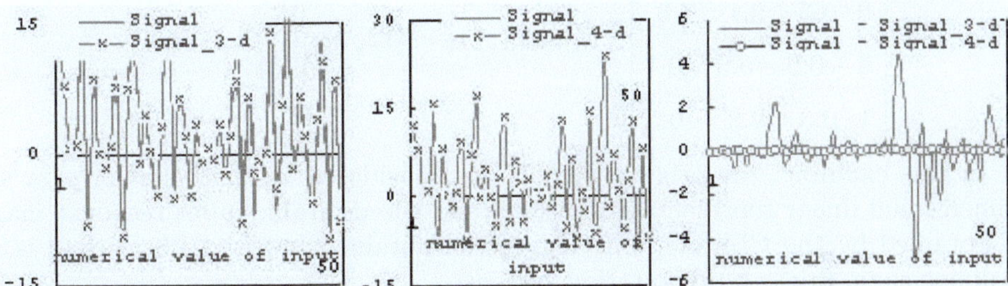

Fig. 8.3. The left is the original input response map and the behavior of a 3-dimensional linear representation system obtained by the CLS method. The middle is the original input response map and the behavior of a 4-dimensional linear representation system obtained by the CLS method. The right is the difference between the original one and the behavior of the 3-dimensional linear representation system obtained by the CLS method or the 4-dimensional linear representation system obtained by the CLS method in Example (8.27).

1) Since the ratio $\frac{5.9}{81.4} = 0.07$ obtained by the square root of $H_{\underline{a}\,(4,40)}^{T}(0,\cdots.3)\,H_{\underline{a}\,(4,40)}(0,\cdots,3)$ is somewhat small, an approximate 3-dimensional linear representation system obtained by the CLS method may be somewhat good.

2) After determining the independent vectors $\underline{a}$, $\underline{S}_l(u_1)\underline{a}$ and $\underline{S}_l(u_2)\underline{a}$ whose numerical value of input are 0, 1 and 2, we will continue the approximate realization algorithm by the CLS method.

Therefore, an approximate 3-dimensional linear representation system $\sigma_1 = ((\boldsymbol{R}^3, F_1), x_1^0, h_1)$ obtained by the CLS method is constructed as follows:

$$F_1(u_1) = \begin{bmatrix} 0 & 1.1 & -0.2 \\ 1 & 0.6 & 0.2 \\ 0 & -1.6 & 1.8 \end{bmatrix}, \quad F_1(u_2) = \begin{bmatrix} 0 & -1 & -0.1 \\ 0 & 0.5 & 0 \\ 1 & 0 & -0.5 \end{bmatrix},$$

$$F_1(u_3) = \begin{bmatrix} -0.5 & 0.3 & -0.09 \\ 0.7 & 0.7 & 0.3 \\ 2.4 & -0.4 & 0.07 \end{bmatrix}, \quad x_1^0 = \boldsymbol{e}_1, \quad h_1 = [12, 7.9, 0.8].$$

For reference, a 4-dimensional linear representation system $\sigma_2 = ((\boldsymbol{R}^4, F_2), x_2^0, h_2)$ obtained by the CLS method can be expressed as follows:

$$F_2(u_1) = \begin{bmatrix} 0 & 1.2 & -0.1 & 0.7 \\ 1 & 0.4 & 0 & 0 \\ 0 & -2 & 1.2 & 0 \\ 0 & 0.2 & 0.3 & 0.8 \end{bmatrix}, \quad F_2(u_2) = \begin{bmatrix} 0 & -1 & -0.1 & -0.8 \\ 0 & 0.5 & 0 & 0.3 \\ 1 & 0 & -0.5 & 0.4 \\ 0 & 0 & 0 & -0.6 \end{bmatrix},$$

$$F_2(u_3) = \begin{bmatrix} 0 & 0.4 & 0 & -0.2 \\ 0 & 0.5 & 0.1 & 1.3 \\ 0 & -0.9 & -0.3 & -0.4 \\ 1 & 0.3 & 0.2 & -0.5 \end{bmatrix}, \quad x_2^0 = \boldsymbol{e}_1, \quad h_2 = [12, 8, 1, 1].$$

In this example, the original signals are considered as the behavior of a 4-dimensional linear representation system and the desirable input response map is obtained by the CLS method. The model obtained by the CLS method is a 3-dimensional linear representation system.

For reference, a 4-dimensional linear representation system is also obtained by the CLS method. The system completely reconstructs the original system.

Just as we somewhat expected, the following table and Fig. 8.3 truly indicate that the 3-dimensional linear representation system obtained by the CLS method is not such a good approximation. For reference, the behavior of the same dimensional linear representation system as the original system is shown. Hence, there does not exist a good approximation for the given system.

dimen-ion	ratio of matrices	mean values of square root for sum of			cosine ① and ②	error ratio
		signal ①	signal by CLS ②	error ③	$\cos\theta$	③/①
$a_{0,1,2}$	0.07	1.1	1.1	0.19	0.985	0.17
$a_{0,1,2,3}$	0	1.1	1.1	0	1	0

Example 8.28. Let the signals be the behavior of the following 5-dimensional linear representation system: $\sigma = ((\boldsymbol{R}^5, F), x^0, h)$,

where

$$F(u_1) = \begin{bmatrix} 0 & 0 & 0.1 & 0.1 & -0.1 \\ 1 & 0 & 0.7 & 0.1 & 0.8 \\ 0 & 0 & -0.1 & 0.5 & 0.2 \\ 0 & 0 & -0.5 & -1.5 & 0.5 \\ 0 & 1 & -0.5 & -1.5 & 0.8 \end{bmatrix}, \quad F(u_2) = \begin{bmatrix} 0 & 0.2 & -1.6 & 0.8 & -0.3 \\ 0 & 0.5 & 0 & -0.2 & 0.2 \\ 1 & -0.5 & 0.2 & 0.5 & 0.2 \\ 0 & -0.5 & 0.2 & 0.7 & 0.3 \\ 0 & 0 & 0.1 & 1.1 & 0.7 \end{bmatrix},$$

$$F(u_3) = \begin{bmatrix} 0 & 0.3 & 0 & -0.5 & 0.3 \\ 0 & 0.8 & 0.5 & -0.2 & 0.8 \\ 0 & 0.1 & 2 & -1 & -0.2 \\ 1 & -1.2 & 0 & -0.8 & 1 \\ 0 & 0 & 0 & 0 & 0.8 \end{bmatrix}, \quad x^0 = \mathbf{e}_1, \quad h = [12, -7, -3, 4, 5].$$

Then the approximate realization problem is solved as follows:

covariance matrix	eigenvalues								
	1	2	3	4	5	6	7	$\cdots$	13
$H^T_{\underline{a}\,(5,40)}(0,\cdots,4)H_{\underline{a}\,(5,40)}(0,\cdots,4)$	23617	18578	14120	1606	437				
$H^T_{\underline{a}\,(6,40)}(0,\cdots,5)H_{\underline{a}\,(6,40)}(0,\cdots,5)$	23918	22217	17175	1896	449	0			
$H^T_{\underline{a}\,(7,40)}(0,\cdots,6)H_{\underline{a}\,(7,40)}(0,\cdots,6)$	56796	22260	17240	2038	457	0	0		
$H^T_{\underline{a}\,(13,40)}(0,\cdots,12)H_{\underline{a}\,(13,40)}(0,\cdots,12)$	168524	104576	47931	4801	1265	0	0	$\cdots$	0
covariance matrix	square root of eigenvalues								
$H^T_{\underline{a}\,(5,40)}(0,\cdots,4)H_{\underline{a}\,(5,40)}(0,\cdots,4)$	154	136	119	40	21				
$H^T_{\underline{a}\,(6,40)}(0,\cdots,5)H_{\underline{a}\,(6,40)}(0,\cdots,5)$	155	149	131	44	21	0			
$H^T_{\underline{a}\,(7,40)}(0,\cdots,6)H_{\underline{a}\,(7,40)}(0,\cdots,6)$	238	149	131	45	21	0	0		
$H^T_{\underline{a}\,(13,40)}(0,\cdots,12)H_{\underline{a}\,(13,40)}(0,\cdots,12)$	410	323	219	22	36	0	0	$\cdots$	0

1) Since the ratio $\frac{21}{154} = 0.14$ obtained by the square root of $H^T_{\underline{a}\,(5,40)}(0,\cdots.4) \times H_{\underline{a}\,(5,40)}(0,\cdots,4)$ is somewhat large, the approximate 4-dimensional linear representation system obtained by the CLS method may not be good.

2) After determining the independent vectors $\underline{a}$, $\underline{S}_l(u_1)\underline{a}$, $\underline{S}_l(u_2)\underline{a}$ and $\underline{S}_l(u_3)\underline{a}$ whose numerical value of input are 0, 1, 2 and 3, we will continue the approximate realization algorithm by the CLS method.

Therefore, an approximate 4-dimensional linear representation system $\sigma_1 = ((\boldsymbol{R}^4, F_1), x_1^0, h_1)$ obtained by the CLS method is constructed as follows:

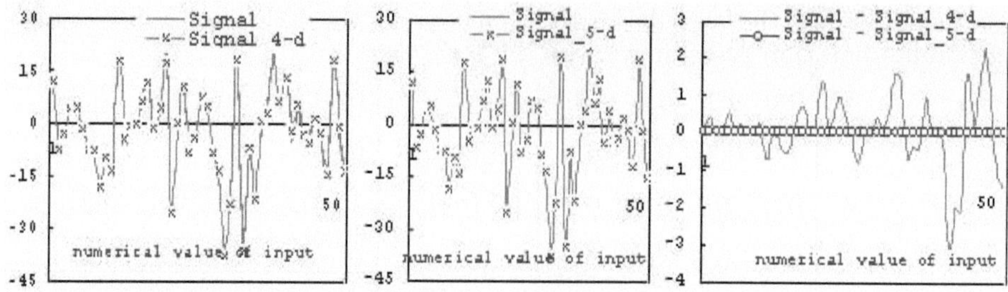

Fig. 8.4. The left is the original input response map and the behavior of a 4-dimensional linear representation system obtained by the CLS method. The middle is the original input response map and the behavior of a 5-dimensional linear representation system obtained by the CLS method. The right is the difference between the original one and the behavior of the 4-dimensional linear representation system obtained by the CLS method or the 5-dimensional linear representation system obtained by the CLS method in Example (8.28).

$$
F_1(u_1) = \begin{bmatrix} 0 & -0.04 & 0.1 & 0.1 \\ 1 & -0.7 & 1 & 1.1 \\ 0 & -0.02 & -0.08 & 0.6 \\ 0 & -0.1 & -0.4 & -1.3 \end{bmatrix}, \quad
F_1(u_2) = \begin{bmatrix} 0 & 0.2 & -1.6 & 0.9 \\ 0 & 0.5 & -0.06 & -0.9 \\ 1 & -0.5 & 0.2 & 0.5 \\ 0 & -0.5 & 0.2 & 0.6 \end{bmatrix},
$$

$$
F_1(u_3) = \begin{bmatrix} 0 & 0.3 & 0 & -0.5 \\ 0 & 0.8 & 0.5 & -0.2 \\ 0 & 0.1 & 2 & -1 \\ 1 & -1.2 & 0 & -0.8 \end{bmatrix}, \quad x_1^0 = \mathbf{e}_1, \quad h_1 = [12, -7.4, -3, 3.9].
$$

For reference, a 5-dimensional linear representation system
$\sigma_2 = ((\mathbf{R}^5, F_2), x_2^0, h_2)$ obtained by the CLS method can be expressed as
follows:

$$
F_2(u_1) = \begin{bmatrix} 0 & 0 & 0.1 & 0.1 & -0.1 \\ 1 & 0 & 0.7 & 0.1 & 0.8 \\ 0 & 0 & -0.1 & 0.5 & 0.2 \\ 0 & 0 & -0.5 & -1.5 & 0.5 \\ 0 & 1 & -0.5 & -1.5 & 0.8 \end{bmatrix}, \quad
F_2(u_2) = \begin{bmatrix} 0 & 0.2 & -1.6 & 0.8 & -0.3 \\ 0 & 0.5 & 0 & -0.2 & 0.2 \\ 1 & -0.5 & 0.2 & 0.5 & 0.2 \\ 0 & -0.5 & 0.2 & 0.7 & 0.3 \\ 0 & 0 & 0.1 & 1.1 & 0.7 \end{bmatrix},
$$

$$
F_2(u_3) = \begin{bmatrix} 0 & 0.3 & 0 & -0.5 & 0.3 \\ 0 & 0.8 & 0.5 & -0.2 & 0.8 \\ 0 & 0.1 & 2 & -1 & -0.2 \\ 1 & -1.2 & 0 & -0.8 & 1 \\ 0 & 0 & 0 & 0 & 0.8 \end{bmatrix}, \quad x_2^0 = \mathbf{e}_1, \quad h_2 = [12, -7, -3, 4, 5].
$$

In this example, the original signals are considered as the behavior of a 5-dimensional linear representation system and the desirable input response map is obtained by the CLS method. The model obtained by the CLS method is a 4-dimensional linear representation system.

For reference, a 5-dimensional linear representation system is also obtained by the CLS method. The system completely reconstructs the original system.

Just as we thought, the following table and Fig. 8.4 truly indicate that the 4-dimensional linear representation system obtained by the CLS method is not a good approximation. For reference, the behavior of the same dimensional linear representation system as the original system is shown. Hence, there does not exist a good approximation for the given system.

dimen-ion	ratio of matrices	mean values of square root for sum of			cosine ① and ②	error ratio
		signal ①	signal by CLS ②	error ③	$\cos\theta$	③/①
$a_{0,1,2,3}$	0.14	1.87	1.86	0.13	0.997	0.07
$a_{0,1,2,3,4}$	0	1.87	1.87	0	1	0

Example 8.29. Let the signals be the behavior of the following 6-dimensional linear representation system: $\sigma = ((\boldsymbol{R}^6, F), x^0, h)$,

$$
\text{where } F(u_1) = \begin{bmatrix} 0 & 0 & 0.4 & 0 & -0.3 & -0.3 \\ 1 & 0 & -0.1 & 0 & -0.1 & 0.3 \\ 0 & 0 & -0.2 & 0 & 0.4 & 0.1 \\ 0 & 0 & 0 & 0 & 0 & 0 \\ 0 & 1 & -0.5 & 0 & -0.8 & 0.1 \\ 0 & 0 & 0.2 & 0 & -0.3 & -0.1 \end{bmatrix}, \quad F(u_2) = \begin{bmatrix} 0 & 0 & 0.9 & 0 & -1.6 & -0.6 \\ 0 & 0 & -0.9 & 0 & 1.9 & 0.5 \\ 1 & 0 & -0.9 & 0 & 1.5 & 0.7 \\ 0 & 0 & 0 & 0 & 0 & 0 \\ 0 & 0 & -1 & 0 & 2.1 & 0.6 \\ 0 & 1 & 0.7 & 0 & -1.2 & -0.5 \end{bmatrix},
$$

$$
F(u_3) = \begin{bmatrix} 0 & 0 & 0.1 & 0 & 0.3 & -0.1 \\ 0 & 0 & 0 & 0 & 0.1 & 0 \\ 0 & 0 & 0 & 0 & 0.1 & -0.01 \\ 1 & 0 & 0 & 0 & 0 & 0 \\ 0 & 0 & 0 & 0 & 0.1 & 0 \\ 0 & 0 & 0.01 & 0 & 0.03 & 1.2 \end{bmatrix}, \quad x^0 = \mathbf{e}_1, \quad h = [12,\ -7,\ -3,\ 4,\ 5,\ 1].
$$

Then the approximate realization problem is solved as follows:

covariance matrix	eigenvalues									
	1	2	3	4	5	6	7	8	$\cdots$	13
$H^T_{\underline{a}\,(6,40)}(0, \cdots, 5) H_{\underline{a}\,(6,40)}(0, \cdots, 5)$	12736	4062	1375	426	194	14				
$H^T_{\underline{a}\,(7,40)}(0, \cdots, 6) H_{\underline{a}\,(7,40)}(0, \cdots, 6)$	12736	4062	1375	426	194	14	0			
$H^T_{\underline{a}\,(8,40)}(0, \cdots, 7) H_{\underline{a}\,(8,40)}(0, \cdots, 7)$	13667	4123	1832	426	200	14	0	0		
$H^T_{\underline{a}\,(13,40)}(0, \cdots, 12) H_{\underline{a}\,(13,40)}(0, \cdots, 12)$	14175	6812	3164	428	262	14	0	0	$\cdots$	0
covariance matrix	square root of eigenvalues									
$H^T_{\underline{a}\,(6,40)}(0, \cdots, 5) H_{\underline{a}\,(6,40)}(0, \cdots, 5)$	113	64	37	21	14	3.7				
$H^T_{\underline{a}\,(7,40)}(0, \cdots, 6) H_{\underline{a}\,(7,40)}(0, \cdots, 6)$	113	64	37	21	14	3.7	0			
$H^T_{\underline{a}\,(8,40)}(0, \cdots, 7) H_{\underline{a}\,(8,40)}(0, \cdots, 7)$	117	64	43	21	14	3.7	0	0		
$H^T_{\underline{a}\,(13,40)}(0, \cdots, 12) H_{\underline{a}\,(13,40)}(0, \cdots, 12)$	119	83	56	21	16	3.7	0	0	$\cdots$	0

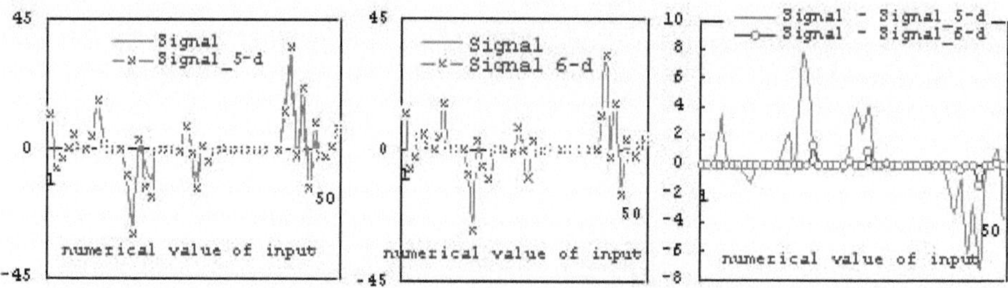

Fig. 8.5. The left is the original input response map and the behavior of a 5-dimensional linear representation system obtained by the CLS method. The middle is the original input response map and the behavior of a 6-dimensional linear representation system obtained by the CLS method. The right is the difference between the original one and the behavior of the 5-dimensional linear representation system obtained by the CLS method or the 6-dimensional linear representation system obtained by the CLS method in Example (8.29).

1) Since the ratio $\frac{3.7}{113} = 0.03$ obtained by the square root of $H^T_{\underline{a}\ (6,40)}(0, \cdots .5)$ $\times H_{\underline{a}\ (6,40)}(0, \cdots , 5)$ is somewhat small, the approximate 5-dimensional linear representation system obtained by the CLS method may be somewhat good.

2) After determining the independent vectors $\underline{a}$, $\underline{S}_l(u_1)\underline{a}$, $\underline{S}_l(u_2)\underline{a}$, $\underline{S}_l(u_3)\underline{a}$ and $\underline{S}_l(u_1|u_1)\underline{a}$ whose numerical value of input are 0, 1, 2, 3 and 4, we will continue the approximate realization algorithm by the CLS method.

Therefore, an approximate 5-dimensional linear representation system $\sigma_1 = ((\boldsymbol{R}^5,\ F_1),\ x_1^0,\ h_1)$ obtained by the CLS method is constructed as follows:

$$F_1(u_1) = \begin{bmatrix} 0 & 0 & 0.5 & 0 & -0.5 \\ 1 & 0 & -0.1 & 0 & -0.09 \\ 0 & 0 & -0.3 & 0 & 0.6 \\ 0 & 0 & -1.3 & 0 & 4.5 \\ 0 & 1 & -0.5 & 0 & -0.8 \end{bmatrix}, F_1(u_2) = \begin{bmatrix} 0 & 1.2 & 1.6 & 0 & -2.7 \\ 0 & -0.3 & -1 & 0 & 2.2 \\ 1 & -1 & -1.5 & 0 & 2.6 \\ 0 & -34.4 & -17.2 & 0 & 28.8 \\ 0 & -0.2 & -1 & 0 & 2.2 \end{bmatrix},$$

$$F_1(u_3) = \begin{bmatrix} 0 & 0 & 0.1 & 0 & 0.3 \\ 0 & 0 & 0 & 0 & 0.1 \\ 0 & 0 & 0 & 0 & 0.1 \\ 1 & 0 & -0.01 & 0 & -0.03 \\ 0 & 0 & 0 & 0 & 0.1 \end{bmatrix}, x_1^0 = \mathbf{e}_1,\ h_1 = [12.1,\ -7,\ -3.1,\ 0.5,\ 5].$$

For reference, a 6-dimensional linear representation system $\sigma_2 = ((\boldsymbol{R}^6,\ F_2),\ x_2^0,\ h_2)$ obtained by the CLS method can be expressed as follows:

$$
F_2(u_1) = \begin{bmatrix} 0 & 0 & 0.4 & 0 & -0.3 & -0.3 \\ 1 & 0 & -0.1 & 0 & -0.1 & 0.3 \\ 0 & 0 & -0.2 & 0 & 0.4 & 0.1 \\ 0 & 0 & 0 & 0 & 0 & 0 \\ 0 & 1 & -0.5 & 0 & -0.8 & 0.1 \\ 0 & 0 & 0.2 & 0 & -0.3 & -0.1 \end{bmatrix}, \ F_2(u_2) = \begin{bmatrix} 0 & 0 & 0.9 & 0 & -1.6 & -0.6 \\ 0 & 0 & -0.9 & 0 & 1.9 & 0.5 \\ 1 & 0 & -0.9 & 0 & 1.5 & 0.7 \\ 0 & 0 & 0 & 0 & 0 & 0 \\ 0 & 0 & -1 & 0 & 2.1 & 0.6 \\ 0 & 1 & 0.7 & 0 & -1.2 & -0.5 \end{bmatrix},
$$

$$
F_2(u_3) = \begin{bmatrix} 0 & 0 & 0.1 & 0 & 0.3 & -0.1 \\ 0 & 0 & 0 & 0 & 0.1 & 0 \\ 0 & 0 & 0 & 0 & 0.1 & -0.01 \\ 1 & 0 & 0 & 0 & 0 & 0 \\ 0 & 0 & 0 & 0 & 0.1 & 0 \\ 0 & 0 & 0.01 & 0 & 0.03 & 1.2 \end{bmatrix}, \ x_2^0 = \mathbf{e}_1, \ h_2 = [12, \ -7, \ -3, \ 4, \ 5, \ 1].
$$

In this example, the original signals are considered as the behavior of a 6-dimensional linear representation system and the desirable input response map is obtained by the CLS method. The model obtained by the CLS method is a 5-dimensional linear representation system.

For reference, a 6-dimensional linear representation system is also obtained by the CLS method. The system nearly reconstructs the original system.

Even though we had a somewhat good expectation, the following table and Fig. 8.5 indicate that the 5-dimensional linear representation system obtained by the CLS method is a bad approximation. The unwanted result is caused by many zero values of the given input response map. For reference, the behavior of the same dimensional linear representation system as the original system is shown. Finally, there does not exist a good approximation for the given system.

dimen-ion	ratio of matrices	mean values of square root for sum of			cosine ① and ②	error ratio
		signal ①	signal by CLS ②	error ③	$\cos\theta$	③/①
$a_{0,1,2,3,4}$	0.03	1.16	1.34	0.34	0.97	0.29
$a_{0,1,2,3,4,5}$	0	1.16	1.16	0	1	0

8.5 Noisy Realization of Linear Representation Systems

In this section, we discuss a noisy realization problem of linear representation systems.

For noise $\{\bar{\gamma}(t) : t \in N\}$ added to an unknown linear representation system a, we will obtain observed data $\{\hat{\gamma}(|\omega|) + \bar{\gamma}(|\omega|) : \omega \in U^*\}$.

For the given data $\{\hat{\gamma}(|\omega|) + \bar{\gamma}(|\omega|) : \omega \in U^*\}$, σ which satisfies $a_\sigma(\omega) \approx \hat{\gamma}(|\omega|) : \omega \in U^*$ is called a noisy realization of a.

Roughly speaking, we can propose the following noisy realization problem: For the given data $\{\hat{\gamma}(|\omega|) + \bar{\gamma}(|\omega|) : \omega \in U^*\}$, find a linear representation system σ which satisfies $a_\sigma(\omega) \approx \hat{\gamma}(|\omega|)$ for any $\omega \in U^*$.

In order to make our discussion simple, we assume that the set Y of outout is the set R of real numbers, namely 1-output.

A situation for noisy realization problem 8.30.

Let the observed object be a linear representation system and added noise to the output. Then we will obtain the data $\{\gamma(t) = \hat{\gamma}(t) + \bar{\gamma}(t) : 0 \leq t \leq \underline{N}\}$ for some integer $\underline{N} \in N$, where $\hat{\gamma}(t)$ is the exact signal which comes from the observed linear representation system and $\bar{\gamma}(t)$ is the noise added at time of observation.

Problem 8.31. Problem statement of a noisy realization for linear representation systems

Let $H_{\underline{a}\ (p,\bar{p})}$ be the measured finite-sized Input/output matrix. Then find the cleaned-up Input/output matrix $\hat{H}_{\underline{a}\ (p,\bar{p})}$ such that $H_{\underline{a}\ (p,\bar{p})} = \hat{H}_{\underline{a}\ (p,\bar{p})} + \bar{H}_{\underline{a}\ (p,\bar{p})}$ holds.

Namely, find out a minimal dimensional linear representation system $\sigma = ((R^n, F_r), g_r, h_r, h^0))$ which realizes $\hat{H}_{\underline{a}\ (p,\bar{p})}$.

Theorem 8.32. *Algorithm of noisy realization for Linear Representation Systems*

Let an input response map $\underline{a}$ be a considered object which is a linear representation system. Then an approximate realization $\sigma = ((R^n, F_s), g_s^0, h_s)$ of $\underline{a}$ is given by the following algorithm:

1) Based on the square root of eigenvalues for a matrix
$H_{\underline{a}\ (n,\bar{p})}(|||\omega_1|||, \cdots, |||\omega_n|||) H_{\underline{a}\ (n,\bar{p})}(|||\omega_1|||, \cdots, |||\omega_n|||)^T$, *determine the value n of rank for the matrix, where $|||\omega_1|||$, $|||\omega_2|||$, $\cdots$ and $|||\omega_n|||$ are suitably selected in order of numerical value of input and*
$\{\underline{S_l}(\omega_i)\underline{a} - \underline{a}; 1 \leq i \leq n,\ \omega_i \in U^*\}$ *is a set of independent vectors.*
Namely, determine the value n of rank for the matrix $H_{\underline{a}\ (n,\bar{p})}(|||\omega_1|||, \cdots, |||\omega_n|||)$ such that a set of the square root of eigenvalues for the covariance matrix composed of relatively small and equally-sized numbers is excluded, where the signal part effected by the set may be the noisy part of the observed data.

2) In order to determine g_s, the CLS method is used as follows:
① In particular, set $g_s(\omega_i) := \mathbf{e}_i$ for $\omega_i \in U$. Namely, $g_s(\omega_1) := \mathbf{e}_1$, $g_s(\omega_2) := \mathbf{e}_2$, $\cdots$, $g_s(\omega_k) := \mathbf{e}_k$ for some $k \in N$.
For $u \in U$ such that $u \notin \{\omega_i; 1 \leq i \leq n\}$ and $\omega_r < u$,
$g_s(u) = \sum_{j=1}^{r} b_{u,j} \underline{S_l}(\omega_j)\underline{a}$ is obtained as follows:

Let a matrix $A_u \in \mathbf{R}^{1 \times (r+1)}$ be $A_u := [b_{u,1}, b_{u,2}, \cdots, b_{u,r}, -1]$.
Choose the coefficients $\{b_{u,j} : 1 \leq j \leq r\}$ such that
$\sum_{j=1}^{r}(\underline{S_l}(\omega_j)\underline{\bar{a}}) \cdot (\underline{S_l}(\omega_j)\underline{\bar{a}}) + (\underline{S_l}(u)\underline{\bar{a}}) \cdot (\underline{S_l}(u)\underline{\bar{a}})$ take a minimum value,
where $\{\underline{S_l}(\omega_j)\underline{\bar{a}} \in \mathbf{R}^{L \times 1} : 0 \leq j \leq r\}$ are given by
the equation $[\underline{S_l}(\omega_1)\underline{\bar{a}}, \underline{S_l}(\omega_2)\underline{\bar{a}}, \cdots, \underline{S_l}(\omega_r)\underline{\bar{a}}, \underline{S_l}(u)\underline{\bar{a}}]^T :=$
$A_u^T[A_u A_u^T]^{-1} A_u H_{\underline{a}\,(r+1,L)}^T (\|\|\omega_1\|\|, \|\|\omega_2\|\|, \cdots, \|\|\omega_r\|\|, \|\|u\|\|)$
and $H_{\underline{a}\,(\|\|u\|\|+1,L)}^T (\|\|\omega_1\|\|, \|\|\omega_2\|\|, \cdots, \|\|\omega_r\|\|, \|\|u\|\|) :=$
$[\underline{S_l}(\omega_1)\underline{a}, \cdots, \underline{S_l}(\omega_r)\underline{a}, \underline{S_l}(u)\underline{a}]$. And $\cdot$ denotes the inner product
of two vectors.

3) In order to obtain F_s, the CLS method is used as follows:
① Let $_i\mathbf{f}_j \in \mathbf{R}^n$ in $F_s(u_i) := [_i\mathbf{f}_1\ _i\mathbf{f}_2\ \cdots\ _i\mathbf{f}_n\ ]$ be $_i\mathbf{f}_j := [_if_{j,1}\ _{i}f_{j,2}\ \cdots\ _if_{j,n}]^T$
for $1 \leq i \leq m$, where $_if_{j,k}$ is given by the following:
$\underline{S_l}(u_i)\underline{S_l}(\omega_j)\underline{a} = \sum_{k=1}^{n}\ _if_{j,k}\underline{S_l}(\omega_k)\underline{a}$, $_if_{j,k} \in \mathbf{R}$ in the sense of
$F(U_{L-p}^*, Y)$.
② For i $(1 \leq i \leq m)$, j $(1 \leq j \leq n)$ and for the maximum number
r $(1 \leq r \leq n)$ such that $\omega_r, \omega_j \in \{\omega_j; 1 \leq j \leq n\}$ and $\|\|\omega_r\|\| < \|\|u_i|\omega_j\|\|$,
let a matrix $_iA_j \in \mathbf{R}^{1 \times (r+1)}$ be $_iA_j := [_if_{j,1}, _i f_{j,2}, \cdots, _i f_{j,r}, -1]$.
Choose the coefficients $_if_{j,k} = 0$ for k $(r+1 \leq k \leq n)$ and $\{_if_{j,k} : 1 \leq k \leq r\}$
such that $\sum_{i=1}^{r}(\underline{S_l}(\omega_i)\underline{\bar{a}}) \cdot (\underline{S_l}(\omega_i)\underline{\bar{a}}) + (\underline{S_l}(u_i|\omega_j)\underline{\bar{a}}) \cdot (\underline{S_l}(u_i|\omega_j)\underline{\bar{a}})$ takes a
minimum value, where $\{\underline{S_l}(\omega_i)\underline{\bar{a}} \in \mathbf{R}^{L \times 1} : 0 \leq i \leq r\}$ and $\underline{S_l}(u_i|\omega_j)\underline{\bar{a}}$ are
given by the following equation:
$[\underline{S_l}(\omega_1)\underline{\bar{a}}, \underline{S_l}(\omega_2)\underline{\bar{a}}, \cdots, \underline{S_l}(\omega_r)\underline{\bar{a}}, \underline{S_l}(u_i|\omega_j)\underline{\bar{a}}]^T :=$
$_iA_j^T[_iA_j\ _iA_j^T]^{-1}\ _iA_j\ H_{\underline{a}\,(n+1,L)}^T(\|\|\omega_1\|\|, \|\|\omega_2\|\|, \cdots, \|\|\omega_r\|\|, \|\|u_i|\omega_j\|\|)$ and
$H_{\underline{a}\,(r+1,L)}^T(\|\|\omega_1\|\|, \|\|\omega_2\|\|, \cdots, \|\|\omega_r\|\|, \|\|u_i|\omega_j\|\|) :=$
$[\underline{S_l}(\omega_1)\underline{a}, \underline{S_l}(\omega_2)\underline{a}, \cdots, \underline{S_l}(\omega_r)\underline{a}, \underline{S_l}(u_i|\omega_j)\underline{a}]$, and $\cdot$ denotes the inner product
of two vectors.

4) In order to determine h_s, the CLS method is used as follows:
① For the first ω_{r_1+1}, $\omega_{r_1} \in \{\omega_i; 1 \leq i \leq n\}$ such that $\omega_{r_1+1} > \omega_{r_1}$ and
$\|\|\omega_{r_1+1}\|\| - \|\|\omega_{r_1}\|\| > 1$ when starting out from ω_1,
set $\underline{S_l}(\lambda_1)\underline{a} := \sum_{i=1}^{r_1} b_{\lambda_1,i}\mathbf{e}_i$ for the obtained equation $\underline{S_l}(\lambda_1)\underline{a} =$
$\sum_{i=1}^{r_1} b_{\lambda_1,i}(\underline{S_l}(\omega_i)\underline{a})$ for λ_1 such that $\|\|\lambda_1\|\| = \|\|\omega_{r_1}\|\| + 1$.
Let a matrix $A_{\lambda_1} \in \mathbf{R}^{1 \times (r_1+1)}$ be $A_{\lambda_1} := [b_{\lambda_1,1}, b_{\lambda_1,2}, \cdots, b_{\lambda_1,r_1}, -1]$.
Choose the coefficients $\{b_{\lambda_1,i} : 1 \leq i \leq r_1\}$ such that
$\sum_{i=1}^{r_1}(\underline{S_l}(\omega_i)\underline{\bar{a}}) \cdot (\underline{S_l}(\omega_i)\underline{\bar{a}}) + (\underline{S_l}(\lambda_1)\underline{\bar{a}}) \cdot (\underline{S_l}(\lambda_1)\underline{\bar{a}})$ take a minimum value,
where $\{\underline{S_l}(\omega_i)\underline{\bar{a}} \in \mathbf{R}^{L \times 1} : 0 \leq i \leq r_1\}$ are given by
the equation $[\underline{S_l}(\omega_1)\underline{\bar{a}}, \underline{S_l}(\omega_2)\underline{\bar{a}}, \cdots, \underline{S_l}(\omega_{r_1})\underline{\bar{a}}, \underline{S_l}(\lambda_1)\underline{\bar{a}}]^T :=$
$A_{\lambda_1}^T[A_{\lambda_1} A_{\lambda_1}^T]^{-1} A_{\lambda_1} H_{\underline{a}\,(r_1+1,L)}^T(\|\|\omega_1\|\|, \|\|\omega_2\|\|, \cdots, \|\|\omega_{r_1}\|\|, \|\|\lambda_1\|\|)$
and $H_{\underline{a}\,(\|\|\lambda_1\|\|+1,L)}^T(\|\|\omega_1\|\|, \|\|\omega_2\|\|, \cdots, \|\|\omega_{r_1}\|\|, \|\|\lambda_1\|\|) :=$
$[\underline{S_l}(\omega_1)\underline{a}, \cdots, \underline{S_l}(\omega_{r_1})\underline{a}, \underline{S_l}(\lambda_1)\underline{a}]$. And $\cdot$ denotes the inner product of two
vectors.
Then let $h_{\lambda_1 s} \in \mathbf{R}^{1 \times r_1}$ be
$h_{\lambda_1 s} := [\underline{a}(\omega_1) - (\underline{\bar{a}}(\omega_1), \underline{a}(\omega_2) - (\underline{\bar{a}}(\omega_2), \cdots, \underline{a}(\omega_{r_1}) - (\underline{\bar{a}}(\omega_{r_1})]$.
② For the first ω_{r_2+1}, $\omega_{r_2} \in \{\omega_i; 1 \leq i \leq n\}$ such that $\omega_{r_2+1} > \omega_{r_2}$ and
$\|\|\omega_{r_2+1}\|\| - \|\|\omega_{r_2}\|\| > 1$ when starting out from ω_{r_1+1},
set $\underline{S_l}(\lambda_2)\underline{a} := \sum_{i=1}^{r_2} b_{\lambda_2,i}\mathbf{e}_i$ for the obtained equation $\underline{S_l}(\lambda_2)\underline{a} =$

$\sum_{i=1}^{r_2} b_{\lambda_2,i}(\underline{S_l}(\omega_i)\underline{a})$ *for* λ_2 *such that* $|||\lambda_2||| = |||\omega_{r_2}||| + 1$.

Let a matrix $A_{\lambda_2} \in \boldsymbol{R}^{1\times(r_2+1)}$ *be* $A_{\lambda_2} := [b_{\lambda_2,1}, b_{\lambda_2,2}, \cdots, b_{\lambda_2,r_2}, -1]$.

Choose the coefficients $\{b_{\lambda_2,i} : 1 \leq i \leq r_2\}$ *such that*

$\sum_{i=1}^{r_2}(\underline{S_l}(\omega_i)\underline{a}) \cdot (\underline{S_l}(\omega_i)\underline{a}) + (\underline{S_l}(\lambda_2)\underline{a}) \cdot (\underline{S_l}(\lambda_2)\underline{a})$ *take a minimum value,*

where $\{\underline{S_l}(\omega_i)\underline{a} \in \boldsymbol{R}^{L\times1} : 0 \leq i \leq r_2\}$ *are given by*

the equation $[\underline{S_l}(\omega_1)\underline{a}, \underline{S_l}(\omega_2)\underline{a}, \cdots, \underline{S_l}(\omega_{r_2})\underline{a}, \underline{S_l}(\lambda_2)\underline{a}]^T :=$

$A_{\lambda_2}^T [A_{\lambda_2} A_{\lambda_2}^T]^{-1} A_{\lambda_2} H_{\underline{a}\ (r_2+1,L)}^T(|||\omega_1|||, |||\omega_2|||, \cdots, |||\omega_{r_2}|||, |||\lambda_2|||)$

and $H_{\underline{a}\ (|||\lambda_2|||+1,L)}^T(|||\omega_1|||, |||\omega_2|||, \cdots, |||\omega_{r_2}|||, |||\lambda_2|||) :=$

$[\underline{S_l}(\omega_1)\underline{a}, \cdots, \underline{S_l}(\omega_{r_2})\underline{a}, \underline{S_l}(\lambda_2)\underline{a}]$. *And* $\cdot$ *denotes the inner product of two*

vectors.

Then let $h_{\lambda_2 s} \in \boldsymbol{R}^{1\times r_2}$ *be*

$h_{\lambda_2 s} := [\underline{a}(\omega_{r_1+1}) - (\bar{\underline{a}}(\omega_{r_1+1}), \underline{a}(\omega_{r_1+2}) - (\bar{\underline{a}}(\omega_{r_1+2}), \cdots, \underline{a}(\omega_{r_2}) - (\bar{\underline{a}}(\omega_{r_2})]$.

$\vdots$

$\textcircled{t}$ *For* $\omega \in U^*$ *such that* $|||\omega||| = |||\omega_n||| + 1$, *let a matrix* $A_\omega \in \boldsymbol{R}^{1\times(n+1)}$

be $A_\omega := [b_{\omega,1}, b_{\omega,2}, \cdots, b_{\omega,n}, -1]$.

Choose the coefficients $\{b_{\omega,i} : 1 \leq i \leq n\}$ *such that*

$\sum_{i=1}^{n}(\underline{S_l}(\omega_i)\underline{a}) \cdot (\underline{S_l}(\omega_i)\underline{a}) + (\underline{S_l}(\omega)\underline{a}) \cdot (\underline{S_l}(\omega)\underline{a})$ *take a minimum value,*

where $\{\underline{S_l}(\omega_i)\underline{a} : 1 \leq i \leq n\}$ *and* $\underline{S_l}(\omega)\underline{a}$ *are*

given by the following equation:

$[\underline{S_l}(\omega_1)\underline{a}, \underline{S_l}(\omega_2)\underline{a}, \cdots, \underline{S_l}(\omega_n)\underline{a}, \underline{S_l}(\omega)\underline{a}]^T :=$

$A_\omega^T [A_\omega A_\omega^T]^{-1} A_\omega H_{\underline{a}\ (n+1,L)}^T(|||\omega_1|||, |||\omega_2|||, \cdots, |||\omega_n|||, |||\omega|||)$

and $H_{\underline{a}\ (n+1,L)}^T(|||\omega_1|||, |||\omega_2|||, \cdots, |||\omega_n|||, |||\omega|||) :=$

$[\underline{S_l}(\omega_1)\underline{a}, \cdots, \underline{S_l}(\omega_n)\underline{a}, \underline{S_l}(\omega)\underline{a}]$, *and* $\cdot$ *denotes the inner product of two*

vectors.

Then let $h_{\omega s}$ *be*

$h_{\omega s} := [\underline{a}(\omega_{r_t+1}) - (\bar{\underline{a}}(\omega_{r_t+1}), \underline{a}(\omega_{r_t+2}) - \bar{\underline{a}}(\omega_{r_t+2}), \cdots, \underline{a}(\omega_n) - \bar{\underline{a}}(\omega_n)]$.

Finally, let $h_s \in \boldsymbol{R}^{1\times n}$ *be* $h_s := [h_{\lambda_1 s}, h_{\lambda_2 s}, \cdots, h_{\omega s}]$.

[proof]. In 1), the number of dimensions is determined by checking what part is the noisy part and by using the ratio of Hankel matrix norm, which imply the noise to signal ratio. According to Theorem (8.23), a linear representation system $\sigma = ((\boldsymbol{R}^n, F_s), g_s, h_s)$ is obtained as follows. In 2), g_s is obtained directly or by using the CLS method for A_u corresponding to the matrix A in Proposition (2.14). In 3), F_s is obtained by using the CLS method for $_iA_j$ corresponding to the matrix A in Proposition (2.14). In 4), h_s is obtained by using the CLS method for A_{λ_1}, A_{λ_2}, A_{λ_ω} corresponding to the matrix A in Proposition (2.14).

Remark : Let S and N be the norm of a signal and noise. Then the selected ratio of matrices in the algorithm may be considered as $\frac{N}{S+N}$.

In the figures of this chapter, we use a notation $Signal\_n\ d$ as an input response map obtained by a n-dimensional linear representation system.

In the examples of this chapter, a notation $H_{\underline{a}\ (r+1,40)}^T(0, \cdots, r)$ is used in place of $H_{\underline{a}\ (r+1,40)}^T(0, 1, 2, \cdots, r-1, r)$.

Example 8.33. Let the signals be the input response map of the following 3-dimensional linear representation system: $\sigma = ((\boldsymbol{R}^3, F), x^0, h)$, where $F(u_1) =$

$$\begin{bmatrix} 0 & 0 & 0 \\ 1 & 0 & -0.1 \\ 0 & 1 & 0 \end{bmatrix}, \quad F(u_2) = \begin{bmatrix} -1 & 0.5 & 0.1 \\ 0 & -1 & 0 \\ 0 & -0.1 & -0.1 \end{bmatrix}, \quad F(u_3) = \begin{bmatrix} 0 & 0.1 & 0 \\ 0 & 0.2 & 0 \\ 0 & -0.01 & 0 \end{bmatrix},$$

$x^0 = \mathbf{e}_1$, $h = [9.3, \ -8.6, \ 1.2]$.

Then the noisy realization problem is solved as follows:

covariance matrix	eigenvalues					
	1	2	3	4	5	6
$H^T_{\underline{a}\ (2,40)}(0,1)\,H_{\underline{a}\ (2,40)}(0,1)$	2029	498				
$H^T_{\underline{a}\ (3,40)}(0,1,2)\,H_{\underline{a}\ (3,40)}(0,1,2)$	3155	640	1.3			
$H^T_{\underline{a}\ (4,40)}(0,\cdots,3)\,H_{\underline{a}\ (4,40)}(0,\cdots,3)$	3155	640	1.6	1.1		
$H^T_{\underline{a}\ (5,40)}(0,\cdots,4)\,H_{\underline{a}\ (5,40)}(0,\cdots,4)$	3163	641	5.4	1.6	1.1	
$H^T_{\underline{a}\ (6,40)}(0,\cdots,5)\,H_{\underline{a}\ (6,40)}(0,\cdots,5)$	5350	807	5.4	1.7	1.5	1.0
covariance matrix	square root of eigenvalues					
$H^T_{\underline{a}\ (2,40)}(0,1)\,H_{\underline{a}\ (2,40)}(0,1)$	45	22				
$H^T_{\underline{a}\ (3,40)}(0,1,2)\,H_{\underline{a}\ (3,40)}(0,1,2)$	56	25	1.1			
$H^T_{\underline{a}\ (4,40)}(0,\cdots,3)\,H_{\underline{a}\ (4,40)}(0,\cdots,3)$	56	25	1.3	1		
$H^T_{\underline{a}\ (5,40)}(0,\cdots,4)\,H_{\underline{a}\ (5,40)}(0,\cdots,4)$	56	25	2.3	1.3	1	
$H^T_{\underline{a}\ (6,40)}(0,\cdots,5)\,H_{\underline{a}\ (6,40)}(0,\cdots,5)$	73	28	2.3	1.3	1.2	1

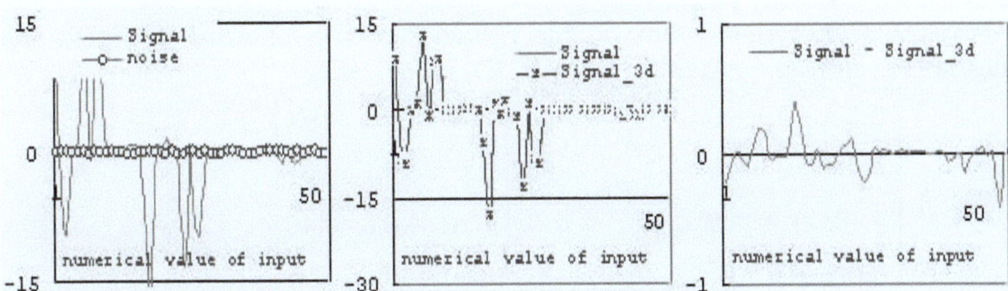

Fig. 8.6. The left is the original input response map and added noise to an original 3-dimensional linear representation system. The middle is the original input response map and the behavior of a 3-dimensional linear representation system obtained by the CLS method. The right is the difference between the original input response map and the behavior of the 3-dimensional linear representation system obtained by the CLS method in Example (8.33).

1) A set $\{1.3,\ 1.2,\ 1\}$ is composed of relatively small and equally-sized numbers in the square root of eigenvalues for $H_{\underline{a}\ (6,40)}^T (0, \cdots, 5) H_{\underline{a}\ (6,40)} (0, \cdots, 5)$.

2) After determining the independent vectors $\underline{a}$, $\underline{S}_l(u_1)\underline{a}$ and $\underline{S}_l(u_1|u_1)\underline{a}$ whose numerical value of input are 0, 1 and 4, we will continue the noisy realization algorithm by the CLS method.

Therefore, a noisy 3-dimensional linear representation system $\sigma_1 = ((\boldsymbol{R}^3,\ F_1),\ x_1^0,\ h_1)$ obtained by the CLS method is constructed as follows:

$$F_1(u_1) = \begin{bmatrix} 0 & 0 & -0.02 \\ 1 & 0 & -0.1 \\ 0 & 1 & -0.2 \end{bmatrix},\quad F_1(u_2) = \begin{bmatrix} -1 & 0.5 & 0.08 \\ 0 & -1 & 0 \\ 0 & 0.08 & -0.2 \end{bmatrix},\quad F_1(u_3) = \begin{bmatrix} 0 & 0.1 & 0 \\ 0 & 0.2 & 0 \\ 0 & -0.07 & 0 \end{bmatrix},$$

$x_1^0 = \mathbf{e}_1,\ h_1 = [9.2,\ -8.4,\ 1.3].$

In this example, the original signals are considered as the behavior of a 3-dimensional linear representation system and the desirable input response map is obtained by the CLS method. The model obtained by the CLS method is a 3-dimensional linear representation system.

Just as we expected, the following table and Fig. 8.6 truly indicate that the 3-dimensional linear representation system obtained by the CLS method is a good noisy realization.

dimen-ion	ratio of matrices	mean values of square root for sum of			cosine ① and ②	error ratio
		signal ①	signal by CLS ②	error ③	$\cos\theta$	③/①
$a_{0,1,4}$	0.02	0.714	0.71	0.02	0.999	0.02

Example 8.34. Let the signals be the behavior of the following 4-dimensional linear representation system: $\sigma = ((\boldsymbol{R}^4, F), x^0, h)$, where

$$F(u_1) = \begin{bmatrix} 0 & 1.2 & -0.1 & -0.7 \\ 1 & 0.4 & 0 & 0 \\ 0 & -2 & 1.2 & 0 \\ 0 & 0.2 & 0.3 & 0.8 \end{bmatrix},\quad F(u_2) = \begin{bmatrix} 0 & -1 & -0.1 & -0.8 \\ 0 & 0.8 & 0 & 0.3 \\ 1 & 0 & -0.5 & 0.4 \\ 0 & 0 & 0 & -0.6 \end{bmatrix},$$

$$F(u_3) = \begin{bmatrix} 0 & 1.4 & 0 & -1.2 \\ 0 & 0.5 & 0.1 & 1.3 \\ 0 & -0.9 & -0.3 & -0.4 \\ 1 & 0.3 & 0.2 & -0.5 \end{bmatrix},\quad x^0 = \mathbf{e}_1,\ h = [12,\ 8,\ 1,\ -2].$$

Then the noisy realization problem is solved as follows:

covariance matrix	eigenvalues						
	1	2	3	4	5	6	7
$H^T_{\underline{a}\ (3,40)}(0,1,2)H_{\underline{a}\ (3,40)}(0,1,2)$	6723	3150	443				
$H^T_{\underline{a}\ (4,40)}(0,\cdots,3)H_{\underline{a}\ (4,40)}(0,\cdots,3)$	7041	5355	3128	232			
$H^T_{\underline{a}\ (5,40)}(0,\cdots,4)H_{\underline{a}\ (5,40)}(0,\cdots,4)$	12745	5450	4112	600	2.9		
$H^T_{\underline{a}\ (6,40)}(0,\cdots,5)H_{\underline{a}\ (6,40)}(0,\cdots,5)$	13000	9761	5430	622	3.6	2.8	
$H^T_{\underline{a}\ (7,40)}(0,\cdots,6)H_{\underline{a}\ (7,40)}(0,\cdots,6)$	21140	10431	5795	636	3.6	3.4	2
covariance matrix	square root of eigenvalues						
$H^T_{\underline{a}\ (3,40)}(0,1,2)H_{\underline{a}\ (3,40)}(0,1,2)$	82	56	21				
$H^T_{\underline{a}\ (4,40)}(0,\cdots,3)H_{\underline{a}\ (4,40)}(0,\cdots,3)$	84	73	56	15			
$H^T_{\underline{a}\ (5,40)}(0,\cdots,4)H_{\underline{a}\ (5,40)}(0,\cdots,4)$	113	74	64	24	1.7		
$H^T_{\underline{a}\ (6,40)}(0,\cdots,5)H_{\underline{a}\ (6,40)}(0,\cdots,5)$	114	99	74	25	1.9	1.7	
$H^T_{\underline{a}\ (7,40)}(0,\cdots,6)H_{\underline{a}\ (7,40)}(0,\cdots,6)$	145	102	76	25	1.9	1.8	1.4

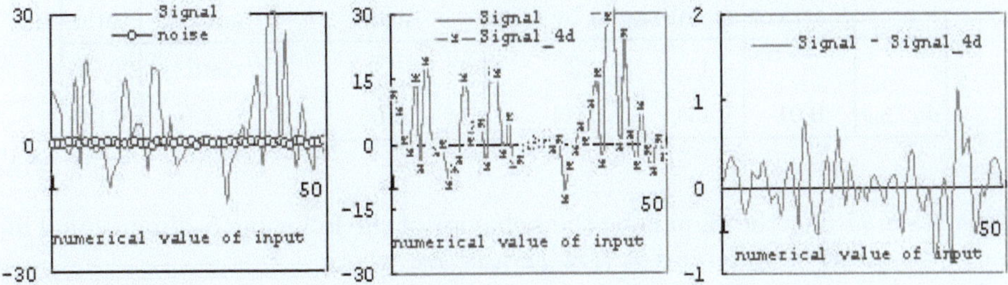

Fig. 8.7. The left is the original input response map and added noise to an original 4-dimensional linear representation system. The middle is the original input response map and the behavior of a 4-dimensional linear representation system obtained by the CLS method. The right is the difference between the original input response map and the behavior of the 4-dimensional linear representation system obtained by the CLS method in Example (8.34).

1) A set $\{1.9,\ 1.8,\ 1.4\}$ is composed of relatively small and equally-sized numbers in the square root of eigenvalues for $H^T_{\underline{a}\ (7,40)}(0,\cdots,6)H_{\underline{a}\ (7,40)}(0,\cdots,6)$.
2) After determining the independent vectors $\underline{a}$, $\underline{S}_l(u_1)\underline{a}$, $\underline{S}_l(u_2)\underline{a}$ and $\underline{S}_l(u_3)\underline{a}$ whose numerical value of input are 0, 1, 2 and 3, we will continue the noisy realization algorithm by the CLS method.

Therefore, a noisy 4-dimensional linear representation system
$\sigma_1 = ((\boldsymbol{R}^4,\ F_1),\ x_1^0,\ h_1)$ obtained by the CLS method is constructed as follows:

$$F_1(u_1) = \begin{bmatrix} 0 & 1.2 & -0.1 & -0.7 \\ 1 & 0.4 & 0 & 0 \\ 0 & -1.9 & 1.1 & 0 \\ 0 & 0.2 & 0.3 & 0.8 \end{bmatrix}, \quad F_1(u_2) = \begin{bmatrix} 0 & -1 & -0.1 & -0.8 \\ 0 & 0.8 & 0 & 0.3 \\ 1 & 0 & -0.5 & 0.4 \\ 0 & 0 & 0 & -0.6 \end{bmatrix},$$

$$F_1(u_3) = \begin{bmatrix} 0 & 1.4 & 0 & -1.2 \\ 0 & 0.5 & 0.1 & 1.3 \\ 0 & -0.9 & -0.3 & -0.4 \\ 1 & 0.3 & 0.2 & -0.5 \end{bmatrix}, \quad x_1^0 = \mathbf{e}_1, \ h_1 = [11.9, \ 7.7, \ 0.7, \ -1.8].$$

In this example, the original signals are considered as the behavior of a 4-dimensional linear representation system and the desirable input response map is obtained by the CLS method. The model obtained by the CLS method is a 4-dimensional linear representation system.

Just as we expected, the following table and Fig. 8.7 truly indicate that the 4-dimensional linear representation system obtained by the CLS method is a good noisy realization.

dimen- ion	ratio of matrices	mean values of square root for sum of			cosine ① and ②	error ratio
		signal	signal by CLS	error	$\cos\theta$	③/①
		①	②	③		
$a_{0,1,2,3}$	0.01	1.434	1.44	0.05	0.999	0.04

Example 8.35. Let the signals be the behavior of the following 5-dimensional linear representation system: $\sigma = ((\mathbf{R}^5, F), x^0, h)$, where

$$F(u_1) = \begin{bmatrix} 0 & 0 & 0.4 & 1.8 & -0.5 \\ 1 & 0 & 1.7 & 0.1 & 1.3 \\ 0 & 0 & -0.1 & 0.5 & 0.2 \\ 0 & 0 & -0.5 & -1.5 & 0.5 \\ 0 & 1 & -0.5 & -1.5 & 0.8 \end{bmatrix}, \quad F(u_2) = \begin{bmatrix} 0 & 0.2 & -1.6 & 1.2 & -0.6 \\ 0 & 0.5 & 0 & -0.2 & 0.2 \\ 1 & -0.5 & 0.2 & 0.5 & 0.7 \\ 0 & -0.5 & 0.2 & 0.7 & 0.3 \\ 0 & 0 & 0.1 & 1.1 & 0.7 \end{bmatrix},$$

$$F(u_3) = \begin{bmatrix} 0 & 1.3 & 0 & -0.5 & -0.1 \\ 0 & 1.8 & 0.5 & -0.3 & 0.8 \\ 0 & 0.1 & 2 & -1 & -0.2 \\ 1 & -1.2 & 0 & -0.8 & 1.5 \\ 0 & 0 & 0 & 0 & 0.8 \end{bmatrix}, \quad x^0 = \mathbf{e}_1, \ h = [12, \ -7, \ -3, \ 4, \ 11].$$

Then the noisy realization problem is solved as follows:

covariance matrix	eigenvalues							
	1	2	3	4	5	6	7	8
$H_{\underline{a}\,(5,40)}^{T}(0,\cdots,4)H_{\underline{a}\,(5,40)}(0,\cdots,4)$	47497	32094	24797	3911	1314			
$H_{\underline{a}\,(6,40)}^{T}(0,\cdots,5)H_{\underline{a}\,(6,40)}(0,\cdots,5)$	47559	38782	31972	4331	1369	4.4		
$H_{\underline{a}\,(7,40)}^{T}(0,\cdots,6)H_{\underline{a}\,(7,40)}(0,\cdots,6)$	125309	44582	33986	7000	1520	5.8	2.9	
$H_{\underline{a}\,(8,40)}^{T}(0,\cdots,7)H_{\underline{a}\,(8,40)}(0,\cdots,7)$	156297	70945	35763	7213	1520	6.1	3	2.9
covariance matrix	square root of eigenvalues							
$H_{\underline{a}\,(5,40)}^{T}(0,\cdots,4)H_{\underline{a}\,(5,40)}(0,\cdots,4)$	218	179	157	63	36			
$H_{\underline{a}\,(6,40)}^{T}(0,\cdots,5)H_{\underline{a}\,(6,40)}(0,\cdots,5)$	218	197	179	66	37	2.1		
$H_{\underline{a}\,(7,40)}^{T}(0,\cdots,6)H_{\underline{a}\,(7,40)}(0,\cdots,6)$	354	211	184	84	39	2.4	1.7	
$H_{\underline{a}\,(8,40)}^{T}(0,\cdots,7)H_{\underline{a}\,(8,40)}(0,\cdots,7)$	395	266	189	85	39	2.5	1.73	1.7

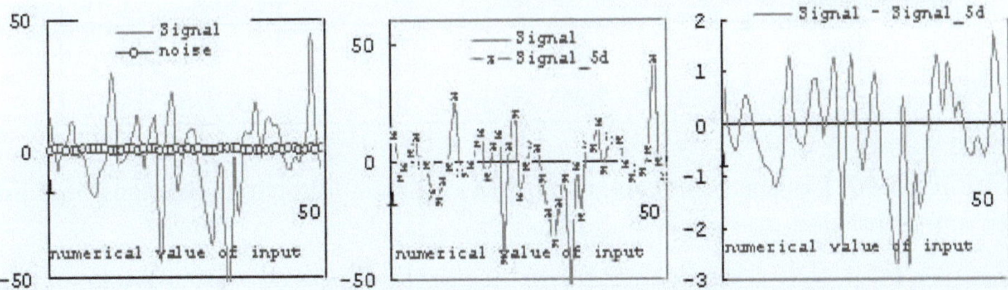

Fig. 8.8. The left is the original input response map and added noise to the original 5-dimensional linear representation system. The middle is the original input response map and the behavior of a 5-dimensional linear representation system obtained by the CLS method. The right is the difference between the original input response map and the behavior of the 5-dimensional linear representation system obtained by the CLS method in Example (8.35).

1) A set $\{2.5,\ 1.73,\ 1.7\}$ is composed of relatively small and equally-sized numbers in the square root of eigenvalues for $H_{\underline{a}\,(8,40)}^{T}(0,\cdots,7)H_{\underline{a}\,(8,40)}(0,\cdots,7)$.
2) After determining the independent vectors $\underline{a}$, $\underline{S}_l(u_1)\underline{a}$, $\underline{S}_l(u_2)\underline{a}$, $\underline{S}_l(u_3)\underline{a}$ and $\underline{S}_l(u_1|u_1)\underline{a}$ whose numerical value of input are 0, 1, 2, 3 and 4, we will continue the noisy realization algorithm by the CLS method.

Therefore, a noisy 5-dimensional linear representation system $\sigma_1 = ((\boldsymbol{R}^5,\ F_1),\ x_1^0,\ h_1)$ obtained by the CLS method is constructed as follows:

$$F_1(u_1) = \begin{bmatrix} 0 & 0 & 0.4 & 1.8 & -0.5 \\ 1 & 0 & 1.7 & 0.1 & 1.3 \\ 0 & 0 & -0.1 & 0.5 & 0.2 \\ 0 & 0 & -0.5 & -1.5 & 0.5 \\ 0 & 1 & -0.5 & -1.5 & 0.8 \end{bmatrix},\ F_1(u_2) = \begin{bmatrix} 0 & 0.2 & -1.6 & 1.2 & -0.6 \\ 0 & 0.5 & 0.01 & -0.2 & 0.2 \\ 1 & -0.5 & 0.2 & 0.5 & 0.7 \\ 0 & -0.5 & 0.2 & 0.7 & 0.3 \\ 0 & -0.02 & 0.11 & 1.1 & 0.7 \end{bmatrix},$$

$$F_1(u_3) = \begin{bmatrix} 0 & 1.33 & -0.01 & -0.5 & -0.1 \\ 0 & 1.8 & 0.5 & -0.2 & 0.8 \\ 0 & 0.1 & 2 & -1 & -0.2 \\ 1 & -1.2 & 0 & -0.8 & 1.5 \\ 0 & -0.02 & 0.01 & 0.02 & 0.8 \end{bmatrix}, \quad x_1^0 = e_1,$$

$h_1 = [11.3, \; -6.7, \; -2.5, \; 3.6, \; 10.5].$

In this example, the original signals are considered as the behavior of a 5-dimensional linear representation system and the desirable input response map is obtained by the CLS method. The model obtained by the CLS method is a 5-dimensional linear representation system.

Just as we expected, the following table and Fig. 8.8 truly indicate that the 5-dimensional linear representation system obtained by the CLS method is a somewhat good noisy realization.

dimen-ion	ratio of matrices	mean values of square root for sum of		error	cosine ① and ②	error ratio
		signal ①	signal by CLS ②	error ③	$\cos\theta$	③/①
$a_{0,1,2,3,4}$	0.01	2.383	2.255	0.14	0.999	0.06

Example 8.36. Let the signals be the behavior of the following 6-dimensional linear representation system: $\sigma = ((R^6, F), x^0, h)$, where

$$F(u_1) = \begin{bmatrix} 0 & 0 & 0.4 & 1.5 & -0.3 & -0.3 \\ 1 & 0 & -0.1 & 0 & -1.1 & 0.8 \\ 0 & 0 & -0.2 & 0 & 0.4 & 0.1 \\ 0 & 0 & 0 & 0 & 0 & 0 \\ 0 & 1 & -0.5 & 0 & -1 & 0.1 \\ 0 & 0 & 0.2 & 0 & -0.3 & -0.1 \end{bmatrix}, \quad F(u_2) = \begin{bmatrix} 0 & 0 & 1.5 & 0 & -1.6 & -0.6 \\ 0 & 0 & -0.9 & 0 & 1.9 & 0.5 \\ 1 & 0 & -0.9 & 0 & 1.5 & 0.7 \\ 0 & 0 & 0 & 0 & 0 & 0 \\ 0 & 0 & -1 & 0 & 2 & 0.6 \\ 0 & 1 & 0.7 & 0 & -1.2 & -0.5 \end{bmatrix},$$

$$F(u_3) = \begin{bmatrix} 0 & 0 & 0.1 & 0 & 0.3 & -0.1 \\ 0 & 0 & 0 & 0 & 0.1 & 0 \\ 0 & 0 & 0 & 0 & 0.1 & 0 \\ 1 & 0 & 0 & 0 & 0 & 0 \\ 0 & 0 & 0 & 0 & 0.1 & 0 \\ 0 & 0 & 0 & 0 & 0 & 0 \end{bmatrix}, \quad x^0 = e_1, \; h = [12, \; -2, \; -6, \; 2, \; 3, \; 1].$$

Then the noisy realization problem is solved as follows:

covariance matrix	eigenvalues								
	1	2	3	4	5	6	7	8	9
$H^T_{\underline{a}\ (6,40)}(0,\cdots,5)H_{\underline{a}\ (6,40)}(0,\cdots,5)$	17835	4575	2269	1495	268	90			
$H^T_{\underline{a}\ (7,40)}(0,\cdots,6)H_{\underline{a}\ (7,40)}(0,\cdots,6)$	17835	4576	2270	1496	249	90	5		
$H^T_{\underline{a}\ (8,40)}(0,\cdots,7)H_{\underline{a}\ (8,40)}(0,\cdots,7)$	18176	4833	2408	1883	275	91	8.7	3.9	
$H^T_{\underline{a}\ (9,40)}(0,\cdots,8)H_{\underline{a}\ (9,40)}(0,\cdots,8)$	22265	12197	3024	2161	278	94	8.7	5	3.6
covariance matrix	square root of eigenvalues								
$H^T_{\underline{a}\ (6,40)}(0,\cdots,5)H_{\underline{a}\ (6,40)}(0,\cdots,5)$	133.5	67.6	48	39	16	9.5			
$H^T_{\underline{a}\ (7,40)}(0,\cdots,6)H_{\underline{a}\ (7,40)}(0,\cdots,6)$	133.5	67.6	47.6	39	15.8	9.5	2.2		
$H^T_{\underline{a}\ (8,40)}(0,\cdots,7)H_{\underline{a}\ (8,40)}(0,\cdots,7)$	135	69.5	49	43	16.6	9.5	2.9	2	
$H^T_{\underline{a}\ (9,40)}(0,\cdots,8)H_{\underline{a}\ (9,40)}(0,\cdots,8)$	149	110	55	46.4	16.7	9.7	2.9	2.2	1.9

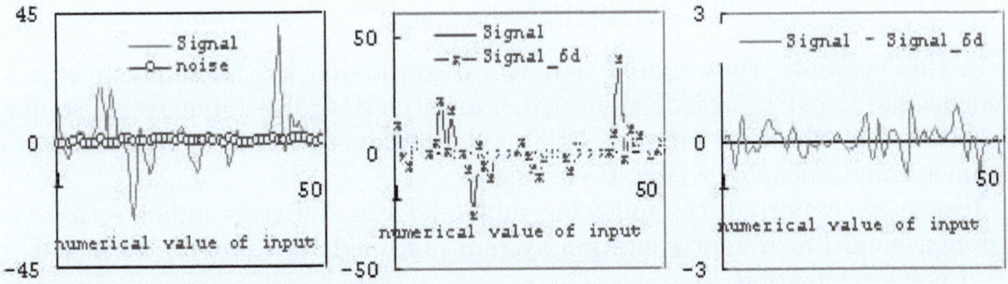

Fig. 8.9. The left is the original input response map and added noise to the original 6-dimensional linear representation system. The middle is the original input response map and the behavior of a 6-dimensional linear representation system obtained by the CLS method. The right is the difference between the original input response map and the behavior of the 6-dimensional linear representation system obtained by the CLS method in Example (8.36).

1) A set $\{2.9,\ 2.2,\ 1.9\}$ is composed of relatively small and equally-sized numbers in the square root of eigenvalues for $H^T_{\underline{a}\ (9,40)}(0,\cdots,8)H_{\underline{a}\ (9,40)}(0,\cdots,8)$.
2) After determining the independent vectors $\underline{a}$, $\underline{S}_l(u_1)\underline{a}$, $\underline{S}_l(u_2)\underline{a}$, $\underline{S}_l(u_3)\underline{a}$, $\underline{S}_l(u_1|u_1)\underline{a}$ and $\underline{S}_l(u_2|u_1)\underline{a}$ whose numerical value of input are 0, 1, 2, 3, 4 and 5, we will continue the noisy realization algorithm by the CLS method.

Therefore, a noisy 6-dimensional linear representation system $\sigma_1 = ((\boldsymbol{R}^6,\ F_1),\ x_1^0,\ h_1)$ obtained by the CLS method is constructed as follows:

$$
F_1(u_1)
$$

$$
= \begin{bmatrix}
0 & 0 & 0.4 & 1.5 & -0.3 & -0.3 \\
1 & 0 & -0.1 & -0.02 & -1.1 & 0.8 \\
0 & 0 & -0.2 & -0.02 & 0.4 & 0.1 \\
0 & 0 & 0.01 & 0.02 & 0.01 & 0.01 \\
0 & 1 & -0.5 & 0 & -1 & 0.1 \\
0 & 0 & 0.2 & -0.02 & -0.4 & -0.1
\end{bmatrix}, \; F_1(u_2) = \begin{bmatrix}
0 & 0 & 1.5 & 0.02 & -1.7 & -0.6 \\
0 & 0 & -0.8 & 0.01 & 1.9 & 0.5 \\
1 & 0 & -0.9 & 0.006 & 1.6 & 0.74 \\
0 & 0 & -0.04 & 0.01 & 0.03 & 0.03 \\
0 & 0 & -0.9 & 0 & 2 & 0.6 \\
0 & 1 & 0.68 & 0.01 & -1.08 & -0.38
\end{bmatrix},
$$

$$
F_1(u_3) = \begin{bmatrix}
0 & 0 & 0.1 & 0 & 0.3 & -0.1 \\
0 & 0.1 & -0.04 & 0.04 & 0.13 & 0 \\
0 & 0.02 & -0.04 & 0.03 & 0.12 & 0 \\
1 & 0 & 0.03 & -0.01 & 0 & -0.01 \\
0 & 0.01 & -0.02 & 0 & 0.12 & 0.01 \\
0 & 0.02 & 0.01 & 0.06 & 0 & 0.01
\end{bmatrix}, \; x_1^0 = \mathbf{e}_1,
$$

$$
h_1 = [11.9, \; -2.1, \; -5.3, \; 1.4, \; 3.3, \; 1.14].
$$

In this example, the original signals are considered as the behavior of a 6-dimensional linear representation system and the desirable input response map is obtained by the CLS method. The model obtained by the CLS method is a 6-dimensional linear representation system.

Just as we expected, the following table and Fig. 8.9 truly indicate that the 6-dimensional linear representation system obtained by the CLS method is a good noisy realization.

dimen-ion	ratio of matrices	mean values of square root for sum of			cosine	error ratio
		signal	signal by CLS	error	① and ②	
		①	②	③	$\cos\theta$	③/①
$a_{0,1,2,3,4,5}$	0.02	1.293	1.273	0.05	0.999	0.04

8.6   Historical Notes and Concluding Remarks

Approximate and noisy realization problems of linear representation systems were proposed from the notion of Hankel matrix norm and the CLS method. The matrix norm is used for determining the number of dimensions of state space and the CLS method is used for determining the parameters of linear representation systems, which are general non-linear systems. Note that there are homogeneous bilinear systems as a subclass of linear representation systems.

Our solution of approximate and noisy realization problems, like we discussed, are the same as with other linear and non-linear systems. In the past, a unified solution of non-linear systems could not be proposed.

In order to insist that our method for approximate and noisy realization is effective for our systems, we gave several examples. As a result of the examples, we have shown that the ratio of Hanke matrix norm implies the degree of approximation. For our noisy realization problem, we have shown that we can determine the number of dimensions of linear representation systems when a set of relatively small and equally-sized numbers of the square root of eigenvalues for a finite-sized Hankel matrix can be found.

However, for an approximate realization problem, we could not fully apply the approximate realization algorithm to the input response maps which have many values near zero, which can be seen in Examples 8.25 and 8.29.

As stated in the Historical notes and concluding remarks of the other chapters, our methods can be roughly summarized as follows:

Intuitively, our several examples for the approximate realization problem show that the smaller the ratio of matrices is, the smaller the error to signal ratio is.

The changing relations among the ratio of matrices and the error to signal ratio are proportial relations.

The several examples suggest that our two features can be expressed as follows:

(1) The ratio of matrices determines a degree of the crossed angle between directions of the approximate signal and the original signal.

(2) The CLS method determines the coefficients of linearly dependent vectors such that the error between the approximate signal and original signal has a minimum value in the sense of the square norm while conseving the crossed angle.

The ratio of 0.01 Hankel matrices ratio ranges from 0.01 to 0.02 for the error to signal ratio.

Intuitively, our several examples for the noisy realization problem show that the smaller the ratio of matrices is, the smaller the error to signal ratio is. The ratio within a 0.02 input/output matrix ratio implies an approximate error to signal ratio of within 6 %.

The several examples suggest that our two features can be expressed as follows:

(1) The ratio of matrices determines a degree of the crossed angle between directions of the obtained signal and the original signal.

(2) The CLS method determines the coefficients of linearly dependent vectors such that the error between the obtained signal and original signal has a minimum value in the sense of the square norm while conseving the crossed angle.

References

Akaike, H.: A New Look at Statistical Model Identification. IEEE Trans. on Automatic Control AC-19, 716–723 (1974)

Bourbaki, N.: Elements de Mathematique, Algebre, Herman, Paris (1958)

Bourbaki, N.: Elements of Mathematics, Theory of Sets (1968); English Translation, Hermann, Paris

Bourbaki, N.: Elements of Mathematics, Algebra Part 1 (1974); English Translation, Hermann, Paris

Chevalley, C.: Fundamental concept of algebra. Academic Press, London (1956)

Chua, L.O.: Introduction to nonlinear network theory. McGraw-Hill, New York (1969)

D'alessandro, D., Isidori, A., Ruberti, A.: Realization and structure theory of bilinear dynamical systems. SIAM J. Contr. 12, 517–534 (1974)

Davison, E.J.: A method for simplifying dynamic systems. IEEE Trans. on Automatic Control AC-11, 93–101 (1966)

Eckart, C., Young, G.: The approximation of one matrix by another of lower rank. Psychometrika 1, 211–218 (1936)

Gantmacher, F.R.: The theory of matrices, vol. 2. Chelsea, NewYork (1959)

Glover, K.: All optimal Hankel-norm approximation of linear multivariable systems and their L^∞ − error bounds. Int. J. Control 39(6), 1115–1193 (1984)

Haber, R., Keviczky, L.: Nonlinear system identification input-output modeling approach, vol. 1. Kluwer Academic Publishers, Dordrecht (1999)

Halmos, P.R.: Finite dimensional vector spaces, D. Van. Nos. Com. (1958)

Hasegawa, Y., Hamada, K., Matsuo, T.: A Realization Algorithm for Discrete-Time Linear Systems. In: Problems in Modern Applied Mathematics, pp. 192–197. World Scientifics, Singapore (2000)

Hasegawa, Y., Matsuo, T.: Realization Theory of discrete-time linear representation systems. SICE Transaction 15(3), 298–305 (1979a) (in Japanese)

Hasegawa, Y., Matsuo, T.: On the discrete time finite-dimensional linear representation systems. SICE Transaction 15(4), 443–450 (1979b) (in Japanese)

Hasegawa, Y., Matsuo, T.: Realization theory of discrete-time Pseudo-Linear Systems. SICE Transaction 28(2), 199–207 (1992) (in Japanese)

Hasegawa, Y., Matsuo, T.: Realization theory of Discrete-Time finite-dimensional Pseudo-Linear systems. SICE Transaction 29(9), 1071–1080 (1993) (in Japanese)

Hasegawa, Y., Matsuo, T.: Realization theory of discrete-time Almost-Linear Systems. SICE Transaction 30(2), 150–157 (1994a) (in Japanese)

Hasegawa, Y., Matsuo, T.: Realization theory of Discrete-Time finite-dimensional Almost-Linear systems. SICE Transaction 30(11), 1325–1333 (1994b) (in Japanese)

Hasegawa, Y., Matsuo, T.: Partial Realization & Real-Time, Partial Realization Theory of Discrete- Time Pseudo-Linear System. SICE Transaction 31(4), 471–480 (1995) (in Japanese)

Hasegawa, Y., Matsuo, T.: Modeling for Hysteresis Characteristic by Affine Dynamical Systems. In: International Conference on Power Electronics, Drivers and Energy Systems, New-Delhi, pp. 1006–1011 (1996a)

Hasegawa, Y., Matsuo, T.: Real-Time, Partial Realization Theory of Discrete-Time Non-Linear Systems. SICE Transaction 32(3), 345–354 (1996b) (in Japanese)

Hasegawa, Y., Matsuo, T.: A relation between Discrete-Time Almost-Linear Systems and "So-called" Linear Systems. SICE Transaction 32(5), 782–784 (1996c) (in Japanese)

Hasegawa, Y., Matsuo, T.: Partial realization and real-time partial realization theory of discretetime almost linear systems. SICE Transaction 32(5), 653–662 (1996d)

Hasegawa, Y., Matsuo, T., Hirano, T.: On the Partial Realization Problem of discrete-time linear representation systems. SICE Transaction 18(2), 152–159 (1982) (in Japanese)

Hasegawa, Y., Niinomi, S., Matsuo, T.: Realization Theory of Continuous-Time Pseudo-Linear Systems. In: IEEE International Symposium on Circuits and Systems, USA, vol. 3, pp. 186–189 (1996)

Hasegawa, Y., Niinomi, S., Matsuo, T.: Realization Theory of Continuous-Time Finite Dimensional Pseudo- Linear Systems. In: Recent Advances in Circuits and Systems, pp. 109–114. World Scientific, Singapore (1998)

Hasegawa, Y., Suzuki, T.: Realization Theory and Design of Digital Images. Lecture Notes in Control and Information Sciences, vol. 342. Springer, Heidelberg (2006)

Horn, R.A., Johnson, C.R.: Matrix Analysis. Cambridge University Press, Cambridge (1985)

Horn, R.A., Johnson, C.R.: Topics in Matrix Analysis. Cambridge University Press, Cambridge (1991)

Isidori, A.: Direct construction of minimal bilinear realization from nonlinear inputoutput maps. IEEE Trans. on Autom. Contr. AC-18, 626–631 (1973)

Realization and modeling in system theory. In: Kaasshocek, M.A., van Schuppen, J.H., Ran, A.C.M. (eds.) Proc. of the Int. Symp. MTNS 1989, Progress in systems and Control Theory, Birkhauser, vol. 1 (1990)

Kalman, R.E.: On the general theory of control systems. In: Proc. 1st IFAC Congress, Moscow, Butterworths London (1960)

Kalman, R.E.: Mathematical description of linear dynamical systems. SIAM J. Contr. 1, 152–192 (1963)

Kalman, R.E.: Algebraic structure of linear dynamical systems. I, The module of $\sum$. Proc. Nat. Acad. Sci (USA) 54, 1503–1508 (1965)

Kalman, R.E.: Algebraic aspects of the theory of dynamical systems. In: Hale, J.K., Lasale, J.P. (eds.) Differential Equations and Dynamical systems, Bologna, Italy, Cremonese, Roma. Lectures on controllability and observability, note for a course held at C.I.M.E, pp. 133–143. Academic Press, New York (1967)

Kalman, R.E.: A theory for the identification of linear systems. In: Brezis, H., Giariet, P.G. (eds.) Proceedings, Colloque Lions (1989)

Kalman, R.E.: Nine Lectures on Identification in Swiss Federal Institute of Technology, Zurich, Switzerland (1997)

Kalman, R.E., Falb, P.L., Arbib, M.A.: Topics in mathematical system theory. McGraw-Hill, New York (1969)

Kung, S.Y., Lin, D.W.: Optimal Hankel-Norm Model Reductions: Multivariable Systems. IEEE tranction Automatic Control AC-26(4) (August 1981)

Matsuo, T.: System theory of linear continuous-time systems using generalized function theory. Res. Rept. of Auto. Contr. Lab. 15 (1968) (in Japanese)

Matsuo, T.: Mathematical theory of linear continuous-time systems. Rept. of Auto. Contr. Lab. 16, 11–17 (1969)

Matsuo, T.: On the realization theory linear infinite dimensional systems. Res. Rept. No. SC-75-9, The Institute of Electrical Engineer of Japan (1975) (in Japanese)

Matsuo, T.: Foundations of mathematical system theory. Journal of SICE 16, 648–654 (1977)

Matsuo, T.: On-line partial realization of a discrete-time non-linear black-box by linear representation systems. In: Proc. of 18th Annual Allerton Con. on Com. Contr. and Computing, Monticello, Illinois (1980)

Matsuo, T.: Realization theory of continuous-time dynamical systems. Lecture Notes in Control and Information Sciences, vol. 32. Springer, Heidelberg (1981)

Matsuo, T., Hasegawa, Y.: On linear representation systems and affine dynamical systems, Res. Rept. No. SC-77-27, The institute of Electrical Engineers of Japan (1977) (in Japanese)

Matsuo, T., Hasegawa, Y.: Realization theory of discrete-time systems. SICE Transaction 152, 178–185 (1979)

Matsuo, T., Hasegawa, Y.: Two-Dimensional Arrays and Finite-Dimensional Commutative Linear Representation Systems. Electronics and Communications in Japan 64-A(5), 11–19 (1981)

Matsuo, T., Hasegawa, Y.: Partial Realization & Real-Time, Partial Realization Theory of Discrete- Time Almost-Linear System. SICE Transaction 32(5), 653–662 (1996) (in Japanese)

Matsuo, T., Hasegawa, Y.: Realization Theory of Discrete-Time Dynamical Systems. Lecture Notes in Control and Information Sciences, vol. 296. Springer, Heidelberg (2003)

Matsuo, T., Hasegawa, Y., Niinomi, S., Togo, M.: Foundations on the Realization Theory of Two-Dimensional Arrays. Electronics and Communications in Japan 64-A(5), 1–10 (1981)

Matsuo, T., Hasegawa, Y., Okada, Y.: The partial Realization Theory of Finite Size Two-Dimensional Arrays. Trans. IECE 64-A(10), 811–818 (1981) (in Japanese)

Matsuo, T., Niinomi, S.: The realization theory of continuous-time affine dynamical systems. SICE Trans. 17(1), 56–63 (1981) (in Japanese)

Mohler, R.R.: Nonlinear Systems. In: Applications to Bilinear Control, vol. II, Prentice-Hall, Englewood Cliffs (1991)

Niinomi, S., Matsuo, T.: Relation between discrete-time linear representation systems and affine dynamical systems. SICE Trans. 17(3), 350–357 (1981) (in Japanese)

Pareigis, B.: Category and functors. Academic Press, New York (1970)

Smale, S.: Differential dynamical systems. Bull. Amer. Soc. 73, 747–817 (1967)

Index

Lecture Notes in Control and Information Sciences

Edited by M. Thoma, M. Morari

Further volumes of this series can be found on our homepage:
springer.com

Vol. 354: Witczak M.:
Modelling and Estimation Strategies for Fault
Diagnosis of Non-Linear Systems
215 p. 2007 [978-3-540-71114-8]

Vol. 353: Bonivento C.; Isidori A.; Marconi L.;
Rossi C. (Eds.)
Advances in Control Theory and Applications
305 p. 2007 [978-3-540-70700-4]

Vol. 352: Chiasson, J.; Loiseau, J.J. (Eds.)
Applications of Time Delay Systems
358 p. 2007 [978-3-540-49555-0]

Vol. 351: Lin, C.; Wang, Q.-G.; Lee, T.H., He, Y.
LMI Approach to Analysis and Control of
Takagi-Sugeno Fuzzy Systems with Time Delay
204 p. 2007 [978-3-540-49552-9]

Vol. 350: Bandyopadhyay, B.; Manjunath, T.C.;
Umapathy, M.
Modeling, Control and Implementation of Smart
Structures 250 p. 2007 [978-3-540-48393-9]

Vol. 349: Rogers, E.T.A.; Galkowski, K.;
Owens, D.H.
Control Systems Theory
and Applications for Linear
Repetitive Processes
482 p. 2007 [978-3-540-42663-9]

Vol. 347: Assawinchaichote, W.; Nguang, K.S.;
Shi P.
Fuzzy Control and Filter Design
for Uncertain Fuzzy Systems
188 p. 2006 [978-3-540-37011-6]

Vol. 346: Tarbouriech, S.; Garcia, G.; Glattfelder,
A.H. (Eds.)
Advanced Strategies in Control Systems
with Input and Output Constraints
480 p. 2006 [978-3-540-37009-3]

Vol. 345: Huang, D.-S.; Li, K.; Irwin, G.W. (Eds.)
Intelligent Computing in Signal Processing
and Pattern Recognition
1179 p. 2006 [978-3-540-37257-8]

Vol. 344: Huang, D.-S.; Li, K.; Irwin, G.W. (Eds.)
Intelligent Control and Automation
1121 p. 2006 [978-3-540-37255-4]

Vol. 341: Commault, C.; Marchand, N. (Eds.)
Positive Systems
448 p. 2006 [978-3-540-34771-2]

Vol. 340: Diehl, M.; Mombaur, K. (Eds.)
Fast Motions in Biomechanics and Robotics
500 p. 2006 [978-3-540-36118-3]

Vol. 339: Alamir, M.
Stabilization of Nonlinear Systems Using
Receding-horizon Control Schemes
325 p. 2006 [978-1-84628-470-0]

Vol. 338: Tokarzewski, J.
Finite Zeros in Discrete Time Control Systems
325 p. 2006 [978-3-540-33464-4]

Vol. 337: Blom, H.; Lygeros, J. (Eds.)
Stochastic Hybrid Systems
395 p. 2006 [978-3-540-33466-8]

Vol. 336: Pettersen, K.Y.; Gravdahl, J.T.;
Nijmeijer, H. (Eds.)
Group Coordination and Cooperative Control
310 p. 2006 [978-3-540-33468-2]

Vol. 335: Kozłowski, K. (Ed.)
Robot Motion and Control
424 p. 2006 [978-1-84628-404-5]

Vol. 334: Edwards, C.; Fossas Colet, E.;
Fridman, L. (Eds.)
Advances in Variable Structure and Sliding Mode
Control
504 p. 2006 [978-3-540-32800-1]

Vol. 333: Banavar, R.N.; Sankaranarayanan, V.
Switched Finite Time Control of a Class of
Underactuated Systems
99 p. 2006 [978-3-540-32799-8]

Vol. 332: Xu, S.; Lam, J.
Robust Control and Filtering of Singular Systems
234 p. 2006 [978-3-540-32797-4]

Vol. 331: Antsaklis, P.J.; Tabuada, P. (Eds.)
Networked Embedded Sensing and Control
367 p. 2006 [978-3-540-32794-3]

Vol. 330: Koumoutsakos, P.; Mezic, I. (Eds.)
Control of Fluid Flow
200 p. 2006 [978-3-540-25140-8]

Vol. 329: Francis, B.A.; Smith, M.C.; Willems,
J.C. (Eds.)
Control of Uncertain Systems: Modelling,
Approximation, and Design
429 p. 2006 [978-3-540-31754-8]

Vol. 328: Loría, A.; Lamnabhi-Lagarrigue, F.;
Panteley, E. (Eds.)
Advanced Topics in Control Systems Theory
305 p. 2006 [978-1-84628-313-0]

Vol. 327: Fournier, J.-D.; Grimm, J.; Leblond, J.;
Partington, J.R. (Eds.)
Harmonic Analysis and Rational Approximation
301 p. 2006 [978-3-540-30922-2]

Vol. 326: Wang, H.-S.; Yung, C.-F.; Chang, F.-R.
H_∞ Control for Nonlinear Descriptor Systems
164 p. 2006 [978-1-84628-289-8]

Vol. 325: Amato, F.
Robust Control of Linear Systems Subject to
Uncertain
Time-Varying Parameters
180 p. 2006 [978-3-540-23950-5]

If you have any concerns about our products,
you can contact us on
ProductSafety@springernature.com

In case Publisher is established outside the EU,
the EU authorized representative is:
Springer Nature Customer Service Center GmbH
Europaplatz 3, 69115 Heidelberg, Germany

Printed by Libri Plureos GmbH
in Hamburg, Germany